Proceedings of the 8th European Congress of Neurosurgery

Barcelona, September 6–11, 1987

Edited by
F. Isamat, A. Jefferson, F. Loew, L. Symon

Volume 2

Spinal Cord and Spine Pathologies
Basic Research in Neurosurgery

Acta Neurochirurgica
Supplementum 43

Springer-Verlag Wien New York

Professor Dr. Fabian Isamat
Servicio de Neurocirugía, Hospital de Bellvitze, University of Barcelona, Barcelona, Spain

Dr. Antony Jefferson
Newport, Dyfed, United Kingdom

Professor Dr. Friedrich Loew
Neurochirurgische Universitätsklinik, Homburg/Saar, Federal Republic of Germany

Professor Lindsay Symon, TD, FRCS
Department of Neurological Surgery, Institute of Neurology, The National Hospital, London, U.K.

With 97 Figures

Softcover reprint of the hardcover 1st edition 1988

Library of Congress Cataloging-in-Publication Data. European Congress of Neurosurgery (8th: 1987: Barcelona, Spain) Proceedings of the 8th European Congress of Neurosurgery: Barcelona, September 6–11, 1987/edited by F. Isamat ... [et al.]. VIII, 213 p. 21 × 27.7 cm.— (Acta neurochirurgica. Supplementum, ISSN 0065-1419;43).ISBN-13:978-3-7091-8980-1 (U.S.:v.2).1.Nervous system—Surgery—Congresses.I. Isamat,F. II.Title.III.Series. [DNLM: 1. Neurosurgery—congresses. W1 AC8661 no.43/WL 368 E89p 1987]. RD593.E95 1987. 617'.48-dc19. 88-20167.

ISSN 0065-1419

ISBN-13:978-3-7091-8980-1 e-ISBN-13:978-3-7091-8978-8
DOI: 10.1007/978-3-7091-8978-8

Preface

The 8th European Congress of Neurosurgery which took place in Barcelona from September 6 to 11, 1987, was an unforgettable experience. Many factors contributed to its success: the splendid ambience of Barcelona, the generous hospitality and warm friendship of the hosts, and the marvellous organization and high scientific standard of papers, workshops and discussions.

For financial reasons it would not have been possible to publish all the papers presented during the congress. In order to preserve as much as possible of the scientific results, the Programme Committee of the European Association of Neurosurgical Societies had already selected before the congress a certain number of abstracts to be published as full papers. These are compiled in two supplement volumes of Acta Neurochirurgica and deal with the following main topics:

Volume 1:
1. Intraoperative and posttraumatic monitoring and brain protection
2. Cerebro-vascular lesions
3. Intracranial tumours
4. Benign intracranial cystic lesions, hydrocephalus, CSF-volumes
5. Central pain syndromes

Volume 2:
6. Spinal cord and spine pathologies
7. Basic research in neurosurgery.

The Editors
F. Isamat, A. Jefferson, F. Loew, L. Symon

Contents

Listed in Current Contents

VI. Spinal Cord and Spine Pathologies

Fornari, M., Pluchino, F., Solero, C. L., Giombini, S., Luccarelli, G., Oliveri, G., Lasio, G.: Microsurgical Treatment of Intramedullary Spinal Cord Tumours . . . 3

Yasui, T., Hakuba, A., Katsuyama, J., Nishimura, S.: Microsurgical Removal of Intramedullary Spinal Cord Tumours: Report of 22 Cases . . . 9

Roosen, N., Dahlhaus, P., Lumenta, Ch. B., Lins, E., Stork, W., Gahlen, D., Bock, W. J.: Magnetic Resonance (MR) Imaging in the Management of Primary and Secondary Syringomyelic Cavities, and of Other Cystic Lesions of the Spinal Cord . . . 13

Powell, M.: Syringomyelia: How MRI Aids Diagnosis and Management . . . 17

Tokuno, H., Hakuba, A., Suzuki, T., Nishimura, Sh.: Operative Treatment of Chiari Malformation with Syringomyelia . . . 22

Bidziński, J.: Pathological Findings in Suboccipital Decompression in 63 Patients with Syringomyelia 26

Bidziński, J.: Late Results of the Surgical Treatment of Syringomyelia . . . 29

Philippon, J., Sangla, S., Lara-Morales, J., Gazengel, J., Rivierez, M., Horn, Y. E.: Treatment of Syringomyelia by Syringo-peritoneal Shunt . . . 32

Alphen, H. A. M. van, Braakman, R., Berfelo, M. W., Broere, G., Bezemer, P. D., Kostense, P. J.: Chemonucleolysis or Discectomy? Results of a Randomized Multicentre Trial in Patients with a Herniated Lumbar Intervertebral Disc . . . 35

Ferrer, E., García-Bach, M., López, L., Isamat, F.: Lumbar Microdiscectomy: Analysis of 100 Consecutive Cases. Its Pitfalls and Final Results . . . 39

Cervellini, P., Curri, D., Bernardi, L., Volpin, L., Benedetti, A.: Computed Tomography After Lumbar Disc Surgery: a Comparison Between Symptomatic and Asymptomatic Patients 44

Ray, Ch. D.: Transfacet Decompression with Dowel Fixation: a New Technique for Lumbar Lateral Spinal Stenosis . . . 48

Gomes, F.: Automated Percutaneous Nucleotomy—Initial Experience in Twenty-Five Cases of Contained Lumbar Disc Herniation . . . 55

Onik, G., Maroon, J., Day, A., Helms, C.: Automated Percutaneous Discectomy: Preliminary Experience 58

Sonntag, V. K. H., Hadley, M. N., Dickman, C. A., Browner, C. M.: Atlas Fractures: Treatment and Long-term Results . . . 63

Spetzler, R. F., Hadley, M. N., Sonntag, V. K. H.: The Transoral Approach to the Anterior Superior Cervical Spine. A Review of 29 Cases . . . 69

Gambacorta, D., Reale, F.: Posterior Acrylic Surgical Fixation of Odontoid Fractures . . . 75

Zygmunt, S. C., Ljunggren, B., Ålund, M., Brattström, H., Säveland, H. G., Holtås, S., Larsson, E. M., Redlund-Johnell, I.: Realignment and Surgical Fixation of Atlanto-axial and Subaxial Dislocations in Rheumatoid Arthritis (RA) Patients . . . 79

Steudel, W. I., Rosenthal, D., Lorenz, R., Merdes, W.: Prognosis and Treatment of Cervical Spine Injuries with Associated Head Trauma . . . 85

Rochkind, S., Barr-Nea, L., Bartal, A., Nissan, M., Lubart, R., Razon, N.: New Methods of Treatment of Severely Injured Sciatic Nerve and Spinal Cord. An Experimental Study 91

Zierski, J., Müller, H., Dralle, D., Wurdinger, T.: Implanted Pump Systems for Treatment of Spasticity 94

VII. Basic Research in Neurosurgery

Shibata, T., Burger, P. C., Kleihues, P.: Ki-67 Immunoperoxidase Stain as Marker for the Histological Grading of Nervous System Tumours . . . 103

Westphal, M., Hänsel, M., Nausch, H., Rohde, E., Köppen, J., Fiola, M., Hölzel, F., Herrmann, H.-D.: Glioma Biology in vitro: Goals and Concepts . . . 107

Casalone, R., Minelli, A., Butti, G., Gaetani, P., Silvani, V., Solero, C. L., Danesino, C.: Gene Dosage Effect in Cells with Monosomy of Chromosome 22 Derived from Human Meningiomas 114
Helseth, E., Unsgaard, G., Dalen, A., Vik, R.: The Effect of Type Beta Transforming Growth Factor on Proliferation of Clonogenic Cells from Human Gliomas 118
Colombatti, M., Bisconti, M., Lorenzi, P., Stevanoni, G., Dipasquale, B., Gerosa, M., Tridente, G.: Human Glioma Cell Lines: Tumour Associated Antigens Distribution and Sensitivity to Antibody-Toxin or Ligand-Toxin Conjugates. A Preliminary Report 121
Gaini, S. M., Riboni, L., Cerri, C., Grimoldi, N., Sganzerla, E. P., Berra, B.: Ganglioside Content and Composition in Human Gliomas 126
Maleci, A., Ausiello, C., Gagliardi, F. M., Guidetti, B., Cassone, A.: The Response of Peripheral Blood Mononuclear Cells of Glioma-Bearing Patients to Stimulation with Microbial Antigen and IL-2: Proliferation and IFN-Gamma Production 130
Imieliński, B. L., Borowska-Lehman, J., Zółtowska, A.: Immunological and Immunohistochemical Studies in Brain Tumours 135
Hirakawa, K., Naruse, Sh., Higuchi, T., Horikawa, Y., Tanaka, Ch., Ebisu, T.: The Investigation of Experimental Brain Tumours Using ^{31}P-MRS and ^{1}H-MRI 140
Rommel, Th., Bodsch, W.: Glucocorticosteroid Treatment of Vasogenic Oedema 145
Shimizu, K., Tsuda, N., Okamoto, Y., Matsui, Y., Miyao, Y., Tamura, K., Yamada, M., Nakatani, S., Ikeda, T., Mogami, H.: Transplant-induced Recovery from 6-OHDA Lesions of the Nigrostriatal Dopamineneurones in Mice 149
Takayama, H., Toya, Sh., Shinozaki, T., Inoue, H., Otani, M., Kohsaka, Sh., Tsukada, Y.: Possible Synapse Formation by Embryonic Cerebellar Tissue Grafted into the Cerebellum of the Weaver Mutant Mouse 154
González-Darder, J. M., Barcia-Salorio, J. L., Barberá, J., Broseta, J.: Intraventricular Transplantation of Omentum for Treatment of Hydrocephalus. An Experimental Study in Dogs 159
Rijen, P. C. van, Verheem, A., Tulleken, C. A. F.: Proton Magnetic Resonance Imaging in Experimental Cerebral Ischaemia 162
Langmoen, I. A., Berg-Johnsen, J.: Intracellular Recordings from Neurones in Rat Cerebral Cortex During Hypoxia 168
Higuchi, T., Naruse, S., Horikawa, Y., Tanaka, C., Ebisu, T., Yamamoto, K., Hirakawa, K.: Pathophysiological Investigation of Experimental Cerebral Ischaemia Using in vivo ^{31}P-NMR Spectroscopy and ^{1}H-MRI 172
Rodriguez y Baena, R., Gaetani, P., Silvani, V., Spanu, G., Marzatico, F.: Effect of Nimodipine on Mitochondrial Respiration in Different Rat Brain Areas After Subarachnoid Haemorrhage 177
Wallenfang, Th., Fries, G., Jantzen, J. P., Bayer, J., Trautmann, F.: Pathomechanism of Brain Oedema in Experimental Intracerebral Mass Haemorrhage 182
Dóczi, T., Joó, F., Szerdahelyi, P., Bodosi, M.: Regulation of Brain Water and Electrolyte Contents: the Opposite Actions of Central Vasopressin and Atrial Natriuretic Factor (ANF) 186
Roda, J. M., Alvarez, F., Garcia-Villalon, A. L., Ruiz, M. R., Gutierrez, M., Garcia Blazquez, M.: An Increment in Unilateral Carotid Blood Flow Produces Cerebral Aneurysms in Rats 189
Moringlane, J. R., Grote, R., Vonnahme, F.-J., Mestres, P., Harbauer, G., Ostertag, C. B.: Occlusion of Experimental Artery Aneurysms by Intrasaccular Injection of Fibrin Sealant 193
Broseta, J., García-March, G., Sánchez-Ledesma, M^{a}. J., Gonçalves, J., Anaya, J., Torregrosa, A., Urbano, J., Collía, F., Ludeña, M^{a}. D., Merchán, M.: Local Pathological Findings in Experimental Dorsal Root Entry Zone Lesions Performed by Mechanical Section, Laser and Radiofrequency. A Comparative Study 198
Barberá, J., Gonzalez, J., Gil, J. L., Sanjuán, M^{a}. A., García, F., Lopez, A.: The Quality and Extension of Nerve Fibre Regeneration in the Centrocentral Anastomosis of the Peripheral Nerve 205
Pallini, R., Fernandez, E., Minciacchi, D., Marchese, E., Sbriccoli, A.: Peripheral Nerve Autografts to the Rat Spinal Cord: a Study of the Origin of Regenerating Fibres Using Fluorescent Double Labelling 210

VI. Spinal Cord and Spine Pathologies

Acta Neurochirurgica, Suppl. 43, 3–8 (1988)

Microsurgical Treatment of Intramedullary Spinal Cord Tumours

M. Fornari, F. Pluchino, C. L. Solero, S. Giombini, G. Luccarelli, G. Oliveri, and G. Lasio

Department of Neurosurgery, Istituto Neurologico "C. Besta", Milano, Italy

Summary

A series of 81 patients operated for intramedullary spinal cord tumours over a span of 12 years (1975–1986) is presented. The mean age of the patients was about 37 years and 9 patients were in the paediatric age (0–15 years). In 29 cases (36%) the tumours were located within the cervical region; in 34 cases the tumours were located in the thoracic region (42%) and in 18 cases the tumours were located within the medullary cone (22%). Ependymoma accounted for 52% of the cases (42 cases), astrocytoma for 20% (16 cases), lipoma for 7% (6 cases), hemangioblastoma for 6% (5 cases), metastatic tumours for 5% (4 cases); in the remaining 8 cases (10%) rare or very rare spinal tumours were found. The 9 patients in the paediatric age and 8 additional cases who had undergone previous surgery were then excluded from this review. In 59% of the patients with ependymoma and in 90% of the patients with astrocytoma severe neurological deficit were already present before the operation.

Total tumour removal was accomplished in 81% of the cases with ependymoma, total or subtotal removal was achieved in 50% of the cases with astrocytoma.

At long term follow-up study fair or good functional results were observed in 19 out of 41 cases of ependymoma (46%) and in 3 cases out of 10 of astrocytoma (30%). The surgical outcome was mostly related to the preoperative neurological conditions and to tumour malignancy grade.

Keywords: Spinal cord neoplasms; astrocytoma; ependymoma; microsurgery; follow-up studies.

Introduction

In spite of the fact that the surgical removal of intramedullary tumours had been pioneered by some neurosurgeons, mainly by Cushing, Gowers, Elsberg, Horrax[1,3,6,9], since 1925, the first large series of patients in which successful removal of intramedullary tumours was performed was reported by Greenwood[7] only about 40 years later. Later on microsurgical techniques were widely introduced in the neurosurgical practice but nevertheless only very few series of treated intramedullary tumours have then been reported[2,4,5,8,10,11,12] and the indications and results of surgery of these lesions appear still extremely controversial. A general agreement has been reached concerning the treatment of ependymoma, whose complete removal is achievable without additional damage to the spinal cord in most of the cases, in spite of the frequent huge longitudinal extent of these tumours. On the other hand the successful removal of astrocytoma has been reported in very few series and it has been mostly related to astrocytoma in childhood, even if holointramedullary[4]; it has been claimed that these tumours should have a well defined plane of cleavage from neural tissue in approximately 50% of adult patients[11]. The aim of this work was to review our series of 81 patients with intramedullary tumours operated during the period from 1975 to 1986 with main regard to the possible factors influencing the long term clinical outcome.

Material and Methods

During the last 12 years (1975–1986), 81 patients with intramedullary spinal cord tumours have been operated at the Istituto Neurologico "C. Besta" of Milano. Fourty-eight patients were males and 34 were females. The age of the patients ranged from 5 to 62 years (mean age about 37 years); 9 patients were in the paediatric age (0–15 years) and were excluded from this study (5 ependymoma and 4 astrocytoma). Patients in adolescent and juvenile age (15–20 years) showed clinical and pathological conditions similar to those of the adult patients and have not therefore been differentiated from them.

The diagnosis was mostly reached by myelography, although the combination of this examination with CT scanning resulted in useful additional information in many cases. During the last few years the early recognition of these lesions and the differential diagnosis from syringomyelia, multiple sclerosis and inflammatory diseases have been impressively improved by the aid of Magnetic Resonance Imaging (MRI) (Fig. 1).

In 29 cases (36%) the tumours were located in the cervical region, in 34 cases the tumours were located in the thoracic region (42%) and in 18 cases (22%) the tumours were located in the lumbosacral region within the conus medullaris often extending towards the cauda equina (14 cases) (see also Table 1). As far as the histological classification is concerned, ependymoma accounted for 52% of the cases

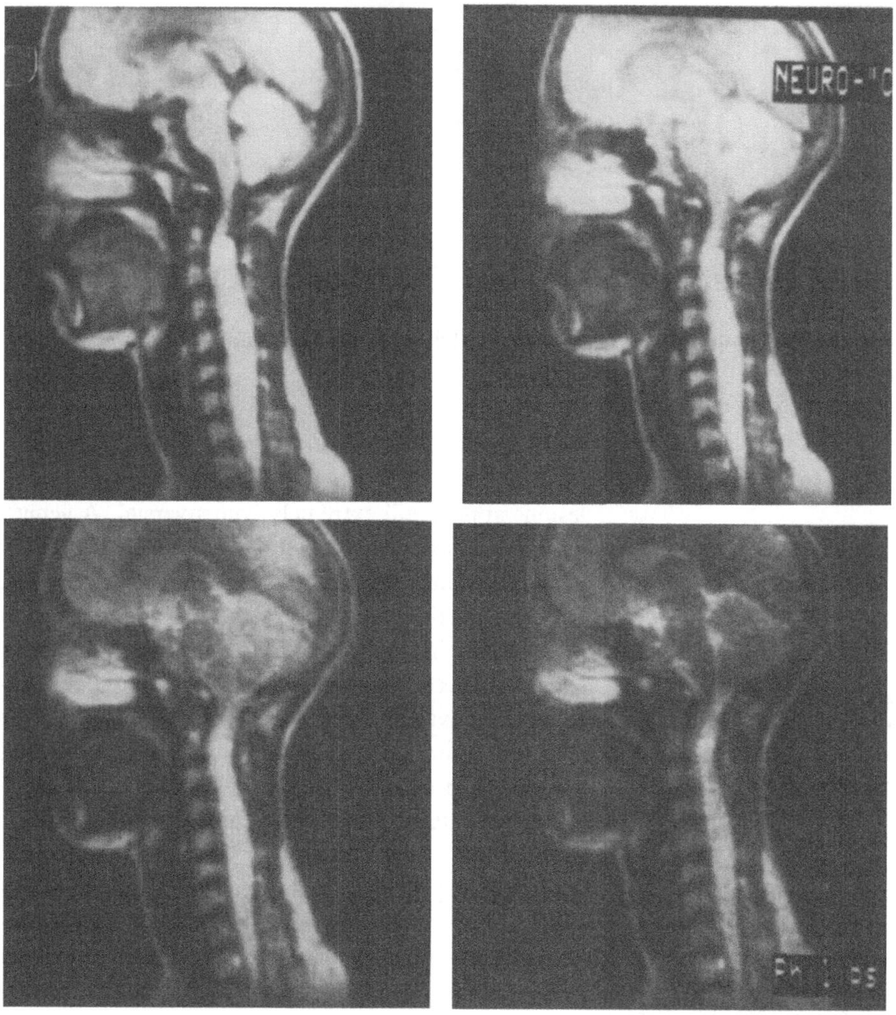

Fig. 1. MRI in a case of cervical intramedullary lipoma. The tumour and its relationship with the spinal cord are clearly outlined. An exophytic portion of the tumour was also extending extramedullary from the dorsal surface of the spinal cord

(42 cases), astrocytoma accounted for 20% of the cases (16 cases), lipoma accounted for 7% of the cases (6 cases), hemangioblastoma accounted for 6% of the cases (5 cases), metastatic tumours accounted for 5% of the cases (4 cases), in the remaining 8 cases rare (epidermoid tumours: 3 cases) or very rare intramedullary tumours were found (10%). In 21 cases the tumours were found extending for 3 to 6 vertebral segments and in 6 cases (7%) for 7 vertebral segments or longer (in two cases holointramedullary astrocytoma were present). Astrocytomas were mostly (69%) found in the thoracic region while ependymoma were found with about the same frequency in the cervical and thoracic region. The lumbosacral region was fairly typical for the epidermoid tumours (3 cases). All the patients have been operated by microsurgical technique; the tumours located in the cervical region were mostly operated with the patient in the sitting

Table 1

		Ependymoma	Astrocytoma	Lipoma	Hemangio-blastoma	Epidermoid	Metastatic	Rare T.
Cervical Region	29 cases	17	5	2	1		1	3
Thoracic Region	34 cases	13	11	4	4		1	1
Lumbosacral Region	18 cases	12				3	2	1
Tot	81 cases	42	16	6	5	3	4	5

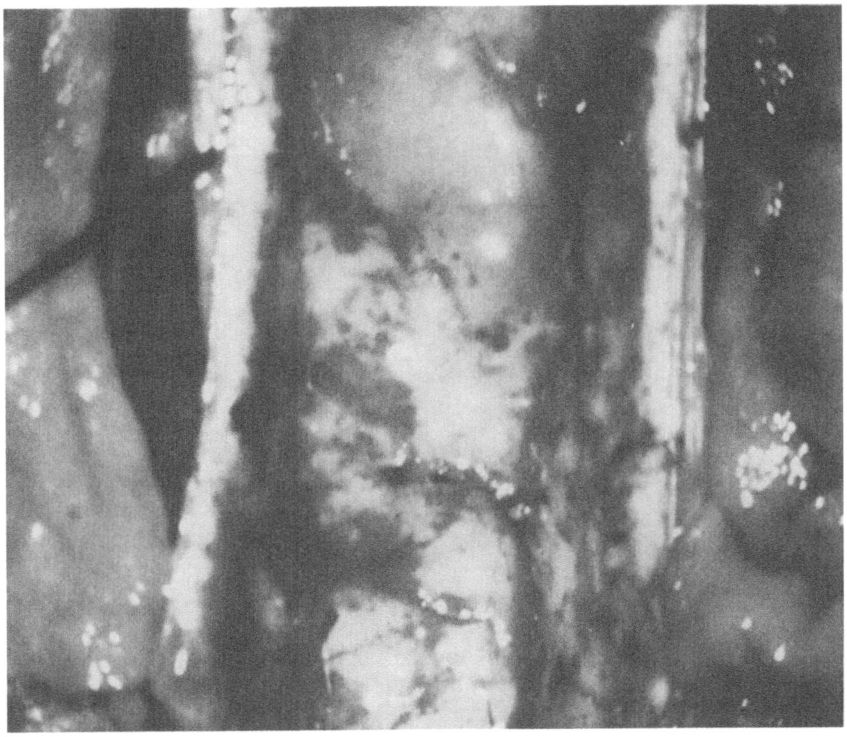

Fig. 2. Operative photograph showing the cavity left within the spinal cord after complete removal of an intramedullary ependymoma. The borders of the myelotomy stay open also after removal of pial stay sutures

position (with ultrasonic cardiac monitor in use and intravenous catheter inserted to prevent air embolism), while in the other cases the patients were carefully adjusted in the prone position in order to obtain a free abdomen and chest and therefore minimizing the venous congestion around the area of surgery.

Results

In consideration of the different clinical and surgical problems which arise when dealing with intramedullary tumours of different type, it looked worthwhile to analyze the results of this series according to the tumour type.

i) Ependymoma (31 cases): ependymoma was the most relevant group in this series (42 patients: 52%). Eleven patients were then excluded from this study because they were in the paediatric age (5 cases) or because they had been previously operated elsewhere (6 cases). Finally 31 patients have been considered: 15 were females and 16 were males and the mean age was 39 years. The tumours were located in the cervical region in 7 cases (23%), in the cervico-thoracic region in 9 cases (29%), in the thoracic region in 6 cases (19%), in the conus medullaris region in 9 cases (29%). The tumour extended for 1–2 spinal segments in 4 patients; for 3 to 5 segments in 15 cases; for more than 5 segments in 12 cases. The first symptom consisted in sensory disturbances (pain, paresthesia or hypoesthesia) in 23 patients (74%), in motor deficits in 5 cases (16%) and in bowel, bladder and sexual organs dysfunction in 3 cases (10%). The mean duration of the clinical history was fairly long, lasting for about 53 months (range: 1 month–15 years). At time of the operation the neurological examination was normal or showed only mild sensory and motor deficits in 13 patients (41%); in 18 cases (59%) severe or devastating neurological deficits (up to paraplegia) were already present before surgery. In one case sudden worsening of symptoms and signs was observed after myelography. Complete tumour removal was achieved in 25 cases (81%) while subtotal removal was accomplished in the remaining 6 cases (19%) (Fig. 2), due to the difficulties encountered in the dissection of the tumour from the surrounding spinal cord tissue. According to the histological examination 25 cases were classified as ependymoma, 2 cases as subependymoma, 4 cases as anaplastic ependymoma or ependymoblastoma. A huge cranial or caudal cystic cavity was encountered in 7 cases. Out of the 13 patients with absent to mild neurological deficits, one presented severe neurological worsening in the immediate postoperative course, which did not clear in spite of adequate rehabilitation therapy. Two other of these patients with anaplastic ependymoma presented clinical progression of the disease because of the rapid regrowth of the tumours within 2 and 3 years respectively. The remaining 10 patients at long-term follow-up study (mean duration 62 months) were doing fairly

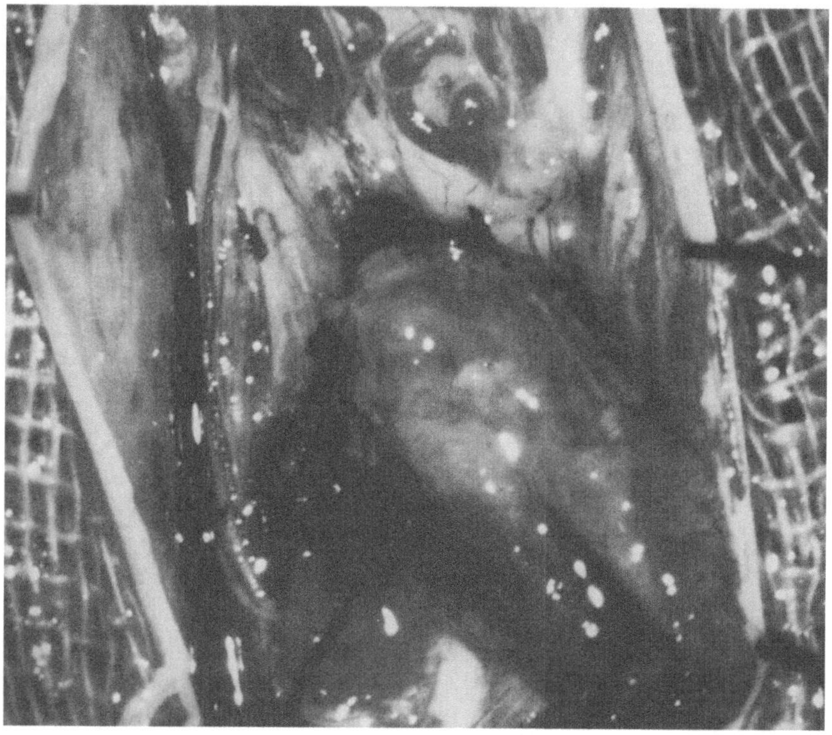

Fig. 3. Operative photograph showing complete removal of an hemangioblastoma. The tumour has been removed in block, after careful dissection from the spinal cord and coagulation of its feeding arteries. Note the extremely dilated veins surrounding the cavity left by tumour removal

well, with normal social life and no recurrence of symptoms. Out of the 18 cases with severe preoperative neurological deficits 5 cases showed an immediate and permanent worsening of deficits, 1 case with anaplastic ependymoma showed progression of the disease because of tumour growth within 2 years. At late follow-up study (mean duration 60 months) 9 cases were unchanged or showed only minor improvement, 3 cases were greatly improved and had returned to fairly normal social life.

ii) Astrocytoma (10 cases): out of the 16 patients (20%) with astrocytoma, 6 were excluded because they were in the paediatric age (4 cases) or because that had been previously operated elsewhere (2 cases). Finally 10 cases have been considered: 5 were females and 5 were males. The tumour was located in the cervical region in 4 cases and in the dorsal region in 6 cases, and it extended for 3 segments in 8 cases, over 5 segments in 2 cases. The initial symptom consisted in motor complaints in 8 cases, in sensory disturbances in 2 cases. The mean duration of the clinical history was about 34 months (range 1 month–7 years) and neurological examination at time of surgery showed mild deficits in only one case while in the remaining 9 cases a severe neurological disability was already present. Complete removal was achieved in 2 cases, subtotal removal in 3 cases while in the remaining 5 cases only partial removal was accomplished (WHO). Grade I–II astrocytoma were found in 7 cases, anaplastic astrocytoma in 2 cases and glioblastoma in 1 case. These last three cases (which received only partial removal) showed devastating progression of the disease and two of them died within 18 months after surgery. Another patient in which subtotal removal had been performed presented severe postoperative worsening after surgery and died because of respiratory insufficiency. At follow-up study out of the remaining 6 cases, 3 cases (2 with partial and 1 with apparent complete removal) after an initial postoperative improvement showed clinical recurrence after 6 months, 2 years and 8 years respectively; the last 3 cases (2 with subtotal and 1 with apparent complete removal) have absent or mild neurological deficits allowing fairly normal social life, 1, 4 and 10 years respectively after surgery.

iii) Hemangioblastoma (5 cases) *and lipoma* (6 cases): hemangioblastoma were located in the thoracic region in 4 cases and in the cervical region in 1 case. The duration of the clinical history ranged from 4 months to 4 years and the neurological examination was strictly related to the duration of the clinical history, showing severe neurological deficit in 2 cases. Complete removal was achieved in all the cases (Fig. 3). At follow-up study (mean duration about 4 years) the two patients with severe preoperative deficits are still showing a complete neurological disability, while the remaining 3 cases showed no variation at all of their deficits which were still causing a minimal or moderate disability.

All the 6 cases operated for intramedullary lipoma (which were seated in the cervical region in 2 cases and in the thoracic region in 4 cases) underwent complete tumour removal and the neurological examination at long-term follow-up study (mean duration about 5 years) appeared strictly consistent with the preoperative neurological conditions. Normal neurological examination and normal social life was observed in half of the patients. No clinical recurrence has been observed.

iiii) Epidermoid (3 cases), *rare* (5 cases) *and metastatic* (4 cases) *tumours:* complete removal was achieved in all the three cases with epidermoid tumours of the conus medullaris region although it couldn't be completely excluded that some portion of the capsule had been left attached to the spinal roots, in spite of all attempts made to remove it. All these patients presented good or fair neurological conditions either preoperatively and postoperatively at follow-up study (mean duration about 4 years) and no clinical recurrence has been observed.

Rare tumours (2 germinoma, 1 lymphoma, 1 neurinoma, 1 choriocarcinoma) and metastatic tumours (3 cases of metastatic melanoma and 1 case of breast carcinoma) accounted for 11% (9 cases) of all intramedullary tumours in this series and their incidence shouldn't be underestimated when dealing with intramedullary tumours but their limited number does not allow any consideration concerning the surgical results which are obviously related to the nature and the stage of the disease and to the preoperative neurological deficits.

Discussion and Conclusions

By the analysis of this series it appeared that in a percentage of cases which can be even higher than 50% the correct diagnosis of intramedullary tumours could not be reached until severe neurological deficits were already present. In the group of ependymoma patients this poor neurological status showed a clear postoperative improvement in only 3 cases out of 18 (16%), furthermore 4 cases (12%) of ependymoma with anaplastic dedifferentiation presented postoperatively a rapid tumour regrowth and progression of the disease and surgery itself caused permanent worsening of deficit in one of the 13 cases with mild preoperative disability. In the group of astrocytoma patients a significative neurological improvement occurred postoperatively in 2 cases out of 9 (22%).

Malignant rapidly growing tumours were found in 3 cases (30%), permanent worsening of deficits occurred after operation in 1 case. As a whole at long term follow-up study fair or good results were obtained in 19 cases of ependymoma (out of 31) and in 3 cases (out of 10) of astrocytoma (all grades).

These observations should lead to the conclusions, which have been already emphasized by other authors[2, 5, 10, 11], that the prognosis and the results of the surgical treatment of intramedullary tumours are mostly related to the early recognition of the lesion in a stage when minor neurological deficits are still present. This was confirmed true, in this series, for every tumour type (ependymoma, astrocytoma, hemangioblastoma etc.) irrespectively for the longitudinal extension and the site of the tumours. The results with malignant ependymoma and astrocytoma have been discouraging because these tumours are not amenable to radical surgery and show a very rapid rate of growth.

Malis reported poor results in his series of operated astrocytoma[10] and Stein referred that postoperative neurological deterioration occurred in about 21% of 31 operated patients[11]. In spite of the encouraging results which have been reported in paediatric series[4] surgery of intramedullary tumours should still be considered as a challenging field for neurosurgeons. Probably the introduction of MRI will allow an early diagnosis of these lesions and the use of new surgical instruments (such as CO_2 surgical laser and ultrasonic surgical aspirator) will further minimize the surgical trauma improving the surgical results.

References

1. Cushing H (1905) The special field of neurological surgery. Bull Hopkins Hosp 16: 77–85
2. De Sousa AL, Kalsbeck JE, Mealey J Jr, Campbell RL, Hockey A (1979) Intraspinal tumours in children: a review of 81 cases. J Neurosurg 51: 437–445
3. Elsberg CA (1925) Tumours of the spinal cord and the symptoms of irritation and compression of the spinal cord and nerve roots: pathology, symptomatology, diagnosis and treatment. Paul B Hoeber, New York
4. Epstein F, Epstein N (1981) Surgical management of holocord intramedullary spinal cord astrocytomas in children: report of three cases. J Neurosurg 54: 829–832
5. Garrido E, Stein BM (1977) Microsurgical removal of intramedullary spinal cord tumours. Surg Neurol 7: 215–219
6. Gowers WR, Horsley V (1963) A case of tumour of the spinal cord: removal, recovery. In: Wilkins RH (ed) Neurosurgical classics XI. J Neurosurg 20: 815–824
7. Greenwood J Jr (1963) Intramedullary tumours of the spinal cord: a follow-up study after total surgical removal. J Neurosurg 20: 665–668

8. Guidetti B, Fortuna A (1967) Surgical treatment of intramedullary hemangioblastoma of the spinal cord. Report of 6 cases. J Neurosurg 27: 530–540

9. Horrax G, Henderson DG (1939) Encapsulated intramedullary tumour involving the whole spinal cord from medulla to conus: complete enucleation with recovery. Surg Gynec Obstet 68: 814–819

10. Malis LI (1978) Intramedullary spinal cord tumours. Clin Neurosurg 25: 512–540

11. Stein BM (1979) Surgery of intramedullary spinal cord tumours. Clin Neurosurg 26: 529–542

12. Yaşargil MG, Antic J, Laciga R, de Preux J, Fideler RW, Boone SC (1976) The microsurgical removal of intramedullary spinal hemangioblastomas: report of twelve cases and a review of the literature. Surg Neurol 6: 141–148

Address for correspondence: Dr. M. Fornari, Reparto di Neurochirurgia, Istituto Neurologico "C. Besta", Via Celoria 11, 20133 – Milano, Italy.

Acta Neurochirurgica, Suppl. 43, 9–12 (1988)

Microsurgical Removal of Intramedullary Spinal Cord Tumours: Report of 22 Cases

T. Yasui, A. Hakuba[1], J. Katsuyama[1], and S. Nishimura[1]

Department of Neurosurgery, Baba Memorial Hospital, Osaka, Japan, [1] Department of Neurosurgery, Osaka City University Medical School, Osaka, Japan

Summary

The management of patients with intramedullary spinal cord tumours is controversial. The authors have treated these tumours in accordance with the principles of radical resection, using microsurgical techniques. There have been 22 cases of intramedullary spinal cord tumours (4 astrocytomas, 3 glioblastoma multiformes, 5 ependymomas, 7 haemangioblastomas, 1 dermoid, 1 lipoma and 1 schwannoma) thus treated at our department during the past 13 years. Sixteen were adults and six were children. The early and late results of their surgical treatment have been analyzed, with follow-up periods ranging from 10 months to 12 years. Of the 22 patients, total removal was performed in 16, subtotal removal in 5 and biopsy in 1. Guidetti's criteria were adopted in assessing the results of surgery. Long-term results were as follows: 8 very good, 5 good, 3 fair, 3 poor and 3 deceased. The 3 latter cases were glioblastoma multiformes. Gross total removal of intramedullary tumours is technically feasible and compatible with neurological recovery. Since the majority of intramedullary tumours are benign and their course may span not only years but decades, long-term follow-up is essential.

Keywords: Intramedullary tumours; spinal cord tumours; microsurgery; radiotherapy; outcome.

1. Introduction

The optimum treatment of patients with intramedullary spinal cord tumours is unclear. In the past, such tumours were often treated with biopsy or subtotal removal followed by irradiation – a therapy usually associated with early tumour recurrence and progressive neurological impairment. In an attempt to improve on the outcome for patients with intramedullary tumours, the authors have treated these tumours in accordance with the principle of radical resection, using microsurgical techniques. We report here our results in 22 consecutive patients who underwent surgery over the last 13 years at the hands of one neurosurgeon (AH), using similar microsurgical techniques.

2. Summary of Cases (Table 1)

2.1. Patient Population

Twenty-two patients underwent this surgery during the last 13 years. There were 10 men and 12 women, ranging in age from 1 to 59 years (median age 30 years). Sixteen were adults, six were children. Signs and symptoms, usually related to tumour location, consisted of backache, paraesthesia, and sensory and motor deficits below the level of the tumour. Unusual symptoms included scoliosis (cases 2, 5), subarachnoid haemorrhage due to rupture of an aneurysm of a tumour-feeding artery (case 17) and a skin dimple (case 20). One case (case 8) had 3 spinal cord tumours, *i.e.*, 1 ependymoma and 2 meningiomas, with no evidence of neurofibromatosis. The mean duration from onset of symptoms to admission to our institute was 42 months; however, there was considerable variation, according to tumour type. Three patients (case 2, 4, 21) had had previous surgery (not radical resection).

2.2. Diagnostic Evaluation

X-ray evaluations of the early cases were via plain film, tomography and iophendylate myelography. After the introduction of computed tomography (CT) to our hospital in 1979, metrizamide myelography and metrizamide CT became the main diagnostic procedures. In 1984 a 0.5 superconductive magnetic resonande imaging (MRI) system was installed in our clinic; from that time MRI became the most useful diagnostic tool. 8 patients (Cases 4, 6, 10, 11, 12, 18, 19, 22) underwent MR examination. This diagnostic procedure can accurately localize spinal cord tumours and cystic cavities above and below the tumours without the need for introducting contrast material into the

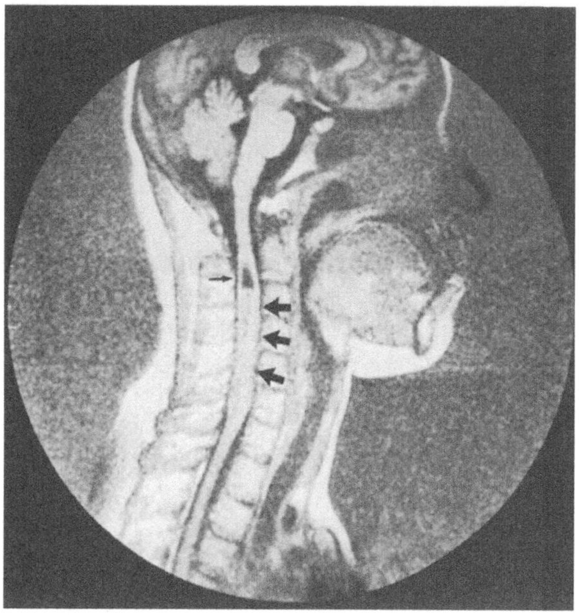

Fig. 1. Magnetic resonance image of case 10. The spinal cord is diffusely widened (large arrows); a cyst is also seen (small arrow)

subarachnoid space (Fig. 1). When haemangioblastoma is suspected, spinal angiography is also indicated.

2.3. Operative Technique

In principle, patients with lesions above the Th 3 level were operated on in the sitting position and those with lesions below the Th 3 level were operated upon in the crouching position. Semi-prone or park-bench positions were also adopted in some cases. A relatively wide laminectomy, using a high-speed airdrill, was performed over the area of suspected solid tumour, with no instrument placed between the lamina and the dura. In the children, osteoplastic laminotomy was performed to prevent postoperative spinal deformity.

The spinal cord was opened in the posterior raphé, using a razor blade or diamond knife, under the higher magnifying power of the microscope.

After the tumour was encountered, raphé dissection was continued in the longitudinal plane over the extent of the tumour. The pia mater at the edge of the spinal cord incision was sewn to the dura mater with monofilament nylon threads (8–0) to gently keep the cord incision open. Recently, we have used sensory-evoked potential monitoring during radical tumour removal and, in actual tumour removal, the Cavitron Ultrasonic Surgical Aspirator (CUSA). Use of the CUSA, however, was limited to internal decompression only after isolation of the tumour from the neural structure; we could not use it in work very close to the ill-defined margin of these tumours for fear of injuring the normal tissue. A Malis bipolar coagulator and cutter system (10–15 output power) was generally used for tumour removal, combined with Systec for irrigation with normal saline solution and simultaneous suction to provide a clear margin of dissection and reduce the thermal effect upon normal tissue. A watertight dural closure was generally possible. If necessary, fascia lata was used as a dural graft.

2.4. Radiotherapy

Radiotherapy was used in 7 cases (cases 1, 3, 5, 6, 7, 8, 9). All the early glioma cases were given postoperative radiotherapy regardless of histological diagnosis and extent of operative procedure. Recently, however, radiotherapy was recommended only for those presumed to have malignant tumours.

2.5. Results

The pathological diagnoses of specimens removed in surgery at our institution are given in Table 1. Reports of intramedullary schwannoma are extremely rare in the literature; we experienced only one case (case 22). In most cases an attempt was made to achieve total removal. Determination of completeness of removal was estimated by the operating surgeons. We have analyzed the early (about one month follow-up) and late (current status) results of this surgical treatment, with follow-up periods ranging from 10 months to 13 years (mean 58 months). Guidetti's criteria were adopted for assessing the surgical results. Each patient's current clinical status was assessed by outpatient neurological examination, communication with referring physicians, or by telephoning the patient. Of the 22 patients, total tumour removal was performed in 16, subtotal removal in 5 and biopsy in 1. Early results were as follows: 5 very good, 5 good, 2 fair, 9 poor, and 1 deceased. Six cases (cases 1, 3, 6, 8, 9, 10) were worsened; all of these are gliomas. However, even the unchanged or aggravated cases gradually improved as time went by; the late results were as follows: 8 very good, 5 good, 3 fair, 3 poor, and 3 deceased. The three latter cases were glioblastoma multiformes; they died 1 month (case 6), 6 months (case 7) and 5.3 years (case 5), respectively, following surgery and postoperative radiotherapy. The operative results in the 16 cases of total removal were assessed. Early results were as follows: 4 very good, 5 good, 2 fair and 5 poor. Late results were better: 7 very good, 5 good, 3 fair and 1 poor.

Table 1. *Summary of 22 Cases*

Case No.	Age (yr.) & Sex	Duration of Symptoms	Location	Pathology	Surgery	Preoperative state*	Results*		Remarks
							Early	Late	
1	17/M	1 yr.	medulla—C2	astrocytoma Grade2	total	fair	poor	good	
2	10/M	7mo.	medulla—Th3	astrocytoma Grade2	total	poor	fair	fair	previous operation
3	15/F	1 yr.	medulla—Th1	astrocytoma Grade2	subtotal	fair	poor	poor	
4	49/M	8yr.	C3—Th2	astrocytoma Grade2	total	good	good	good	previous operation
5	1/F	6mo.	medulla—Th2	glioblastoma multiforme	subtotal	poor	poor	death	
6	48/F	6yr.	C1—Th1	glioblastoma multiforme	biopsy	fair	death	—	
7	13/M	1 mo.	Th9—Th12	glioblastoma multiforme	subtotal	poor	poor	death	
8	14/M	1 yr. 6mo.	C4—Th2	ependymoma	total	fair	poor	good	multiple tumors
9	28/F	1 yr. 4mo.	C6—conus	ependymoma	total	fair	poor	fair	holocord tumor
10	38/F	3yr.	C3—C6	ependymoma	total	good	fair	very good	
11	52/M	7yr.	C1—C6	ependymoma	total	poor	poor	poor	
12	44/M	5yr.	C2—C5	ependymoma	total	poor	poor	fair	
13	26/M	4yr.	Th8	hemangioblastoma	total	fair	very good	very good	
14	29/F	4yr.	Th11, 12	hemangioblastoma	total	fair	very good	very good	
15	27/F	1 yr.	Th5—Th7	hemangioblastoma	total	poor	good	very good	
16	28/F	8mo.	C5, 6	hemangioblastoma	total	poor	good	very good	
17	37/M	?	medulla—C1	hemangioblastoma	subtotal	good	very good	very good	SAH
18	37/F	5mo.	medulla—C2	hemangioblastoma	total	good	very good	very good	
19	39/F	6yr.	medulla—C1	hemangioblastoma	total	fair	good	good	
20	1/F	1 yr.	conus	dermoid	total	very good	very good	very good	dermal sinus
21	54/F	20yr.	C6—Th6	lipoma	subtotal	poor	poor	poor	previous operation
22	59/M	3yr.	Th9—Th12	schwannoma	total	good	good	good	

* In assessing the pre-and postoperative states. Guidetti's classification was used.
yr. : years. mo. : months. SAH : subarachnoid hemorrhage

3. Discussion

It goes without saying, that the principle of treatment of the neoplasm is radical removal. However, in the case of intramedullary spinal cord tumours, this principle is not always correct, since the radical procedure may severely damage the normal cord and impart irreversible neurological deficits postoperatively.

Therefore, radiotherapy following a decompressive procedure used to be the recommended and time-honored treatment for most intramedullary tumours. However, the commitment to surgery as the primary treatment for such tumours was reaffirmed by Greenwood in his report[4] on the successful removal of 10 intramedullary tumours. The tumours were removed without mortality and without significant morbidity. He stressed the use of microinstrumentation and magnified vision in accomplishing the successful removal of these lesions. He also pointed out the difference between the astrocytoma and the ependymoma, the former not being reliably associated with a plane between tumour and spinal cord, whereas the latter was invariably associated with an identifiable interface between tumour and spinal cord. Subsequently there have been a number of reports emphasizing the appropriateness of microsurgery and the excellent results obtained by radical removal of these tumours[2, 3, 4, 6, 7].

We have treated such tumours in accordance with the principle of radical resection, using microsurgical techniques. A number of surgical authorities agree that ependymomas or other encapsulated tumours are potentially curable and that an aggressive approach at removing such tumours is appropriate[3, 4, 5, 6].

Nevertheless, these same authors believe that attempts to remove infiltrating tumours are not indicated. For example, Malis[6] questions the operability of spinal cord astrocytomas. Guidetti *et al.*[5] believe that radical removal of spinal cord astrocytomas is precluded, except for a few cases in which a clear cleavage plane can be found. In a relatively small series, Stein[7] achieved total removal of 9 of 11 intramedullary tumours. Of the 5 astrocytomas in his series, total removal was accomplished in 3. Even in the astrocytomas, he often observed a well-defined plane between neoplasm and spinal cord. We also performed total removal of astrocytomas in 3 of 4 cases. However, the intraoperative resectable plane is not a true cleavage between tumour and spinal cord. Histological examination of specimens

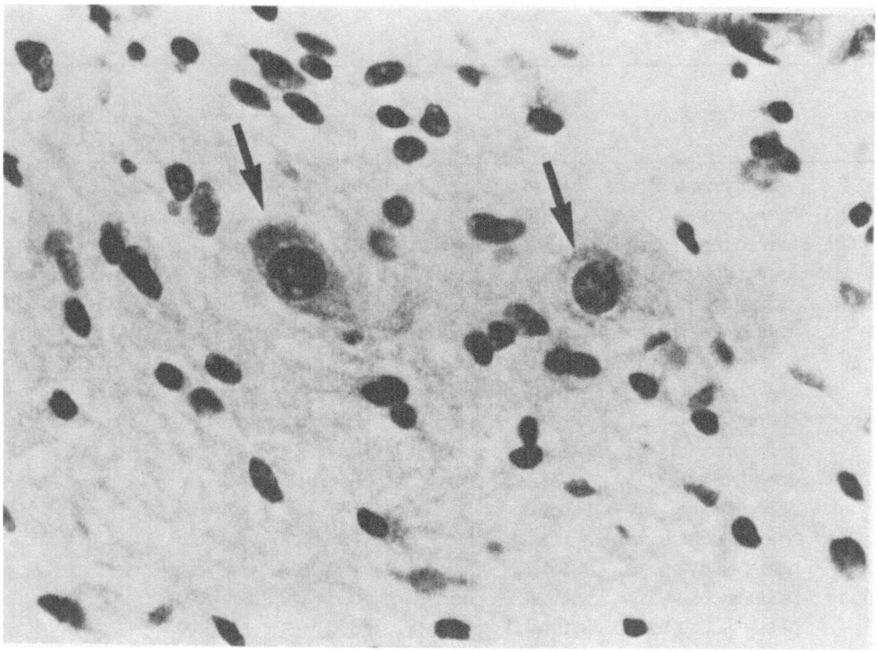

Fig. 2. Case 2, astrocytoma. Photomicrograph of specimen from an astrocytoma totally removed from the cervical spinal cord. The infiltrating nature of the tumour is characterized by the astrocytic tumour cells surrounding the anterior horn cells (arrows)

showed the inclusion of anterior horn cells surrounded by tumour (Fig. 2). Haemangioblastomas are usually well-demarcated, with no infiltration, and are regularly resectable. We have experienced 7 haemangioblastomas; with one exception (case 17), they were totally removed. The capsule of one dermoid (case 20) was lipomatous and partially adherent to the spinal cord. We performed total removal of the tumour with a gliotic layer 100 μ in thickness surrounding the capsule. Because lipomas are tightly adherent to the spinal cord and are not DNA-producing neoplasms, many authors have recommended subtotal removal as the treatment of choice[6]. We agree completely with this opinion. Intramedullary schwannomas are extremely rare. We experienced one case, located subpially, which was totally removed.

The effectiveness of postoperative radiotherapy remains controversial and indications therefore are empirical[1]. However, radiotherapy has been recommended as primary therapy for ependymomas and subependymomas.

References

1. Chigasaki H, Pennybacker JB (1968) A long follow-up study of 128 cases of intramedullary spinal cord tumours. Neurol Med Chir (Tokyo) 10: 25–66
2. Cooper PR, Epstein F (1985) Radical resection of intramedullary spinal cord tumours in adults. J Neurosurg 63: 492–499
3. Fischer G, Mansuy L (1980) Total removal of intramedullary ependymomas: follow-up study of 16 cases. Surg Neurol 14: 243–249
4. Greenwood J Jr (1963) Intramedullary tumours of spinal cord. A follow-up study after total surgical removal. J Neurosurg 20: 665–668
5. Guidetti B, Mercuri S, Vagnozzi R (1981) Long-term results of the surgical treatment of 129 intramedullary spinal gliomas. J Neurosurg 54: 323–330
6. Malis LI (1978) Intramedullary spinal cord tumours. Clin Neurosurg 25: 512–539
7. Stein BM (1979) Surgery of intramedullary spinal cord tumours. Clin Neurosurg 26: 529–542

Address for correspondence: T. Yasui, Department of Neurosurgery, Baba Memorial Hospital, 244 Higashi 4, Hamadera-Funao-Cho, Sakai, Osaka 592, Japan.

Acta Neurochirurgica, Suppl. 43, 13–16 (1988)

Magnetic Resonance (MR) Imaging in the Management of Primary and Secondary Syringomyelic Cavities, and of Other Cystic Lesions of the Spinal Cord

N. Roosen[1], **P. Dahlhaus**[1], **Ch. B. Lumenta**[1], **E. Lins**[1], **W. Stork**[2], **D. Gahlen**[2], and **W. J. Bock**[1]

[1] Neurosurgical University Clinic, Düsseldorf, and [2] Institute of Roentgenology "Grafenberger Allee", Düsseldorf, Federal Republic of Germany

Summary

During the past three years 27 patients with medullary and/or brainstem cavities were examined with MRI at the Institute of Roentgenology, and we report the radiological MRI correlative findings. Five patients were operated: postoperative MRI examinations were compared with preoperative studies.

Syringomyelia was present in 21 cases, syringobulbia in 2, and syringobulbomyelia in 4. Additional pathology such as the Arnold Chiari malformation or hydrocephalus was found at the craniospinal junction in 8 patients. Two instances of associated intraspinal tumours were noted too. In all cases the syringomyelic cavities involved the cervical cord and in most of them the lesion extended down to the thoracic or even lumbar level. Several patients were shown to have multilocular cavities.

In the patients with syringomyelia a syringoarachnoidal shunting procedure was performed. This resulted in cyst collapse as proved by postoperative MRI examinations; neurological symptomatology was ameliorated too.

Conventional radiological imaging with computed tomography (CT) and myelography was not as efficient as MRI in imaging syringomyelia. The extent of intramedullary cavities as well as the presence of associated pathology is best appreciated on sagittal MRI. The use of special surface coils clearly results in better image quality due to higher resolution.

According to our experience MRI should be the primary examination in syringomyelia; myelography and CT are obsolete in the management of syringomyelia; the insertion of a syringoarachnoidal shunt is a convenient surgical procedure, resulting in clinical amelioration in most cases.

Keywords: Syringomyelia; syringobulbia; MRI; syringoarachnoidal shunt; results.

Introduction

The diagnostic work-up of patients with spinal cord disease has been influenced considerably by the advent of magnetic resonance (MR) imaging[6]. Particularly spinal cord cavities are detected more frequently than previously, and their evaluation is facilitated by MR to a large extent[10].

In this paper patients with spinal cord cavities diagnosed with MR, are reviewed. The MR findings are discussed, and attention is given to those patients, who have been studied both pre- and postoperatively.

Methods

A retrospective search of the patient files of the Institute of Roentgenology "Grafenberger Allee" was done. All patients, in whom spinal cord cavities had been diagnosed by MR, were identified, and their clinical, radiological, and MR data were analyzed. Because several hospitals and many practicing physicians are referring patients for MR to the Institute, only a minority of cases have been seen primarily at the Neurosurgical University Clinic of Düsseldorf.

The MR studies were performed with a Diasonics MT/S^R imager, which has a superconductive magnet operating at 0.35 T. All studies were obtained with spin-echo pulse sequence (SE) using different repetition and echo-delay times (TR resp. TE). The earlier investigations (1984, at the beginning of our experience) were done with the head and body coils for upper cervical resp. cervicothoracolumbar imaging, but soon surface coils were available, which clearly improved imaging results. Slice thickness also could be reduced in order to achieve spinal cord MR scans of higher quality.

Results

This series comprises 27 patients, in whom MR had revealed spinal cord cavities. There were 14 women and 13 men. The average age was 49 years (range 16 to 73 years).

In six patients there was an associated intrinsic tumour of the spinal cord, while one woman had been operated for a lower thoracic meningioma 16 years earlier. The associated intrinsic spinal cord tumours were identified at operation and were low grade astrocytic gliomas, ependymomas, and one haemangioblastoma. In four cases the tumours were located at the caudal end of the spinal cord cavities, *i.e.* in the lower thoracic or in the thoracolumbar region; the two instances of cervical tumours were cystic gliomas.

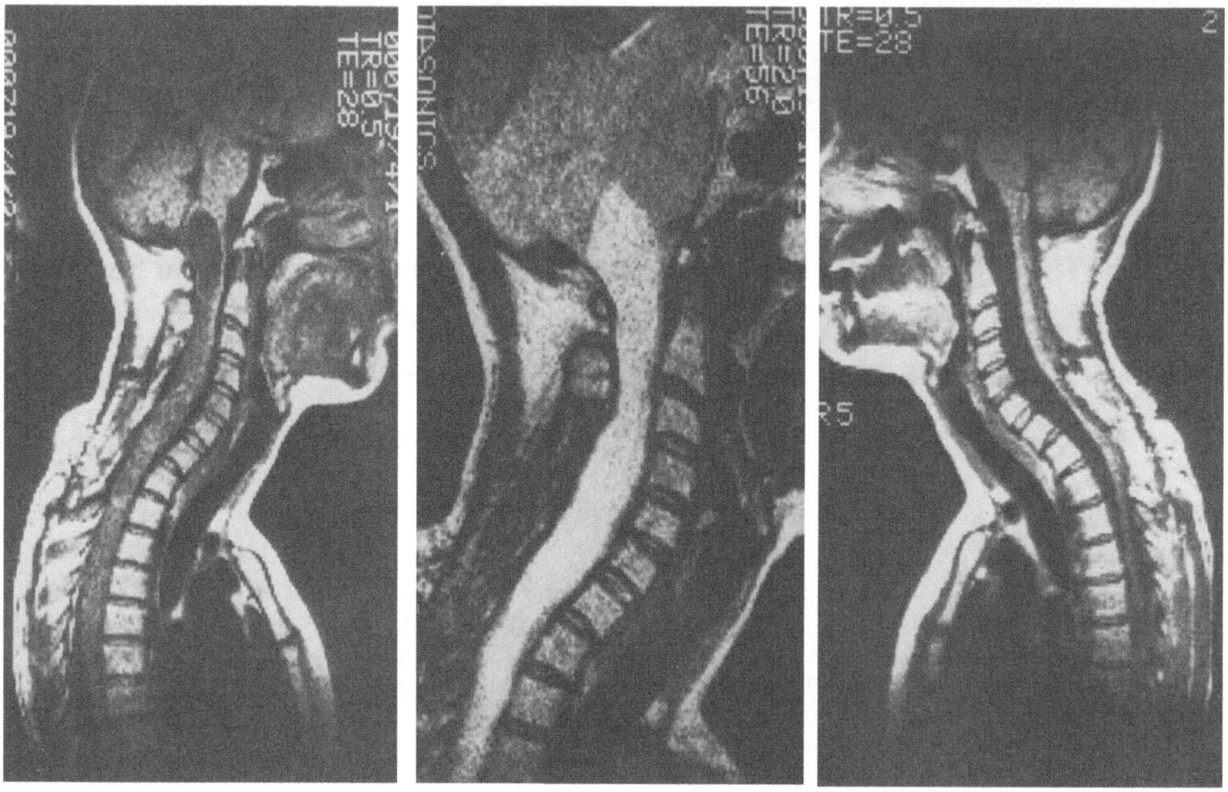

A B C

Fig. 1 A–C. Case 1: A and B: Preoperative MR shows a large, partially cystic lesion, extending from the lower brain stem to approximately the Th_2 level. The upper part of the lesion, which is very hypointense on SE 500/28 (A) and slightly less hyperintense than the remaining part of the lesion on SE 2000/56 (B), was demonstrated intraoperatively to be a cyst. The lower part of the lesion was gliotic with a nodule of recurring ependymoma at the C_6–C_7 level. C: The postoperative MR (SE 500/28) reveals the thin atrophic spinal cord, which shows a slight hyperintensity suggesting some gliotic reaction. The cyst remained collapsed

Although in most cases one single large cavity was seen in the spinal cord, we could detect multiple cavities in eight patients. However it was difficult, to decide on the basis of MR if there were complete septa between these cavities or if these were communicating cavities.

The majority of spinal cord cavities were seen involving the cervical medulla, the upper end usually being localized at the C_1–C_3 level. In two cases the cavity extended upwards within the lower brain stem, so called syringobulbia. Only in one case the cranial end of the syrinx was found as low as the upper thoracic cord (Th_1). Seven cavities were wholly confined to the cervical cord, whereas in 18 patients the syrinx could be traced down to the thoracic cord, regularly ending at the upper or middle thoracic level.

Hydrocephalus was found in only five patients. A type I Arnold-Chiari malformation was present in 11 cases: the prolongation of the cerebellar tonsils beneath the foramen magnum was identified five times.

A total of 11 patients have been operated at the Neurosurgical Clinic of the University Hospital Düsseldorf. In all cases a syringo-subarachnoidal shunt has been inserted after laminectomy. These included three tumour patients, in whom only a partial tumour removal had been done, and eight non-neoplastic syringomyelic cavities. In nine patients MR showed the syrinx to be collapsed postoperatively; in two the diameter remained the same. The immediate clinical postoperative results did not seem to be related to the MR findings. One of the two patients with unchanged syrinx diameter had a stable neurological condition, whereas the other one showed slight improvement. Among the patients with a reduced diameter of the spinal cord cavity, there were two without improvement and two with some neurological deterioration. The results of long-term follow-up (6 months–> 1 year) are not yet all available, and will constitute the subject of a subsequent report. However, a preliminary survey revealed

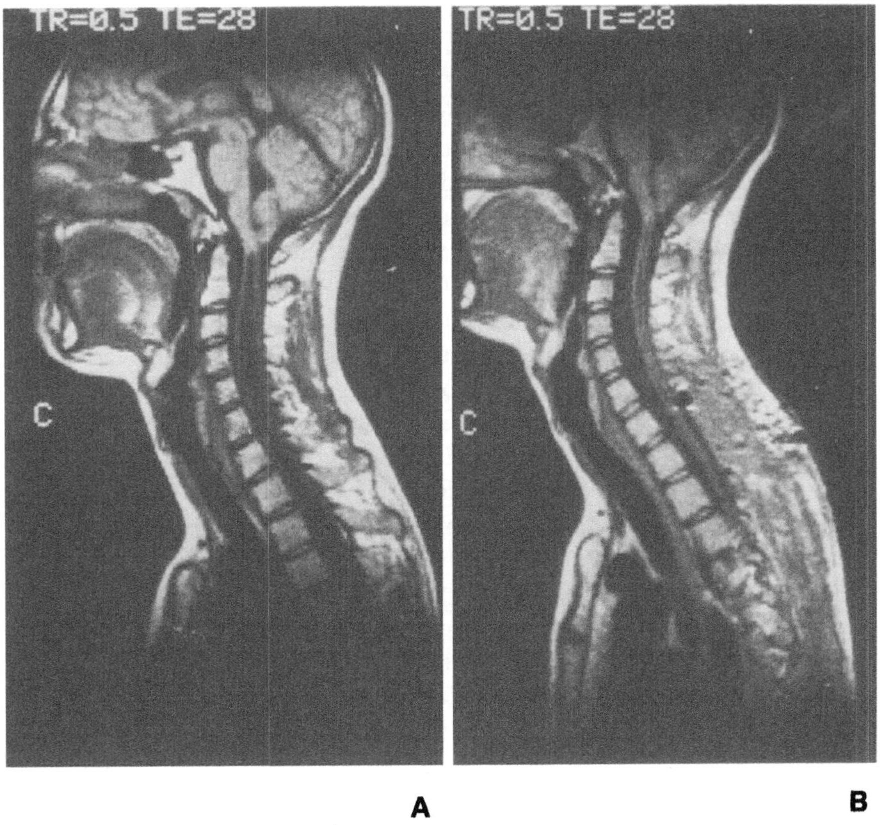

Fig. 2 A, B. Case 2: A: Preoperative MR (SE 500/28) clearly shows syringomyelia from the C_1 level down to the thoracic region. There is no evidence of hydrocephalus, but the cerebellar tonsils are found below the foramen magnum level. B: The postoperative MR visualizes the laminectomy site, the level of syringo-subarachnoidal shunting, and the collapsed syrinx

both a persisting reduction of syrinx diameter as well as a return to the preoperative syrinx diameter.

Illustrative case reports

Case 1: A 38-year-old woman had been operated for an intramedullary tumour, extending from C_4 to Th_2. Neuropathological examination had disclosed an ependymoma. Four years later an MR investigation was done because of a newly progressive syndrome of upper cervical cord and lower brain stem dysfunction (Fig. 1 A and B). On SE 500/28 imaging a hypointense thickening of the spinal cord and lower brain stem can be seen, which remains slightly hypointense on SE 2,000/56 imaging too. The spinal cord at the C_2 to C_7 level is also thicker than normal, and somewhat hypointense (SE 500/28). SE 2,000/56 shows a bright intensity. The lesion at the level of the foramen Monroi was surgically proven to be a cyst, whereas the remaining cervical cord was gliotic, but contained an area of recurrent tumour measuring approximately 4 cm. Six months later a routine follow-up MR was obtained (Fig. 1 C). The cyst could not be seen anymore. The spinal cord was thin and atrophic. A slight hyperintensity, which was heterogeneous and diffuse, can possibly be correlated with gliotic scar tissue present after operation.

Case 2: A 28-year-old man was admitted because of a slowly progressive myelopathic syndrome. Computed tomography (CT) and myelography suggested syringomyelia but were not conclusive. Therefore, MR was obtained (Fig. 2 A). A large cavity was seen within the spinal cord, extending from C_1 level towards the upper thoracic cord. There was no evidence of a connection between this cavity and the ventricular system. No septa were detected within the syrinx. A cervicothoracic laminectomy was performed, and a syringo-subarachnoidal shunt inserted. Short-term follow-up revealed no further progression of symptoms, but no amelioration either. The syrinx, however, was demonstrated to be collapsed (Fig. 2 B).

Discussion

Although the occurrence of cavities within the spinal cord and/or brain stem has been considered as relatively rare[4], modern neuroimaging, in particular MR, seems to disclose this pathological condition more frequently[10]. Intramedullary cavities may develop according to different pathogenetic mechanisms and may present in various ways (a detailed review of these hypotheses is not the subject of this work, and the interested reader is referred to the literature, *e.g.*[1, 3]). Some cases are associated with trauma, haemorrhage, arachnoiditis, of tumours, both intrinsic as well as extrinsic tumours of the spinal cord[8]; whereas in other patients there is an idiopathic syringomyelia[8]. Sometimes a dilatation of the central canal is found (hydromyelia), and on the other hand, cavities may exist which extend lateral or parallel to the central canal (syringomyelia)[9]. These cavities may not always be separated clearly, and the term syringohydromyelia is often used.

Until recently, myelography and CT-myelography have been the investigations of choice for the radiological work-up of patients suspected of having syringomyelia[5]. Central cord enhancement is a specific and sensitive sign, and change in cord diameter is specific too[5]. Nevertheless, definite conclusions about the presence or absence of syringomyelia, and about the longitudinal extent of the disease may be difficult to reach; the situation in spinal cord gliomas is even worse.

With the use of MR, however, exactly sagittal slices of the spinal cord may be obtained[6]. This allows us to evaluate the full extent of cystic lesions of the spinal cord, and to look for additional pathology such as tumours[6,10]. Our series demonstrates that we can define the level of syrinx formation, its diameter, the atrophy of the spinal cord, the gliotic reaction, the presence of septa within the syrinx, and the obliteration of the subarachnoid space. Additional pathology such as tumours or malformations, may be identified, eventually by using paramagnetic substances such as Gd-DTPA[7].

Different operative procedures have been proposed to treat syringomyelia, according to the hypothesized pathogenetic mechanism of syrinx formation[1]. The postoperative situation of these patients can be assessed perfectly by MR[3], which is demonstrated in our series too. Results of long-term follow-up by MR after syringosubarachnoidal shunting are not yet available in the literature, but some communications at recent meetings seem to suggest that not rarely the syrinx diameter tends to increase again 6 to 12 months after operation. Preliminary findings in our own series are suggestive of this evolution after syringosubarachnoidal shunting too. However, this morphological result is probably not correlated with clinical findings. A complete evaluation of clinical results of surgery for syringomyelia is outside the scope of this report. The clinical course of syringomyelia may be extremely variable[2], and therefore, a large scale study with cooperation between several neurosurgical and neurological centers is mandatory to evaluate surgical therapy. It is a great promise for medicine, that, according to our experience and literature findings, MR now offers ideal imaging possibilities for such an enterprise.

References

1. Aboulker J (1979) La syringomyélie et les liquides rachidiens. Neurochirurgie 25 [Suppl] 1: 1–144
2. Anderson NE, Willoughby EW, Wrightson P (1985) The natural history and the influence of surgical treatment in syringomyelia. Acta Neurol Scand 71: 472–479
3. Barkovich AJ, Sherman JL, Citrin CM, Wippold FJ II (1987) MR of postoperative syringomyelia. AJNR 8: 319–327
4. Barnett HJM, Foster JB, Hudgson P (1973) Syringomyelia. Major Probl Neurol 1. WB Saunders Co, Philadelphia
5. Gates PC, Fox AJ, Barnett HJM (1986) CT metrizamide myelography in syringomyelia: sensitivity and specificity. Neurology 36: 1245–1248
6. Han JS, Kaufman B, El Yousef SJ, Benson JE, Bonstelle CT, Alfidi RJ, Haaga JR, Yeung H, Huss RG (1983) NMR imaging of the spine. AJR 141: 1137–1145
7. Kiwit JCW, Lanksch WR, Fritsch H, Lins E, Stork W, Roosen N, Schirmer M, Bock WJ, Marguth F: Magnetic resonance tomography of solid spinal cord tumours with extensive secondary syringomyelia. Adv Neurosurg 16: 211–215
8. Kiwit JCW, Schober R, Gräfin Vitzthum HG, Wechsler W (1986) Neuropathological aspects of the syringomyelic complex. In: Voth D, Glees P, Lorber J (eds) Spina bifida – neural tube defects. Walter de Gruyter, Berlin New York, pp 73–80
9. Larroche JC (1984) Malformations of the nervous system: syringomyelia and syringobulbia. In: Adams HJ, Corsellis J, Duchin EW (eds) Greenfield's neuropathology, 4th ed. Wiley & Sons, New York, pp 396–402
10. Sherman JL, Barkovich AJ, Citrin CM (1987) The MR appearance of syringomyelia: new observations. AJR 148: 381–391

Address for correspondence: Dr. N. Roosen, Neurochirurgische Klinik der Universität Düsseldorf, Moorenstrasse 5, D-4000 Düsseldorf 1, Federal Republic of Germany.

Acta Neurochirurgica, Suppl. 43, 17–21 (1988)

Syringomyelia: How MRI Aids Diagnosis and Management

M. Powell

The National Hospitals for Nervous Diseases, London, U.K.

Summary

MRI is an invaluable tool in the diagnosis and assessment of treatment of syringomyelia. It also confirms the theory of the pathogenesis of the disease and suggests why some operations on the condition fail.

A series of patients with Chiari I, with and without syringomyelia, have been studied with MRI both prior to surgery and following treatment. In four cases, with Chiari I alone and minor symptoms only, management has been conservative. No progression of symptoms or signs have been seen, MRI allowing syringomyelia to be excluded, both at diagnosis or development during follow-up. Patients with Chiari I, six with and one without syringomyelia, were split into two groups; those with symptoms and signs of cervico medullary compression (CMC) and those with pure cord amyotrophy with upper limb weakness and numbness of typical "cape type". In three CMC patients, foramen magnum decompression (FMD) with fascia lata grafting and high syringotomy, and in two FMD without syringotomy, were carried out. The procedure improved pain, CMC symptomatology and clinical signs, but had only minor effects on amyotrophic symptoms. Some improvement was seen after surgery in distention of the syrinx on MRI studies, but there was no progression in the syrinx cavities. In two patients with pure amyotrophic the syrinx cavities. In two patients with pure amyotrophic symptoms, primary syringoperitoneal shunting (SPS) was carried out, and a further two patients with primary FMD subsequently went on to SPS for amyotrophic symptoms and signs. A Spetzler type system was used in the high thoracic region in all four cases. Brief symptomatic clinical improvement resulted from SPS in three of the four cases but this was not sustained and in one, symptoms developed in the segment above which remains distended on MRI.

MRI shows that below the shunt tubing in the cord, the cavity collapses, but above the shunt, the cavity remains distended. It not only supercedes CT myelography by being non-invasive and having superior imaging, but also demonstrates loculi and segmental decompression caused by surgery. Possible mechanisms are discussed.

Keywords: Syringomyelia; Chiari malformation; MRI; treatment; results.

Introduction

Advances in myelography of the last two decades and subsequently in CT myelography have considerably improved the accuracy of the radiological diagnosis of syringomyelia. However, the invasive nature of the investigation has not allowed the same progress in management and follow-up as it is not acceptable to repeatedly submit patients to CT myelography.

MRI can replace CT myelography, with better images of the spinal cord and greater detail of cavitation especially in the saggittal plane. Moreover, being non-invasive it can be used both in the initial diagnosis and in follow-up studies, both to assess progress and the effect of surgical treatment.

Although MRI aids in the management of both spinal tumours and the late effects of trauma, this paper concentrates on syrinx associated with craniocervical abnormalities, in particular cerebellar tonsillar herniation, the Chiari type I abnormality – hydrosyringomyelia. The signs and symptoms of Chiari I and syrinx are astonishingly varied. They fall into two main groups: those with either bulbar or cervico-medullary compression, or both, and those with symptoms or signs of cord cavitation, *i.e.* suspended sensory loss, limb muscle wasting and limb dysaesthetic pain.

The management of Chiari I and syrinx falls into two major categories; 1) foramen magnum decompression (FMD) with or without various procedures in the vicinity or, 2) syrinx shunting. Most authorities tend to recommend one procedure or the other, however, it seems that the former best treats the symptoms of bulbar or cervico-medullary compression, whilst the latter treats the symptoms of cord cavitation.

This paper discusses the short term results of management based on this thesis, and in particular, how MRI helps in the decision making process.

Management and Discussion

Patients

Two series of patients are presented in which MRI was essential either in diagnosis, management or both.

Table 1

Age/sex	Symptoms	Length	Objective neurology
82 F	Patchy glove and stocking sensory loss. Low back pain	2 years	None
36 F	Hemisensory loss to pain and light touch. Headache	8 months	None
43 F	Neck pain and headache	15 months	None
48 F	Retrobulbar neuritis. Personality change	3 years	Abnormal VER's ↑ plantars

Conservative Management Group

Four women aged between 32–48 presented with vague non-specific, symptomatology to the neurological services. During the course of investigation the Chiari type I malformation was discovered in all four and the patients were referred for neurosurgical advice. In all four, signs of cervico medullary compression and cord cavitation were absent. The clinical details are found in Table 1.

In these patients, MRI confirmed the diagnosis of mild tonsillar herniation (Fig. 1) but excluded cord cavitation. All were advised that although they had a minor abnormality, the symptoms were not obviously referable to it and that conservative treatment was indicated. During the follow-up period no change has been noted in symptomatology other than for the better and there has been no change in their neurological signs if present.

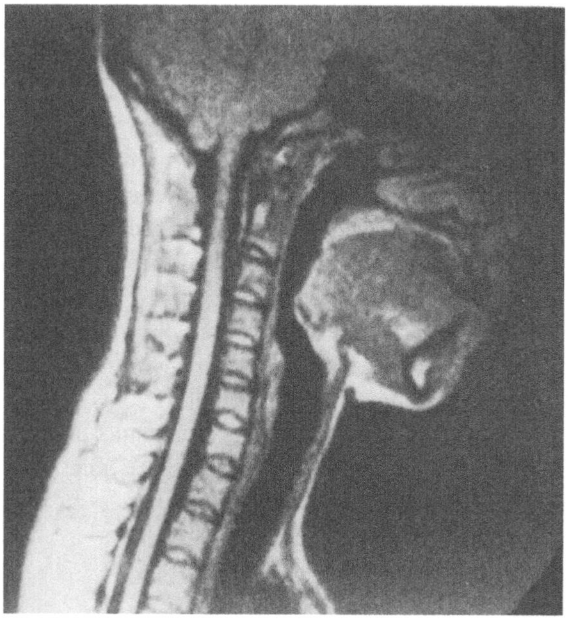

Fig. 1. MRI showing mild tonsillar herniation but no syrinx in patient with nonspecific symptoms and no abnormal signs

Mild tonsillar herniation is now seen as a fairly frequent incidental finding on MRI (15–20% personal communication B. Kendall & J. Stevens). Neurosurgeons must be wary of attributing vague symptoms to physical abnormalities and in particular acting on them.

Operated Group

There were seven patients with Chiari I abnormality, six of whom have syringomyelia. All had neurological signs either referable to craniocervical compression or the cord cavitation. All were investigated with MRI. Patient details are found in Table 2.

Management was directed towards symptoms: those with bulbar or long tract signs underwent FM decompression first, regardless of cord cavitation symptoms, and those with "pure" cavitation symptoms underwent primary syringo-peritoneal (SP) shunting.

In patients 1 and 2, the major symptoms related to the cervico-medullary compression (Fig. 2 a), and the initial management was with FM decompression, dural grafting with fascia lata and high cervical syringotomy. Patient 1 has remained stable symptomatically with mild improvement in tone and reflex changes on examination. Follow-up MRI shows some collapse of the syrinx (Fig. 2 b). In patient 2 following surgery, the symptoms of bulbar dysfunction disappeared and his long tract signs and symptoms improved. Nevertheless his cavitation symptoms progressed. Follow-up MRI (Fig. 3 a) showed continuing distention of his syrinx and he underwent syringo-peritoneal shunting from the mid-thoracic region. An improvement was noted in some of his lower segment signs, but he developed new symptoms in an adjacent higher cord level.

A third MRI (Fig. 3 b) showed collapse of the syrinx cavity below the shunt tubing, but distention above. There was no change on four months follow-up. A second SP shunt was placed into this higher cavity allowing a minor neurological improvement.

Table 2

Patient	Age	Sex	Symptom	Signs	Duration	Operation
1	34	F	Scoliosis Stiff Legs Right hemisensory loss	Right hemisensory loss ↑ Tone lower limbs ↓ Power legs	14 years	1. FMD
2	19	M	Difficulty swallowing and speaking Numb weak right hand Numb left trunk	Nystagmus Cape Sensory loss r. hand Paralysed intrinsic r. hand Sensory loss left trunk Brisk reflexes, ↑ Tone lower limbs	1 year	1. FMD 2. SP shunt 3. SP shunt
3	22	M	Weak numb Rt. hand	Cape distribution sensory loss Rt. arm. Poor intrinsics rt. hand Normal lower limbs	1 year	1. SP shunt 2. Revision of shunt 3. Removal of shunt
4	36	F	Pain in spine Numb trunk	Rt. thoracic abdominal sensory loss	18 months	1. SP shunt
5	61	F	Swallowing difficulty Weak numb arms Weak legs	Nystagmus. Swallowing diff. Cape suspended sensory Rt. arm ↑ Tone lower limbs	6 years	1. FMD 2. SP shunt
6	33	F	Hoarse Voice Swallowing Difficulty	Nystagmus. Absent gag Paralysed cords Disordered Swallowing and ↑ limb tone	9 months	1. FMD
7	67	F	Hoarse Voice. Neck pain	Poor gag. Disordered swallowing. Painful Neck	5 years	1. FMD

FMD = Foramen Magnum Decompression, SP Shunt = Syringo Peritoneal Shunt.

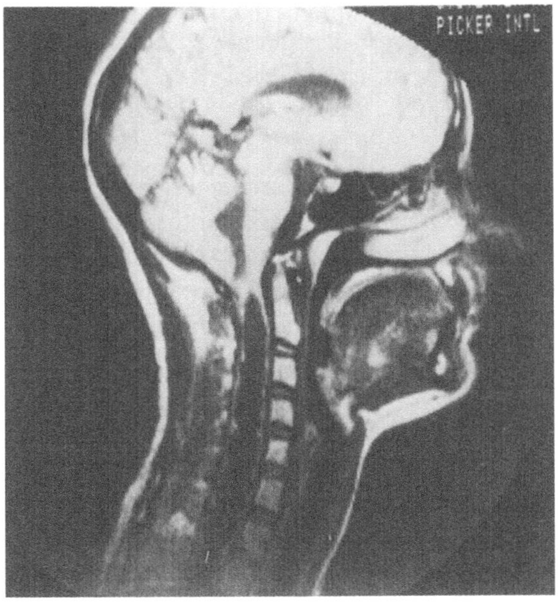

Fig. 2a. MRI of Chiari I and syrinx before Foramen Magnum Decompression

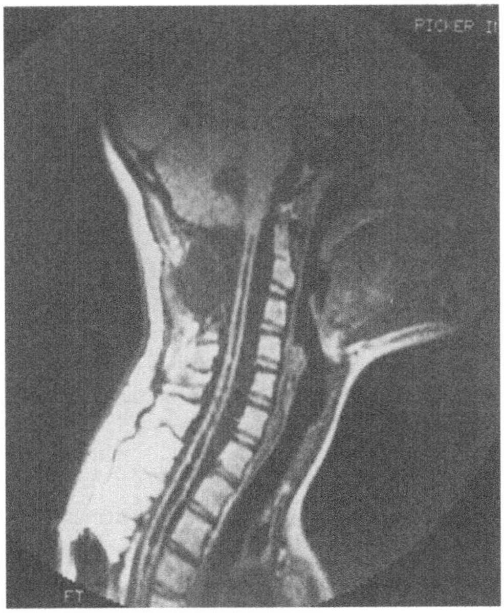

Fig. 2b. MRI of Fig. 2a following surgery. Syrinx collapsed and neurological deterioration halted

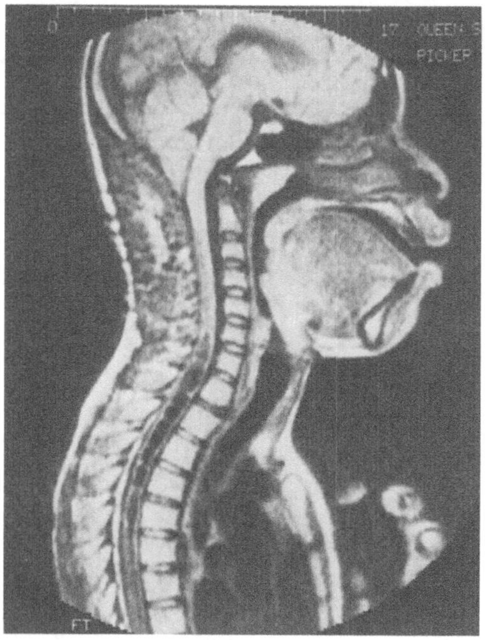

Fig. 3 a. MRI of severe Chiari I and syrinx following Foramen Magnum Decompression. Note continuing distention of syrinx cavity

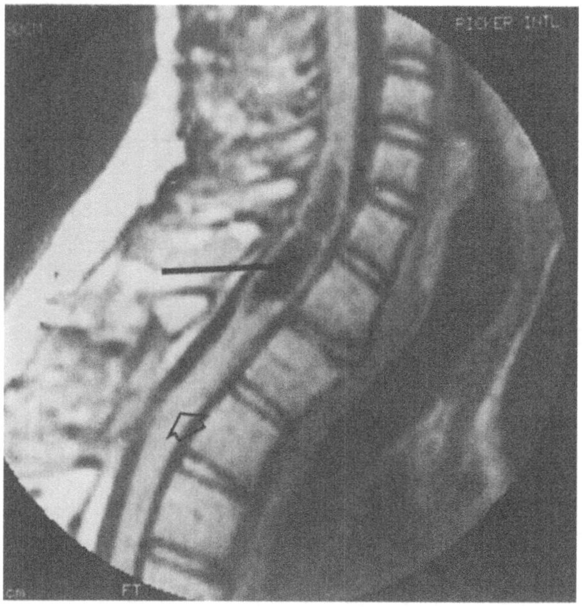

Fig. 3 b. MRI of Fig. 3 a following syringo-peritoneal shunt. Note collapse of cavity below shunt tubing (black arrow) but continuing presence of cavity above (black line)

Patients 3 and 4 both had cavitation symptoms with neither symptoms nor signs referable to the cervicomedullary compression. Consequently they both underwent primary SP shunting (as suggested by Barbaro *et al.* 1984). Patient 3 had an immediate improvement in symptoms and signs, but developed late blockage of the shunt tubing. Following revision he developed a low grade infection and the shunt tubing had to be removed. A long recovery period followed but he had regained preoperative neurological status when he recently developed signs of an ascending syrinx cavity. Patient 4 whose symptoms were mild has not changed following SP shunting apart from a minor wound discomfort. Follow-up MRI shows collapse of the syrinx below the tubing, but continuing mild distention above.

Patient 5 had complex symptomatology including bulbar symptoms despite foramen magnum decompression and subsequent SP shunting. She has remained neurologically stable, but her assessment is complicated by depression made worse by major and insurmountable social problems. Her follow-up MRI shows collapse below the tubing and a continuing cavity above it, similar to patient 4.

Patient 6 presented with signs of bulbar dysfunction but surprisingly no long tract or cavitation signs despite a major syrinx cavity from the C 3 region downwards demonstrated by MRI. Her early postoperative assessment shows little change in her complete glossopharyngeal and vagal palsy, although she is symptomatically otherwise well.

Patient 7 had tonsillar herniation alone, with bulbar signs referrable to this, as well as neck ache and mild increase in tone and reflexes. There were no cavitation symptoms and MRI did not show a syrinx. Her foramen magnum decompression has resulted in early relief of most of her symptoms, but the assessment here is short (4 months).

Comment

Although this paper does not set out to analyse in detail the results of surgery, it appears that the improvement following either surgical approach, though present, is subtle despite an improved radiological picture. MRI demonstrates that by the time symptoms become evident, major structural neural damage has already occurred.

MRI has proved most useful in assessing the results of surgery, particularly in those patients in whom shunt placement has been made. The most surprising finding has been the consistent discovery of collapse of the syrinx below the shunt tubing with continuing distention of the cavity above. This has been seen on the three of the patients who have undergone shunting and had follow-up MRI.

The mechanisms of shunt action suggested by this picture is that the pressure wave is still being transmitted from the foramen magnum region downwards along the syrinx cavity. As Williams had pointed out, syrinx development results from minor low pressure differences rather than from major ones. The action of the shunt would seem, then, to divert the small pressure waves of almost insignificant difference from the cavity below the tubing allowing collapse, but because of continued presence of the pressure wave above, the distention may continue.

Whether this suggests that the syrinx shunting should be carried out at the highest possible level remains to be seen. Certainly with micro-surgery, placement of tubing can be problem-free with minimal neural trauma. Most surgeons who perform syrinx shunting are also aware of the problems caused by septae in the cavity.

Conclusion

1) In syringomyelia MRI is the definitive test in diagnosis and follow-up.

2) There is no one surgical "answer" to Chiari I and syringomyelia.

3) Individual management decisions must be made for individual patients.

References

1. Barbaro NM, Wilson CB, Gutin PH, Edwards MSB (1984) Surgical treatment or syringomyelia, favourable results from syringo peritoneal shunting. J Neurosurg 61: 531–538
2. Logue V, Rice Edwards M (1981) Syringomyelia and its surgical treatment – an analysis of 75 patients. J Neurol Neurosurg Psychiatry 24: 273–284
3. Williams B (1986) Progress in syringomyelia. Neurol Res 8: 130–145

Address for correspondence: Mr. M. Powell, F.R.C.S., Consultant Neurosurgeon, The National Hospitals for Nervous Diseases, Maida Vale Hospital, London, W9 ITL, U.K.

Acta Neurochirurgica, Suppl. 43, 22–25 (1988)

Operative Treatment of Chiari Malformation with Syringomyelia

H. Tokuno, A. Hakuba[1], T. Suzuki, and Sh. Nishimura[1]

Department of Neurosurgery, Tane General Hospital and [1] Osaka City University Medical School, Osaka, Japan

Summary

The authors have operated on and reviewed 38 patients with Chiari malformation. By the introduction of our new operative method, Gardner's operation combined with expansive suboccipital cranioplasty, the operative results were improved.

Keywords: Chiari malformation; syringomyelia; Gardner's operation; suboccipital cranioplasty; treatment; results.

Introduction

Various operative methods for Chiari malformation and syringomyelia have been reported, and their treatment is controversial even now. The surgical treatment of Chiari malformation and syringomyelia has been advanced following the introduction of Gardner's theory and, later, of Williams' theory.

In the last 10 years, we have operated upon 38 patients with Chiari malformation initially by Gardner's operation with microsurgical technique and since 1984, by Gardner's operation combined with "expansive suboccipital cranioplasty". This our new operative method and operative results are detailed in this paper.

Clinical Material

Thirty-eight patients were treated during the last 10 years between 1976 and 1986. The age distribution ranged from 2 to 65 years (mean 36 years). There were 16 males and 22 females. The Chiari malformation of this series was of the so-called adult type which was not associated with meningomyelocoele, but in three cases initial symptoms appeared in infancy or in the childhood. Thirty-one patients (81%) were combined with syringomyelia. In 4 out of them syringomyelia was far advanced and in one case syringobulbia was also present. Hydrocephalus was noted in 4 of 38 cases. The patient's age at which initial symptoms appeared ranged from 3 months to 61 years (mean 27.7 years). The average duration of symptoms before admission was 8.1 years, with a range from 8 months to 20 years.

Symptoms

1) Initial symptoms

Twenty-two patients first complained of pain: headache in 11 (28%), cervical pain or shoulder stiffness in 5 (13%), and shoulder girdle or arm pain in 6 (15%). Four patients (11%) experienced vertigo. Sensory symptoms initially occurred in 12 patients (31%), while motor weakness occurred in 5 (13%) including finger stiffness in 2, weakness of upper extremities in 1, and lower extremities in 2. Three patients had ocular symptoms including blurred vision, visual blackness, and oscillopsia. Two children showed large head with hydrocephalus.

2) Symptoms presenting on admission

Pain was the major complaint. Eighteen patients (47%) experienced headache and cervical pain, while 7 patients (18%) had shoulder girdle or arm pain. Three had a sensation of truncal constriction, sometimes with pain which was so severe that their sleep was disturbed. Sensory symptoms occurred in 21 patients (55%). These included numbness or tingling, in upper as well as lower extremities, and loss of temperature sensation. Gait disturbance was seen 18 (47%). Weakness of upper extremities was noticed in 10 (26%), associated with muscle atrophy in 5 patients. There were other complaints: dysarthria in 2, dysphagia in 1, micturition disturbance in 2, oscillopsia in 3, and mental deterioration or retardation in 2.

Signs

The cranial nerves were involved in 25 patients (66%). Nystagmus was seen in 12 patients (31%), 10 were horizontal, and 2 were rotatory downbeat. Facial sen-

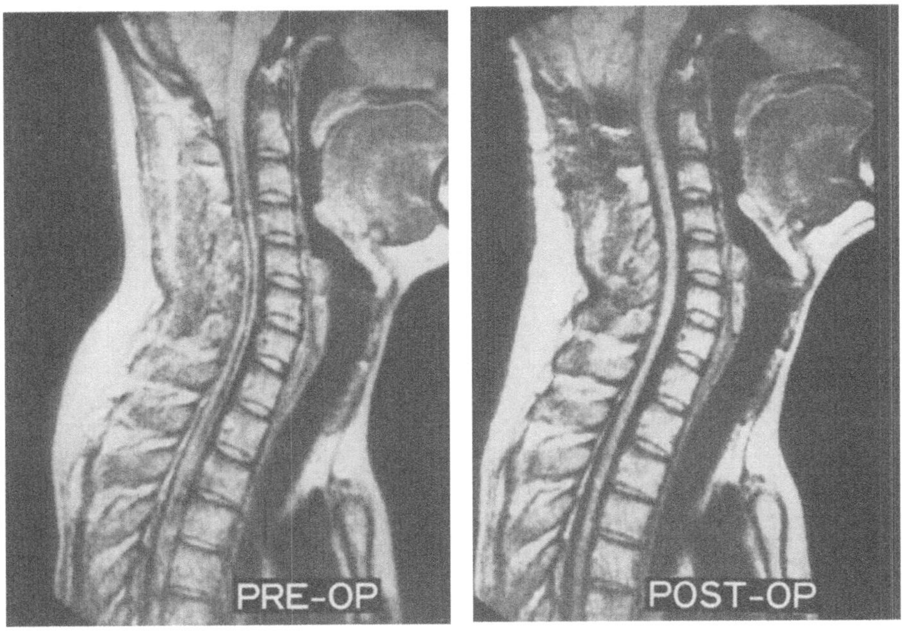

Fig. 1. Left: Preoperative magnetic resonance image (TR 6000 msec, TE 40 msec) showing tonsilar herniation and syrinx. Right: Image obtained 4 months postoperatively (TR 600 msec, TE 40 msec) showing disappearance of syrinx following Gardner's operation combined with expansive suboccipital cranioplasty

sory disturbance in the trigeminal nerve distribution was seen in 10 (26%) while onion skin pattern was seen in 6 (15%). Soft palate dysfunction was seen in 6 (15%) and dysarthria in one patient. Accessory nerve involvement was seen in 4 (10%), and hypoglossal nerve involvement in one patient. Motor weakness was present in 25 patients (65%). Thirty-two patients had a dissociated sensory loss, which varied in distribution; the most extensive was from the dermatome C 2 to L 4, while the least was from C 2–4. Deep sensory disturbance was found in 18 patients (47%). As for the cerebellar signs, limb ataxia was seen in 3 patients, and truncal ataxia in 3 patients. Increased intracranial pressure was found in 4 patients (10%).

Various bony anomalies especially in the craniovertebral junction were observed in 33 patients (86%): these comprised basilar impression in 17 (44%), abnormal spinal curvature in 13 (34%), atlanto-occipital assimilation in 11 (28%), block vertebrae of Klippel-Feil syndrome in 7 (18%), enlarged spinal canal in 6 (15%), atlantoaxial dislocation in 6 (15%), spina bifida of C 1 and/or C 2 in 2, hypoplasia of odontoid process in 2, and narrow spinal canal in 2. On the other hand, 12 patients (32%) were judged to have narrow posterior cranial fossa. The suboccipital bone was flattened in 8 cases, or thinned in 10 cases. The internal occipital protuberance seemed to be located lower than the inion in lateral skull x-ray in 9 cases.

Neuroradiological Investigation

We performed myelography with Pantopaque in 11 patients and gas myelography in 3. Computed tomography was done in 14 cases using CT/T 8,800 or Siemens Somatom II. Metrizamide myelography and metrizamide CT were performed in 21 cases and vertebral angiography in 38 cases. MRI was performed in 5 patients preoperatively and 9 patients pre- and postoperatively. Postoperative MRI showed disappearance of the syrinx in 4 patients, while it became smaller in 5 patients (Fig. 1).

Operative Treatment

The authors' operative approaches are as follows; for the patient with hydrocephalus, ventriculo-peritoneal shunt is the procedure of first choice. Suboccipital decompression and central canal plugging (Gardner's operation) is performed for the patients without hydrocephalus or for the patients with hydrocephalus who have not improved after shunt operation. If the syrinx is very large, syringostomy is added using a silastic tube.

Decompressive suboccipital craniotomy was performed in 3 patients in our early series. Gardner's operation was done in 14 patients, 2 of which had not been improved after shunt operation. One patient was improved after Gardner's operation, but became suddenly tetraparetic after coughing, which presumably displaced the muscle fragments used for central canal plugging; she recovered after terminal ventriculostomy was performed. Two cases without syringomyelia and with irreducible atlantoaxial dislocation were operated on by anterior decompression and fusion.

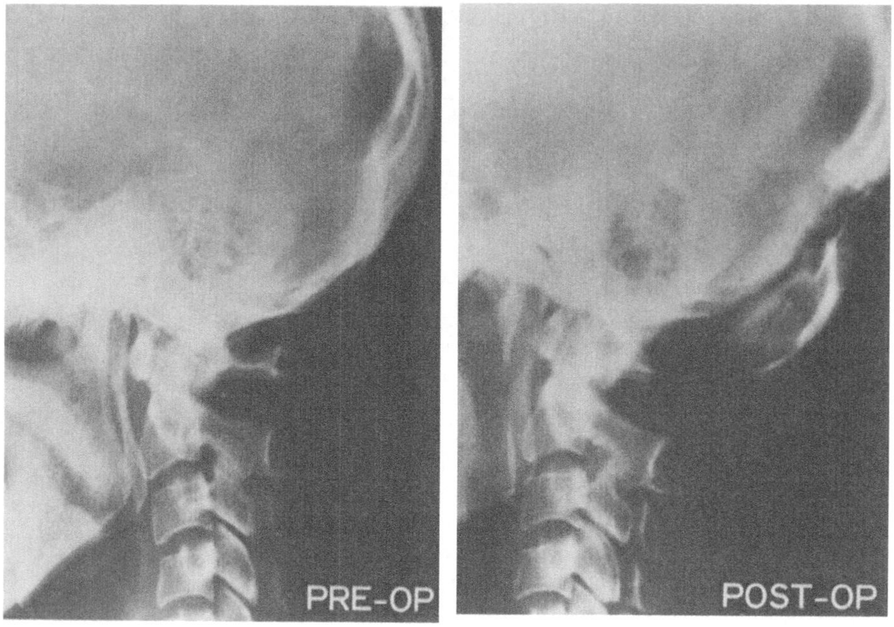

Fig. 2. Expansive suboccipital cranioplasty: Left: Preoperative lateral skull x-ray showing thinned suboccipital bone and narrow posterior fossa. Right: Postoperatively x-ray showing expansion of the posterior fossa following suboccipital cranioplasty. The bone flap is made from the preserved suboccipital bone and iliac bone graft

Our new operative method

Gardner's operation and "expansive suboccipital cranioplasty" was accomplished on the recent 17 patients. One patient showing reducible atlantoaxialdislocation was operated on by this method in addition to occipital-C 2 fusion (Table 1).

Our operative method: After suboccipital osteoplastic craniotomy and C 1 or C 1-2 laminectomy, the dura mater is opened in Y shape. The arachnoid membrane covering over the tonsil is opened extensively. The central canal is plugged by insertion of muscle pieces and Dacron balls utilizing an angled microdissector. The last Dacron ball is transfixed to the pia of obex using interrupted sutures with 10-0 nylon so that it cannot be dislodged. Additionally they are covered with a small piece of either Bemsheets® surgical cotton sheet or oxycel with fibrin glue. The open arachnoid membrane is carefully closed with interrupted 10-0 Dermalon sutures in order to prevent arachnoidal adhesion to the cerebellar surface. Dural plasty is performed with fascia lata graft trapezoid-shaped about 5 cm, 2 cm in each side and 5 cm in height in the region of the foramen magnum. In order to expand the posterior fossa, a larger suboccipital bone flap is made from the preserved suboccipital bone and an iliac bone graft 5 × 3 × 0.5 cm in size (Fig. 2).

Table 1. *Operative Treatment (38 Cases)*

Ventriculoperitoneal shunt		4
Suboccipital craniotomy with laminectomy		3
Gardner's operation		14
+ syringostomy	4	
+ terminal ventriculostomy	1	
+ IV ventricle-cisternal shunt	1	
Gardner's operation		
+ expansive suboccipital cranioplasty		17
+ occipital-C 2 fusion	1	
Anterior decompression and fusion		2

Operative Results

We divided the postoperative state of the patients into 3 groups as follows; "good recovery", *i.e.* complaints have improved or disappeared, or the signs improved even with the same complaints; "no change", *i.e.* signs have not changed; and "poor", *i.e.* complaints and signs became worse.

We followed up 38 patients from 1 year to 10 years (3.4 years in average). Postoperative results showed "good recovery" in 71%, "no change" in 26%, and "poor" in 3%. There was only one poor case in whom postoperative swan neck deformity advanced.

The frequency with which the various neurological symptoms and signs recovered was as follows; headache (89%), pain of extremities (85%) and truncal pain (75%) were all greatly improved; gait disturbance improved symptomatically in 83%, and tetraparesis signs improved in 91% while motor weakness of upper extremities recovered symptomatically in 45%, and objectively in 61%; sensory deficits showed recovery in 87%. Complete recovery of sensory disturbance was rare following Gardner's operation, but 9 cases were observed to show recovery of sensation following our new method. Cerebellar signs had 83% recovery rate, while nystagmus had only 33%. Before the introduction of expansive suboccipital cranioplasty, postoperative good recovery in the Gardner's operation was 67%, thereafter it was improved up to 82%.

The swan neck deformity, developed in 2 patients who had extensive laminectomy, and in one of them it progressed. Muscle fragments used to plug the central canal became dislodged when the patient coughed in one case which was mentioned previously. A 54 year old female with extensive syringomyelia with basilar impression who had been showing symptoms such as oscillopsia for 20 years, developed transient lower cranial nerve palsy after Gardner's operation and expansive suboccipital cranioplasty. Postoperative MRI showed that a significant identation of the ventral portion of the medulla oblongata by basilar impression was still present. Surgical manipulation of the medulla oblongata might cause aggravation of medullary function.

Discussion

Since we started to use expansive suboccipital cranioplasty the patients seemed to show improvement in their symptoms more rapidly. Patients with pain have completely recovered within 2 or 3 weeks; other symptoms also seem to have almost disappeared in a short period. But in some patients, new transient symptoms and signs were observed postoperatively, *e.g.* a patients with thoracic girdle pain was found to have developed numbness in another thoracic dermatome. In one severe case, already mentioned, lower cranial nerve palsy appeared, but it was transient and disappeared during the follow-up period. We have no patient who became permanently worse postoperatively since the introduction of this operative method.

By using this new operative method, we believe we obtain a full decompression of the posterior fossa and normal CSF flow is maintained. 32% of our patients had a narrow posterior fossa either diagnosed radiologically or observed directly intraoperatively. This finding has led us to believe that there is disproportion between the suboccipital cavity and the infratentorial content; indeed, even if the suboccipital space seems to be normal, the infratentorial component may be relatively too large. It is regarded as important that the suboccipital structure is remodelled to a normal physiological one with operation. Some authors reported[2] that cranioplasty may be needed in suboccipital craniotomy for a child with cerebellar tumour. Measurement of posterior cranial fossa volume of each patient is difficult, and we have not measured it in all cases. One author[4] has recently reported a measurement method in which CT scan and computed planimeter are utilized in facial spasm patients. It may be a good method for the evaluation of a bony cavity.

Logue and Edward (1981)[1] reported their 75 patients with syringomyelia and recommended simple decompression because of the preservation of arachnoid membrane. Indeed, some patients will recover by simple decompression and 3 patients in our series did so. It will not be necessary to plug the central canal in all patients. But one of ours, operated on by Gardner's operation, got worse when muscle pieces used for central canal plugging fell out. We believe that central canal plugging is essential for patients with extensive syringomyelia or with a long history because the elasticity of the central canal may be lost in such patients and the canal may remain open even although CSF flow will have been improved by simple decompression.

Other authors[3] have used the syrinx shunt operation without decompression of the posterior fossa which we think is resonable for the large syrinx; we have used syringostomy or syrinx shunt for several patients with large syringomyelia. However, such cases were so infrequent that we have only rarely encountered patients with such a large syrinx that we did not hesitate to incise the cord for syringostomy. When the use of MRI becomes fully developed and widespread patients will be diagnosed at an earlier stage.

As the follow-up period is relatively short, we must follow our patients for longer and re-evaluate the advantages of our operative method in future.

References

1. Logue V, Edwards MR (1981) Syringomyelia and its surgical treatment – an analysis of 75 patients. J Neurol Neurosurg Psychiatry 44: 273–284
2. McLone DG (1985) Cerebellar astrocytomas. Wilkins & Rengachary 83: 754–757
3. Tator CH, Meguro K, Rowed DW (1982) Favorable results with syringosubarachnoid shunts for treatment of syringomyelia. J Neurosurg 56: 517–523
4. Yamamoto Y, Kondo A, Hanakita J *et al* (1987) Measurement and clinical significance of the posterior cranial fossa volume of patients with hemifacial spasm. Neurol Surg 15: 243–248 (in Japanese)

Address for correspondence: H. Tokuno, M.D., Department of Neurosurgery, Tane General Hospital, Osaka, Japan.

Acta Neurochirurgica, Suppl. 43, 26–28 (1988)

Pathological Findings in Suboccipital Decompression in 63 Patients with Syringomyelia

J. Bidziński

Department of Neurosurgery, Academy of Medicine, Warsaw, Poland

Summary

Out of 97 patients with syringomyelia treated surgically, in 63 of them suboccipital decompression was performed. Various kinds of congenital or early acquired abnormalities were found to be the possible cause of the disease. Simple Arnold-Chiari malformation was found in 29 patients. In next 16 cases the Arnold-Chiari malformation was accompanied by partial occlusion of the foramen of Magendie with tiny, translucid membrane, that covered lower 1/2 or 3/4 of the foramen. In another 9 cases of Arnold-Chiari malformation, foramen of Magendie was totally occluded with thick, grayish membrane. In 7 cases lowered cerebellar tonsils were firmly attached to the medulla and joined by adhesions between themselves. In 1 case total atresia of foramen of Magendie was present only. In 1 case no pathological changes were found.

The above mentioned findings point out, that in practically all the cases of syringomyelia various kinds of gross pathology, that leads to difficulties in normal outflow of the CSF from the IVth ventricle to cisterns can be found. From that point of view the suboccipital decompression seems to be the most reasonable surgical procedure.

Keywords: Syringomyelia; Arnold-Chiari malformation; suboccipital decompression.

Introduction

The pathogenesis of syringomyelia is the subject of controversy for many years. Observations by Chiari[2] suggested pure mechanical distension, of the central canal the spinal cord called hydromyelia, that accompanied congenital abnormalities of the hind-brain. His observations seem to have been forgotten for many years. Syringomyelia was and sometimes still is known as untreatable degenerative disease. There were as well other theories of pathogenesis (vascular, infectious), that did not change in any way the consensus on the possibility of treatment of the disease.

The observations of Gardner[3] and his hydrodynamic theory of syringo-hydromyelia brought an increased interest in the pathogenesis of syringomyelia. More frequently applied surgical treatment gave us considerable data confirming that the real causes of disease are congenital or early acquired malformations in the atlanto-occipital region and the hind-brain. They obstruct partly the free circulation of cerebrospinal fluid at the IVth ventricle – foramen magnum level. Although pathogenesis of syringo-hydromyelia seem to have been clarified, the pathomechanism of the central canal distension is still controversial (Gardner[3], Williams[7, 8], Ball and Dayan[1]). In suboccipital craniectomy and decompression various kinds of pathology were found by several authors.

Clinical Material and Results

97 patients with syringomyelia were submitted to surgical therapy in last 12 years. Various surgical methods were applied, but in 63 patients suboccipital craniectomy and decompression was performed in the usual way. This group of patients was further analyzed.

Initial diagnosis in these patients consisted of plain x-rays of the skull and upper cervical spine and ascending positive contrast myelography in the supine position. Most of the patients were examined and treated before CT and NMR diagnosis was introduced and therefore the results of these investigations are not discussed here.

Plain x-rays of the skull revealed pathological changes which are summarised in Table 1.

This examination showed malformations only in 27% of the patients, what is comparable with obser-

Table 1. *Pathology in Plain X-Ray of the Skull*

Platybasia	6 patients
Basilar impression	6 patients
Platybasia and basilar impression	2 patients
Dysplasia of C 1	3 patients
Normal	46 patients

Table 2. *Pathology in Ascending Contrast Myelography*

Arnold-Chiari malformation	13 patients
Arnold-Chiary malformation and block at the level of foramen magnum	30 patients
Slight cervical cord dilatation without pathology at the level of foramen magnum	10 patients
Normal myelography	6 patients
Not performed*	4 patients

* Not performed due to severe kyphoscoliosis of the thoracis spine or intolerence of the patient to head down position.

Table 3. *Pathological Findings in Suboccipital Decompression*

Simple Arnold-Chiari malformation	29 patients
Arnold-Chiari malformation and partial occlusion of foramen of Magendi	16 patients
Arnold-Chiari malformation and total atresia of foramen of Magendie	9 patients
Arnold-Chiari malformation and severe intertonsillar adhesions	7 patients
Atresia of foramen of Magendie	1 patient
No changes	1 patient

vations of others. The most important examination was ascending positive contrast myelography in the supine position, to visualize anatomical relationships at the level of the foramen magnum, especially to find out the Arnold-Chiari malformation. The results of that examination are shown in Table 2.

This examination in 43 patients (out of 59 in which the procedure could be performed) revealed the Arnold-Chiari malformation, in most cases with partial or complete block at the level of the foramen magnum. This investigation was in 73% diagnostically important for the final decision.

The suboccipital craniectomy and decompression consisted of the removal of the arch of C_1, posterior margin of the foramen magnum and part of the occipital squama. The dura was incised in a Y-shape. After the revision of the visible field and improvement of possible malformations, the dura was left open and sutures of the muscles and skin was made. In suboccipital craniectomy several single or double malformations were found as shown in Table 3.

In 16 patients the foramen of Magendie was partly occluded in its lower part by a tiny, translucent membrane, that narrowed its diameter to 1/2 or 1/4 of normal size. Sometimes only in the upper part of the foramen a small opening of 2 mm diameter was left. That tiny membrane made a kind of valve and with rising intraventricular pressure (respiration or Queckenstaedt test) bulged like a "sail" and only through the upper part of the foramen of Magendie did the CSF pour out.

In cases with complete atresia of the foramen of Magendie, it was closed by a thick, grayish membrane, in which in some cases both PICA were involved. That caused considerable difficulty when opening the foramen.

In 7 cases the cerebellar tonsils were joined with very strong adhesions between themselves and with the medulla.

The surgical findings in suboccipital decompression thus revealed, that in 96% of the patients the Arnold-Chiari malformation was present, in half of the patients accompanied by partial or total occlusion of the foramen of Magendie. This points out that Arnold-Chiari malformation is practically a constant finding in syringomyelic patients, even in the cases in which ascending myelography failed to show this malformation.

All of the above mentioned changes lead to disturbances in free outflow of the CSF from the IVth ventricle and cistern and seem to confirm the opinion, that they are responsible for the development of syringo-hydromyelia.

These observations also indicate, that formerly used diagnostic procedures with positive contrast myelography did not reveal in all the cases the presence of malformations. Perhaps this problem will be finally solved with NMR investigation.

Discussion

Gardner[3] in 68 (92%) out of 74 operated patients found Arnold-Chiari malformation and in 30 of them – atresia of the Foramen of Magendie. Newton[6] observed that kind of malformation in 88% of surgically treated patients, Maeder and all[5] – in 5 out of 6 patients. Logue[4] found pathological changes in all 35 operated patients, among them in 31 – Arnold-Chiari malformation and in 2 – occlusion of the Foramen of Magendie. In my personal observations pathological changes were found in all but one out of 63 patients.

The final conclusion is, that syringo-hydromyelia is accompanied in over 90% of the cases by Arnold-Chiari malformation. In about 40% of the cases the foramen of Magendie is partially or totally occluded. The difficulties in free CSF circulation at the IVth ventricle – foramen magnum level play an important role in the development of syringo-hydromyelia. From that point of view the suboccipital craniectomy and decompression to make the CSF pathways free seem to be the most rational surgical procedure in the treatment of syringomyelia.

References

1. Ball MJ, Dayan AD (1972) Pathogenesis of syringomyelia. Lancet 2: 799–801
2. Chiari H (1888) Über die Pathogenese der sogenannten Syringomyelie. Heilkunde 9: 307–336
3. Gardner WJ (1965) Hydrodynamic mechanism of syringomyelia. J Neurol Neurosurg Psychiatry 28: 247–259
4. Logue V (1971) Syringomyelia. Clin Radiol 22: 2–16
5. Maeder RP, Mumenthaler MM, Markwalder H (1970) Symptomatische zervikale Syringomyelie. Deutsch Med Wschr 95: 164–168
6. Newton EJ (1969) Syringomyelia as the manifestation of the defective forth ventricle drainage. Ann Roy Coll Surg England 44: 194–200
7. Williams B (1978) A critical appraisal of posterior fossa surgery for communicating syringomyelia. Brain 101: 223–250
8. Williams B, Fahy G (1983) A critical appraisal of "Terminal ventriculostomy" for the treatment of syringomyelia. J Neurosurg 58: 188–197

Address for correspondence: Prof. Dr. Jerzy Bidziński, Department of Neurosurgery, Academy of Medicine. Banacha Str. 1 a, 02-097 Warzawa, Poland

Acta Neurochirurgica, Suppl. 43, 29–31 (1988)

Late Results of the Surgical Treatment of Syringomyelia

J. Bidziński

Department of Neurosurgery, Academy of Medicine, Warsaw, Poland

Summary

97 patients with syringomyelia were treated surgically during the last 12 years. Only those patients in which the disease showed rapid progression in the last $^1/_2$–1 year were operated. Of these 97 patients 54 were followed for 5 years or longer.

28 patients were treated by suboccipital decompression, 17 by terminal ventriculostomy, 4 by myelotomy and syringo-subarachnoid shunt; 5 had a combination of two procedures. Very good results were obtained in 12 patients (22%), satisfactory – in 31 (58%) and poor results – in 11 (20%). Very good results were more often achieved after suboccipital decompression, but each method had as well excellent as unsatisfactory results, and the number of the patients is too small as to allow statistically significant conclusions.

It is almost impossible to predict the result of the surgical treatment in syringomyelia, but good results were obtained most often in young patients with relatively short duration of the disease. However, it is not the rule. Satisfactory results can even be obtained in older patients with long lasting disease and poor neurological status. The best method of surgical treatment is still under discussion.

Keywords: Syringomyelia; operative treatment; late results.

Introduction

The surgical treatment of syringomyelia has a relatively long history. Abbe[1] was the first, who after laminectomy punctured the syringomyelitic cyst. After temporary improvement the patient died 6 months later. However the controversial opinions on the pathogenesis of syringomyelia and dominating opinions, that it is a incurable degenerative disease, did not encourage neurosurgeons to try surgical treatment at that time.

Chiari[4] noticed that syringomyelia is usually hydromyelia, that accompanies congenital malformations in the cerebellomedullary region. However his observations seem to have been forgotten for many years. The publications of Gardner[6] and his hydrodynamic theory of syringomyelia brought increasing interest in the surgical treatment of this disease. Actually, there are many publications on the surgical treatment of syringomyelia, various surgical procedures are applied and with encouraging results.

There are two main difficulties in the evaluation of the results of surgical treatment. First, the disease itself is chronic, lasts for some decades usually with very slow progression and has sometimes periods of spontaneous remission. The second, from the very beginning of the disease there is irreversible damage of the nervous tissue within the spinal cord, so one can hardly suppose, that the signs and symptoms could be completely abolished by surgery. Usually the patients are referred for surgical treatment at a rather late stage of the disease. To evaluate the results of surgical treatment of syringomyelia it is necessary to apply constant therapeutic criteria and to have very long follow-up.

Clinical Material and Results

Of 97 patients treated surgically for syringomyelia during last 12 years, in 54 the follow-up period is over 5 years, mean – 6.8 years as shown on Table 1.

The most important indication for surgical treatment was the rapid progression of the disease in the last $^1/_2$ to 1 year before surgery. If there were any doubts about the progression of the disease, after diagnostic

Table 1. *Follow-up Period*

Follow-up	No. of patients
10 years or more	6
9 years	3
8 years	6
7 years	9
6 years	12
5 years	18
Mean—6.8 years	

Table 2. *Late Results of Surgical Treatment*

Result	No. of patients	%
Very good	12	22
Good	31	58
Poor	11	20

examinations the patient was dismissed for further evaluation in the outpatient department. In next few months final decision about the operative treatment was done. The patients with stable neurological status were not operated upon. The age of the patient and poor neurological status were not contraindications for surgery.

For evaluation of the results the following outcome scale was used:

1) Very good results – marked improvement and then stable in the follow-up. In some patients treated in the beginning of the disease improvement was practically clinically complete.

2) Good results – slight postoperative improvement and than stable.

3) Poor results – temporary improvement or no improvement and further deterioration.

According to that outcome scale, the results of surgical treatment are presented in Table 2.

Various methods of surgical treatment were applied and in case of poor result of the first operation, another procedure was performed, only if the patient asked.

The results related to the methods of surgical treatment are shown on the Table 3.

The number of patients in each group is too small to state definitively which method of surgical treatment is the best. Comparing to two largest groups, suboccipital decompression and terminal ventriculostomy, it seems, that very good results are obtained more often after suboccipital decompression. However for each method of surgical treatment of syringomyelia there is a group of patients with poor results. So the problem of choosing the best surgical procedure needs further observations.

The most important predictor for the results of surgical treatment is the clinical status of the patient, and the age of the patient and the duration of disease seem to be the most important factors. That shows (at least partly) Table 4.

It is understandable that more very good results were obtained in the group of younger patients, in which the disease began early and the duration was short. The neurological deficit in these cases was relatively small. Even when the disease was in an advanced stage in spite of its short duration, the results were better in that group of patients. However the prognosis before the operation is unpredictable and in some cases very good results were obtained in older patients with long lasting disease.

Table 3. *Surgical Procedure and Results*

Procedure	Result			Total
	Very good	Good	Poor	
1. Suboccipital decompression	8 29% →	14 50% →	6 21% →	28 100%
2. Terminal ventriculostomy	2 12% →	13 76% →	2 12% →	17 100%
3. Myelotomy, syringo-subarachnoid shunt	1	1	2	4
4. Terminal ventricul. and myelotomy		1		1
5. Suboccipital dec. and terminal ventriculostomy	1	2	1	4

Table 4. *Results and the Age of Patients and Duration of Disease*

	Very good	Good	Poor
Mean age of the patients	28 y	40 y	44 y
Mean time of the duration of disease	4 y	11 y	13 y
Mean age of the patients of the beginning of disease	24 y	30 y	32 y

Discussion

It is difficult to find in the literature comparable data, mostly due to the various methods of evaluation of results, the various methods of surgical procedure and the various follow-up periods. Gardner[6] followed for 6 years after suboccipital decompression 16 patients out of 74 treated surgically and found very good results in 8 (50%), good in 5 and further deterioration in 3 (19%). Love and Olafson[9] in their observations for 1 to 10 years after suboccipital decompression found very good result in 10 (30%) and good result in 12 (36%) of 33 surgically treated patients. After terminal ventriculostomy performed as single procedure Gardner and coworkers[7] observed 12 patients for a rather short time of 2.5 years and found very good results in 10 of them. However Williams and Fahy[11] did not notice such good results.

Tator and coworkers[10] followed 20 patients for an average of 5 years after syringo-subarachnoid shunt and obtained very good results in 11 (55%), good – in 4 (20%) and further deterioration in 5 (25%).

Faulhauer and Loew[5] followed 23 patients for an average of 4 years after various surgical procedures in syringomyelia and observed very good results in 4 (17%), satisfactory (stable state) – in 12 (52%) and deterioration in 7 (31%). It seems, that after every method of surgical treatment and long enough follow-up period, about 20–25% of patients show poor results and further progression of the disease. It may suggest, that in groups of patients with comparable pathologic changes and with the same pathogenesis, the mechanism for creating the syringo-hydromyelic cavity may be different.

For the evaluation of the real value of surgical treatment in syringomyelia it is necessary to compare the results with the results of the conservative treatment, that means practically in patients not treated. It is difficult, because there are so few publications about the natural progression of the disease. Boman and Iivanainen[2] observed 55 patients with syringomyelia for 2–45 years. They found, that in 50% of patients there were periods of spontaneous arrest of deterioration lasting up to 10 years. Hertel[8] noticed very slow deterioration or no deterioration in 211 out of 323 patients with syringomyelia. Caraceni and coworkers[3] followed patients with syringomyelia who were not submitted to surgery for 1 to 5 years and found slow deterioration in 63% of patients and no deterioration in 37%.

Surgical treatment of syringo-hydromyelia should be performed in the early stages of the disease and in patients deteriorating in the last months preceeding surgery. Perhaps only after surgical treatment can one achieve the improvement of neurological status that in some early treated cases can lead to clinical recovery.

It is not quite clear why in comparable group of patients treated surgically in about 20–25% of patients the surgical treatment does not give the expected results, independently of the surgical method applied. Further study of the mechanisms of the disease and of the best method of surgical treatment are needed.

References

1. Abbe R (1892) Syringomyelia, operation, exploration of cord, withdrawal of the fluid. J Nerv Ment Dis 17: 512–520
2. Boman R, Iivanainen M (1967) Prognosis in syringomyelia. Acta Neurol Scand 43: 61–68
3. Caraceni TA, Celano J, Borghi P (1978) Evaluation of the results obtained from the surgical treatment of patients affected by syringomyelia associated with Chiari Type I malformation. Acta Neurochir (Wien) 44: 257–263
4. Chiari H (1888) Ueber die Pathogenese des sogenannten Syringomyelie. Heilkunde 9: 307–336
5. Faulhauer K, Loew F (1978) The surgical treatment of syringomyelia. Long term results. Acta Neurochir (Wien) 44: 215–222
6. Gardner WJ (1965) Hydrodynamic mechanism of syringomyelia. J Neurol Neurosurg Psychiatry 28: 247–259
7. Gardner WJ, Bell HS, Poolos PN, Dalu DF, Steinberg Marta (1977) Terminal ventriculostomy for syringomyelia. J Neurosurg 46: 609–617
8. Hertel (1973) according to: Faulhauer and Loew (1978)
9. Love JG, Olafson RA (1966) Syringomyelia. A look at surgical therapy. J Neurosurg 24: 714–718
10. Tator ChH, Kottoo Megur, Rowed DW (1982) Favorable results with syringo-subarachnoid shunts for treatment of syringomyelia. J Neurosurg 56: 517–523
11. Williams B, Fahy G (1983) A critical appraisal of "Terminal ventriculostomy" for the treatment of syringomyelia. J Neurosurg 58: 188–197

Address for correspondence: Prof. Dr. Jerzy Bidziński, Department of Neurosurgery, Academy of Medicine, Banacha Str. 1 a, 02-097 Waszawa, Poland.

Acta Neurochirurgica, Suppl. 43, 32–34 (1988)

Treatment of Syringomyelia by Syringo-peritoneal Shunt

J. Philippon, S. Sangla, J. Lara-Morales, J. Gazengel, M. Rivierez, and **Y. E. Horn**

Service de Neurochirurgie, Hôpital de la Salpétrière, Université de Paris, Paris, France

Summary

Syringomyelic cavitation of various origins (excluding tumour) has been treated by syringo-peritoneal shunting (SPS) in 10 patients. The main clinical and radiological symptoms are reviewed and postoperative results evaluated; among the symptoms, pain is the most likely to improve after surgery. But the global outcome in the majority of cases remains unchanged, some patients showing neurological deterioration. Although MRI is of fundamental importance in diagnosis, the significance of the postoperative appearances on MRI has still to be elucidated. Other studies are certainly necessary to define the exact place of SPS in the treatment of syringomyelia.

Keywords: Syringomelia; syringo-peritoneal shunting; outcome.

Surgical treatment of syringomyelia has varied widely. Unsatisfactory results observed after one type of operation have led to trial of different surgical procedures. Furthermore, the natural course of the disease, highly variable from one patient to another, partly explains difficulties in evaluating long term results.

The most recent technique, syringo-peritoneal shunting (SPS) first described by Philips and Kindt[6] (1981) has been evaluated by some authors[2, 4, 5, 8], but the number of patients remains limited. On the other hand, the increased availability of MRI may lead to a better understanding both of the preoperative conditions as well as of the postoperative results.

Clinical Material and Methods

Patient Population

We reviewed 10 patients treated by SPS between January 1983 and December 1986, having excluded cases with cystic spinal cord tumours. There were 7 males and 3 females, ranging from 32 to 67 years with a mean age of 43 years. The etiology was as followed: 3 patients had posttraumatic cavitation, 6 idiopathic syringomyelia (among them 4 with Arnold Chiari malformation); one was secondary to meningitis. In the 4 patients with non-idiopathic syringomyelia, symptoms of cord cavitation appeared several years after the initial injury (respectively 6, 11, and 34 years in the posttraumatic cases, 6 years in the post meningitic case). The symptoms of immediate paraplegia had markedly regressed in 2 cases but remained important in one.

Symptoms and Signs

The preoperative evolution varied between 9 months and 20 years (more than 10 years in 2 patients and less than one year in 2 others). The main symptoms and signs are summarized in Table 1. Motor weakness in the upper limbs, pyramidal tract disturbance, and areflexia were quite constantly found. Six patients had pain in the neck and upper limbs, associated sensory disturbances, trophic disorders and urinary dysfunction. Signs of cranial nerve involvement were noted in 5 cases.

Radiological Evaluation

Depending upon the technical possibilities and progress in neuroradiological procedures, preoperative radiological evaluation has varied from case to case. Air myelography was performed in 5 cases: widening of the cord and flattening according to the position of the head was found in 2 cases. Computerized tomography scanning showed a centro-spinal hypodensity in 4 cases out of 6; after injection of metrizamide in 5 patients, it was possible to visualize hyperdensity within the cord in 4 cases at 6 hours, and 24 hours; central cavitation could not be proved by this technique in the last one. Cerebellar ectopia was identified in all cases with Arnold Chiari malformations. MRI has been used in the last 5 cases and has permitted a complete evaluation of the extension of the cavity and of the status of the foramen magnum.

Surgical Procedures

Before a syringo-peritoneal shunt was performed, 2 patients (cases 4, 5) had decompression of the foramen magnum; in one (case 9) ventriculo-atrial shunt, foramen magnum decompression and cysto-subarachnoid shunting were successively performed.

Surgical Technique

All patients have been operated upon in the lateral position. The spinal cord dura is exposed after a two-level laminectomy at the inferior level of the cavitation as shown on the preoperative neuroradiological studies – usually mid or upper thoracic level. After opening the meninges, a small incision is made in the thinnest part of the posterior columns (generally between the midline and the emergence of the posterior root). A lumbar peritoneal James shunt

Table 1. *Signs and Symptoms*

Case no.	1	2	3	4	5	6	7	8	9	10	Total
Pain	+	+		+		+	+			+	6
Dissociated sensory loss	+	+		+	+			+	+	+	7
Pyramidal tract signs	+	+	+	+	+	+	+	+	+		9
Motor weakness	+	+	+	+	+	+	+	+	+		9
Areflexia	+	+	+	+	+	+	+	+	+		9
Bladder dysfunction	+	+			+	+	+		+		6
Cranial nerves involvement					+		+	+	+	+	5
Trophic disorders		+	+	+	+		+		+		6

(low pressure: manufactured by Codman) is then inserted into the syrinx cavity. The catheter is fixed to the dura, which is closed. The distal end is directed toward the abdominal cavity after an incision has been made on the left side of the patient; the extremity is introduced into the peritoneal cavity and pushed for a length of about 20 cm. The tube is fixed securely where it enters the peritoneum.

Postoperative Results

Mean postoperative follow-up was 17 months (minimum 2 months, maximum 34). Results may be evaluated from two different perspectives: one concerned with the changes in each of the symptoms in our 10 patients, the other considering the global results in every patient.

Outcome by Symptom

Among the 6 patients with preoperative pain, 5 were immediately pain free; but, in one case, pain recurred 2 weeks later: re-exploration of the shunting system did not reveal any failure in the CSF drainage: one patient's pain was unchanged.

Weakness of the upper limb was improved in 5 out of 9 patients and was unaffected in 4. Of the 9 patients with spasticity, 4 showed improvement; (in one case, a sudden deterioration in leg motility after an accidental fall led to the discovery of a malfunction in the shunt system: it's revision caused only a slight improvement). In the 5 other cases, a transient deterioration of motor function of the lower limb was observed which remained important in 2 cases. Posterior column disturbances were noted postoperatively in half of our cases. They consisted in unilateral or bilateral ataxia (3 cases) or severe shooting pain (2 cases). A secondary improvement occurred in every patient, leaving only a mild disability in one. Trophic disorders, bladder dysfunction, and areflexia were unchanged after operation.

Global Outcome

If one considers the general condition of each patient, after operation, they may be divided into 3 groups:

– good results, with an improvement of the main preoperative signs.

– stabilisation, if the overall condition was not modified (this corresponds in general to partial improvement of some symptoms but worsening of others).

– poor results, with deterioration of the initial status.

In 2 cases, where pain was the most important complaint, the result may be considered as good. In 5 cases, there was no significant modification. Some signs have been improved (mainly pain, cranial nerves involvement and leg stiffness), but other factors led us to consider the final result as no better than the preoperative state, *e.g.* the development of posterior column deficit even when transient, improvement that was transitory or only moderate; 2 patients showed marked increase in the deficit of the legs, one immediately after operation, the other several months later.

Discussion

The evaluation of surgical treatment of syringomyelia is difficult.

It is generally admitted that posttraumatic syringomyelia is the most likely to improve after SPS[7]; the 2 best results in our series were in this category. This is also the conclusion of Suzuki *et al.*[8], with their 17 cases. Idiopathic cases seem to benefit less.

Pain is often relieved by SPS. In our series, 5 out 6 cases were rapidly pain free, even if in one patient the pain returned 2 weeks later. This high proportion of improvement has been observed by Suzuki in 20 out 23 patients, and in the series of Lesoin-Petit[4]. The comparison is more difficult in the work of Barbaro and Wilson[2]: 11 out 15 patients improved after surgery; but different surgical procedures (SPS in 15, syringosubarachnoid shunt in 19, techniques other than shunting in 14) make comparison difficult.

The effects on motor functions remain uncertain. Although the arms may become stronger, transient or persistent deterioration in the motility of the legs was encountered in our series. These results are therefore not as good as those of Suzuki *et al.* (65% improvement of motor signs); on the other hand, Barbaro reported worsening in 3 out of 4 patients with spastic quadriparesis after operation.

In the same way, the relative frequency of posterior column signs, even if some are transitory, does not lead us to be as optimistic as others[4, 5, 6, 8] concerning the general outcome.

MRI in the diagnosis and assessment of syringomyelia is certainly a fundamental advance in the last few years. It can provide without risk the main points for surgical decision; presence of the cavitation within the cord, its extent and size at different levels, presence or absence of hydrocephalus or of a cranio-vertebral malformation. Its value in understanding postsurgical evolution remains at the present time unclear. We have been able to perform a postoperative MRI in 7 patients: in 5, there was a collapse of the central cavity, among them 2 patients were improved, one was unchanged, and 2 others had an increased deficit in their lower limbs.

In 2 other patients, no significant modification of cavitation was observed: the patients remained unchanged, except for increase in posterior column signs. Therefore it seems that there is no obvious correlation between the change in size of the syrinx cavity and the postoperative clinical status. Further evaluation is certainly necessary.

The assessment of the effect of the surgical treatment of syringomyelia is not easy; many factors may explain variability in their evaluation.

The natural history of the disease which varies from one patient to another makes comparisons difficult[1], but it certainly does not account for all the differences observed in different series of patients. Syringomyelia has different etiologies and the better prognosis of post-traumatic cases may explain the better results of some series. Another fact, which has been underlined recently by Faulhauer and Loew[3] is related to length of the postoperative follow-up; immediate positive results may decline with time. Even taking into account all those considerations we cannot share completely the optimism of others authors considering the efficiency and the blandness of the SPS in syringomyelia.

In our study, we have not compared the results of this technique to other types of procedures, but they did not look to us fundamentally different. However this comparison remain debatable (too many variable factors have to be taken into account). It is certain that deterioration may occur, even with those techniques which were considered as relatively benign such as SPS. This fact deserves some comment.

Surgical injury of an already fragile cord may be possible: however a direct lesion due to myelotomy seems improbably (the incision is limited to the zone of introduction of the catheter and performed in an area where the nervous tissue is extremely thin). Another mechanism of deterioration could be the emptying of a cavity accustomed to a chronic pressure, thus modifying the anatomical and physiological properties of the surrounding tissue (a fact which may be observed in some chronic hydrocephalus after any kind of CSF substraction). It is clear that we do not know the pressure inside the cavitation and that the choice of any form of drainage remains purely hypothetical. Further evaluation is certainly needed.

References

1. Anderson NE, Willoughby WE, Whrightson P (1985) The natural history and the influence of surgical treatment in syringomyelia. Acta Neurol Scand 71: 472–479
2. Barbaro NM, Wilson CB (1984) Surgical treatment of syringomyelia. J Neurosurg 61: 531–538
3. Faulhauer K, Loew F (1978) The surgical treatment of syringomyelia. Long-term results. Acta Neurochir (Wien) 44: 215–222
4. Lesoin F, Petit H, Thomas CE, Viaud C, Baleriaux D, Johin M (1986) Use of the syringoperitoneal shunt in treatment of syringomyelia. Surg Neurol 25: 131–136
5. Pecker J, Javalet A, Boutlelis A (1983) La derivation syringo-péritonéale. Neurochirurgie 29: 171–173
6. Philips TW, Kindt GW (1981) Syringoperitoneal shunt for syringomyelia: a preliminary report. Surg Neurol 16: 462–466
7. Roullet E, Philippon J, Marteau R, Lhermitte F (1985) Syringomyélobulbie tardive compliquant un traumatisme médullaire amélioration après dérivation syringo-péritonéale. Rev Neurol 141: 417–419
8. Suzuki M, Davis C, Lindsay S, Gentilli F (1985) Syringoperitoneal shunt for treatment of cord cavitation. J Neurol Neurosurg Psychiatry 48: 620–627
9. Williams B, Fahy G (1983) A critical appraisal of terminal ventriculostomy for treatment of syringomyelia. J Neurosurg 58: 188–197

Address for correspondence: Prof. Dr. J. Philippon, Service de Neurochirurgie, Hôpital de la Salpétrière, F-75634 Paris Cedex 13, France.

Acta Neurochirurgica, Suppl. 43, 35–38 (1988)

Chemonucleolysis or Discectomy? Results of a Randomized Multicentre Trial in Patients with a Herniated Lumbar Intervertebral Disc

(A Preliminary Report)

H. A. M. van Alphen[1], **R. Braakman**[2], **M. W. Berfelo**[3], **G. Broere**[1], **P. D. Bezemer**[4], and **P. J. Kostense**[4]

[1] Department of Neurosurgery, Vrije Universiteit, Amsterdam, [2] Department of Neurosurgery, Erasmus University, Rotterdam, [3] Department of Neurosurgery, De Wever Ziekenhuis, Heerlen, [4] Department of Medical Statistics, Vrije Universiteit, Amsterdam, The Netherlands

Summary

A randomized trial was carried out to compare the results of discectomy and chemonucleolysis in 151 patients, suffering from lumbar disc herniation L 4–L 5 or L 5–S 1. All patients fulfilled strict entry criteria. Seventy eight patients underwent surgical discectomy, 73 were treated with chemonucleolysis.

Within one year of follow-up 18 patients (25%) required surgery after chemonucleolysis, whereas 2 patients (3%) in the surgery group underwent re-operation.

Increase of radicular pain within 30 days after treatment was encountered in 16 patients (22%) in the chemonucleolysis group, as compared to none in the surgery group.

The results of open surgery after preceding chemonucleolysis was successful in 44% and unsuccessful in 56%. The final result of chemonucleolysis, including second treatment, was still slightly less favourable (73%) the result of primary surgery (79%).

Keywords: Sciatica; lumbar disc herniation; discectomy; chemonucleolysis.

Introduction

Over the past 20 years treatment of lumbar disc herniation by chemonucleolysis has become an alternative to open surgical treatment[4]. The results of chemonucleolysis and surgical discectomy in patients with lumbar disc herniation (LDH) have been compared on a larger scale in non-randomized studies[3, 5, 6, 7]. There are, till now, only two small sized, randomized trials comparing discectomy and chemonucleolysis, with conflicting results[1, 2]. This stimulated us to conduct a randomized trial with the aim to establish whether chemonucleolysis is a good alternative to operative treatment of LDH.

In this paper we report the main results of this trial in terms of differences between both groups.

Material and Methods

The study population consists of patients between 18 and 45 years of age, suffering from a unilateral or bilateral radicular syndrome, due to a LDH at one of the levels L 4–L 5 or L 5–S 1, referred to the neurosurgical centres Vrije Universiteit Amsterdam, Erasmus Universiteit Rotterdam or De Wever Ziekenhuis Heerlen, between March 1984 and April 1986. The LDH was proven indisputably by myelography, sometimes supplemented by computerized tomography. Further inclusion criteria were that conservative treatment, consisting of 2 weeks of bedrest followed by physiotherapy, had not given satisfying results and that consent to participate in the trial, after extensive verbal and written information, was given. All patients, fulfilling these criteria, participated in the study.

Excluded were patients with severe paresis, neurogenic micturition disturbances or anaesthesia in the sacral segments – who all underwent emergency surgery –, those who had been operated or subjected to chemonucleolysis for a LDH before, those in whom bony compression was considered to be the most likely cause of the radicular syndrome, and those patients who were pregnant. Suspicion of a sequestered or extruded disc was no reason for exclusion from the study.

Two subgroups of patients were distinguished: one with radicular complaints existing 6 months or shorter and the other with complaints longer than 6 months. Together with the three centres this resulted in 6 strata. In each stratum patients were randomized into approximately two equal groups, balanced over time (Table 1).

Management. Chemonucleolysis was performed under general anaesthesia and radiographic control with the patient in lateral position. Discography was performed, using 1–2 ml water-soluble contrast medium, for verification of the proper position of the point of the needle and to check whether the disc was degenerated or perforated. After an interval of at least 15 minutes 2 ml (4,000 units) of chymopapain (Chymodiactin®, Smith Laboratories, Inc.) was then injected. The patients were discharged 4–7 days after chemonucleolysis.

Discectomy was performed under general anaesthesia by means of an interlaminar approach. The herniated or sequestrated disc was removed and the intervertebral disc space involved was emptied.

Table 1. *Numbers of Patients, Randomized in the Various Strata*

Duration of complaints	Amsterdam		Rotterdam		Heerlen		Total
	⩽ 6 m	> 6 m	⩽ 6 m	> 6 m	⩽ 6 m	> 6 m	
Open surgery	17	20	14	6	13	8	78
Chemonucleolysis	16	19	12	7	11	8	73
Total	33	39	26	13	24	16	151

Table 2. *Some Characteristics of Patients Who Entered the Trial, as Compared with Those Who Refused Consent. Of 247 Eligible Patients 151 (61%) Gave Their Informed Consent*

Duration of complaints	Patients in trial				Consent refused	
	⩽ 6 months		> 6 months		⩽ 6 months	> 6 months
Randomization	surgery	chemo	surgery	chemo		
Total number	44	39	34	34	44	52
Male/female	30/14	28/11	20/14	21/13	31/13	30/22
Mean age	33	34	34	34	34	34
Occupation* (independent/paid employment)	4/35	10/23	6/27	3/29	7/31	10/32

* Not all patients could be classified on the item "occupation".

Subsequently a check was carried out whether the root had a free course in the spinal canal and as far as visible in the intervertebral foramen. The procedure was not performed with the use of the operating microscope. The patients were discharged 7–10 days after surgery. After both procedures the patients were treated with physiotherapy.

Evaluation criteria. The patients were evaluated 12 months after the initial treatment. The outcome was assessed by the doctor and the patient. The doctor's opinion was scored in 4 categories (see Table 3), of which the first two were considered to be satisfactory and the last two unsatisfactory. The patient was asked to complete a questionaire. The question: "are you satisfied with the result of the treatment", was graded in 4 categories. The outcome "second treatment" was added as a fifth category (see Table 4). If the answer was "yes" or "largely", the treatment was considered to be successful. In the other categories the result was classified as unsuccessful. Failures of the treatment were:

- failure of puncturing or injecting the disc;
- persistance or increase of the radicular syndrome for more than 2 months;
- recurrence of radicular syndrome within 12 months after treatment.

Those patients were subjected to a second treatment, which, in all cases, was open surgery. In this paper, second treatment after unsuccessful operation or chemonucleolysis will be marked as an endpoint of the study, regardless the result of this treatment.

Statistical methods. Because no clear indications for heterogeneity between the six strata were found, comparisons between the two groups of patients – chemonucleolysis and open surgery – were made after pooling over strata. Rates were compared by means of Fisher's exact test for fourfold tables, at a two-sided significance level of 5%.

Results

One hundred and fifty one patients were included in the study, of whom 78 underwent discectomy by open surgery, and 73 chemonucleolysis. In the same period 96 other patients in the three participating centres fulfilled the criteria for inclusion, but refused to participate in the trial. Of them 35 preferred open surgery, 41 preferred chemonucleolysis, and 20 refused for other reasons. A survey of various characteristics of all eligible patients is given in Table 2. In the group who refused consent the percentage of patients with radic-

Table 3. *Doctor's Opinion 12 Months After Initial Treatment according to the Following Score:*
1. Pain disappeared and neurological function normal or normalized;
2. Pain lessened and neurological function normal(ized) *or* pain disappeared but neurological function not normalized;
3. Pain not lessened *or* pain not disappeared and neurological function not normalized;
4. Second treatment (surgery) before completion of one-year follow-up

Doctor's opinion	1	2	3	4	Total
Open surgery	32	34	10	2	78
Chemonucleolysis	26	20	9	18	73
Total	58	54	19	20	151

ular complaints existing longer than 6 months was slightly higher, as compared to the participants in the trial (54% and 45%, respectively). However, this difference is not statistically significant. In other respects the two groups were comparable. In the group of patients, included in the study, there also appeared to be no difference in results between the three participating centres. Therefore, the various numbers of patients from the three centres are pooled.

Table 3 shows the judgement of the doctor on the result of treatment, scored in 4 categories at 12-months follow-up. The most important difference between both groups is the number of patients who had to be subjected to a second treatment before completion of one-year follow-up. An operation after a failed chemonucleolysis had to be carried out in 18 patients (25%), a reoperation after failed open surgery in 2 patients (3%), a difference which is statistically significant ($p < 0.0001$). In the group of patients with failed chemonucleolysis, 2 patients are included, in whom the puncture of the intervertebral disc involved was unsuccessful. In all other cases the radicular signs or symptoms persisted for more than 2 months or recurred within 12 months after treatment. If the scores 1 and 2 are considered to be a successful result and the scores 3 and 4 unsuccessful, then the success rate in the surgery group is 86% and in the chemonucleolysis group 62%, as judged by the doctor.

Table 4. *Patient's Satisfaction 12 Months After Initial Treatment.* Answer to the question: "are you satisfied with the result of the treatment?" could be 1: yes, 2: largely, 3: scantly, 4: no. 5: Second treatment (surgery) before completion of one-year follow-up

Patient's answer	1	2	3	4	5	Total
Open surgery	44	16	10	5	2	77*
Chemonucleolysis	31	13	6	5	18	73
Total	75	29	16	10	20	150*

* 1 questionaire not returned.

Table 5. *Patient's Satisfaction 12 Months After Initial Treatment in those Patients who underwent Second Treatment.* Answer to the question: "are you satisfied with the final result of treatment?" could be 1: yes, 2: largely, 3: scantly, 4: no

Patient's answer	1	2	3	4	Total
Open surgery failed		1		1	2
Puncture of the disc failed	2				2
Chemonucleolysis failed	3	4	2	7	16
Total	5	5	2	8	20

In Table 4 the opinion about the result of treatment, as expressed by the patient, is shown. In category 5 again the numbers of patients who needed a second treatment, are given. Considering categories 1 and 2 in this table as being a successful result, then the success rate, according to the patient's opinion, is 78% in the surgery group and 60% in the chemonucleolysis group. Categories 1 to 4 on this scale do not coincide completely with categories 1 to 3 of the doctor's judgement.

In Table 5 the outcome 12 months after initial treatment is given for those patients in whom first treatment failed within this follow-up period. In 16 patients (10 male, 6 female) discectomy was performed after failure of chemonucleolysis. The average age of these patients was 37,6 years, as compared to 34 years in the whole population in the study. The outcome in these 16 patients was favourable in 7 (44%), unfavourable in the other 9 (56%). Two patients in whom chemonucleolysis could not be performed because of failure of puncturing the disc are classified separately. They also were subjected to discectomy.

The surgical procedure was complicated in 3 cases. One patient showed an increase of neurological deficit due to a partial lesion of the nerve root involved. In one case a CSF leakage occurred during the operation, which had no further consequences. One patient suffered a urinary infection postoperatively. In the chemonucleolysis group 16 patients mentioned an increase of radicular pain within 30 days following treatment. Seven of them underwent subsequent surgery within 3 months after initial treatment. The discomfort following surgery, on the other hand, was not counted. One patient showed a slight allergic erythema after chemonucleolysis. No other allergic reactions were seen. Of the two patients in whom puncture of the intervertebral disc for chemonucleolysis failed, one had surgical discectomy during the same session under general anaesthesia. After this procedure, this patient showed up with a partial cauda equina lesion, lasting for some weeks. It is not clear whether this lesion was caused by the repeated puncturing or by the subsequent surgical intervention.

Discussion

Is chemonucleolysis a good alternative treatment in patients with a lumbar herniated intervertebral disc? The only way to answer this question properly is by means of a randomized trial. In the past, only two studies have been carried out in this way. Crawshaw and coworkers compared 27 surgically treated patients with 25 patients, who underwent chemonucleolysis, after a one-year follow-up. Unsuccessful result after surgery was seen in 12% (n = 3) of the patients, as compared to 54% (n = 13) after chemonucleolysis. Twelve of the chemonucleolysis patients and 1 surgical case underwent an operation as second treatment[1]. However, this series is too small for statistical analysis. The same goes for the study by Ejeskär *et al.*, in which the results in 15 patients after chemonucleolysis and in 14 patients after surgery are compared. Eight patients who had chemonucleolysis as initial treatment had to be subjected to surgery subsequently, because of failure of the chemonucleolysis[2].

In the present study an adequate number of patients were available, all fulfilling the very strict entry criteria, and followed for one year after treatment. This series permits well-considered conclusions. The study indicates that surgical discectomy is significantly superior to chemonucleolysis in terms of therapeutic efficacy of the two procedures. Twenty-five percent of the patients in the chemonucleolysis group needed a second treatment versus 3% in the surgery group.

The outcome one year after discectomy was, according to the patients, favourable in 78% of the patients and after chemonucleolysis in 60%. The result of discectomy, carried out after failure of chemonucleolysis, was favourable in 44%, unfavourable in 56%. Final outcome in the chemonucleolysis group, including discectomy after failure of chemonucleolysis was favourable in 73% of the patients. In the discectomy group, in which only 2 patients required a re-operation, final outcome was favourable in 79%.

In the chemonucleolysis group, 16 patients (22%) mentioned an aggravation of radicular pain immediately after the procedure. In 9 out of these an improvement was seen after several weeks to months. In 7 patients an operation was judged to be necessary, because of the gravity of the complaints. In the surgery group no increase of radicular pain was mentioned.

In one of the neurosurgical centres, participating in this study (Amsterdam), yearly almost 400 procedures are carried out in patients complaining of a lumbar radicular syndrome. From the figures of this study, it appeared that in the same centre yearly about 60 patients were eligible for participation in this trial, fulfilling the criteria for inclusion, as mentioned above, including those patients, who refused consent. This means that, according to these criteria, only 15% of the patients, complaining of a lumbar radicular syndrome, are considered to suffer from a real herniated disc. This percentage might misrepresent the real number of LDH in a large population, because of patient selection in a university hospital, and the very strict inclusion criteria with respect to age and localization of the herniated disc in this study. Nevertheless, also according to others' experience (McCulloch, personal communication), the herniated disc can be held responsible for a lumbar radicular syndrome only in a limited percentage of the cases. Chemonucleolysis, despite its limitations mentioned above, being considered as the next step after failure of conservative treatment, should, therefore, be restricted to a very selected group out of all patients with a lumbar radicular syndrome.

Acknowledgements

This study was supported in part by a grant from Ortomed BV, Medische Techniek, Schiedam, The Netherlands.

We thank A. Staal and M. Vermeulen for their valuable advices. We are indebted to Mrs. B. A. A. Koehorst and Mrs. A. Ramcharan for their excellent technical assistance and for preparing this manuscript.

References

1. Crawshaw C, Frazer AM, Merriam WF, Mulholland RC, Webb JK (1984) A comparison of surgery and chemonucleolysis in the treatment of sciatica. Spine 9 (2): 195–198
2. Ejeskär A, Nachemson A, Herberts P, Lysell E, Andersson G, Irstam L, Peterson LE (1983) Surgery versus chemonucleolysis for herniated lumbar discs. A prospective study with random assignment. Clin Orthop 174: 236–242
3. Leavitt F, Garron DC, Whisler WW, d'Angelo CM (1980) A comparison of patients treated by chymopapain and laminectomy for low back pain using a multidimensional pain scale. Clin Orthop 146: 136–143
4. LeBlanc FE (1986) Sciatica – management by chemonucleolysis versus surgical discectomy. Neurosurg Rev 9: 103–107
5. Maroon JC, Abla A (1985) Microdiscectomy versus chemonucleolysis. Neurosurgery 16 (5): 644–649
6. Weinstein J, Spratt KF, Lehmann T, McNeill T, Hejna W (1986) Lumbar disc herniation. J Bone Joint Surg (Am) 68 A (1): 43–54
7. Postacchini F, Lami R, Massobrio M (1987) Chemonucleolysis versus surgery in lumbar disc herniations: correlation of the results to preoperative clinical pattern and size of the herniation. Spine 12 (2): 87–96

Address for correspondence: H. A. M. van Alphen, M.D., Ph.D., Department of Neurosurgery, Free University Hospital, De Boelelaan 1117, 1081 HV Amsterdam, The Netherlands.

Acta Neurochirurgica, Suppl. 43, 39–43 (1988)

Lumbar Microdiscectomy: Analysis of 100 Consecutive Cases. Its Pitfalls and Final Results

E. Ferrer, M. García-Bach, L. López, and **F. Isamat**

Department of Neurosurgery, Hospital de Bellvitge, Hospitalet del Llobregat, Barcelona, Spain

Summary

100 consecutive microsurgically operated patients suffering lumbar prolapsed disc are analyzed in the present work. Three different groups have been distinguished. Under and over forty years of age, and patients operated on after chemonucleolysis failure. General results after all treatments have been 76% good result, 17% improvement and 7% worsening. Special remark should be given about the chemonucleolysis group, that has a highest worsening percentage which raises 25% of the group patients.

Keywords: Lumbar microdiscectomy; chemonucleolysis; low back pain.

Introduction

Many series have been published that analyse the role of the microsurgical technique applied to herniated disc surgery, first proposed and described by Caspar and Loew 1977. Our group of patients operated on under microsurgery is relatively small, but has meant the confirmation of the advantage that the above procedure offers over the classical technique and even on occasions over chemonucleolysis. In this way we study those patients that have not achieved satisfactory results in the operation, in order to give them a second chance to review and try to completely resolve their symptoms. We have also included here the group of patients previously treated with chymopapain who did not experience the expected improvement, and were later operated on microsurgically.

Material and Methods

100 consecutive cases are presented operated on by the same surgical team and using a similar technique. The minimum follow-up period has been six months, which is orientative of the final results achieved. However, in the present work, early results are sufficiently significant in the majority of the cases in order to reach some interesting conclusions.

We have distributed the above 100 cases in three groups. Group A consist of those cases whose age is under 40 and inclues 46 patients. Group B consists of 42 patients over 40. Group C is formed by those patients who had previously been treated by us using chemonucleolysis and whose progress was sufficiently unsatisfactory to require microsurgery. It consists of 12 cases corresponding to 20% of the discolysis performed[2]. The other ones who had been treated by chemonucleolysis were also operated on using however the classical technique (9% of the chemonucleolysis performed).

The distribution by sex (Fig. 1) was 70 males and 30 females. The average age was 41, with a range that oscilated between 19 and 74 years of age. We have seen 48 cases involving the left side, 43 involving the right side and 9 exclusively central. These include 8 cases of foraminal or extreme-lateral herniated discs. Level distribution (Fig. 2) was: 45 patients operated on level L 5–S 1, 44 on L 4–L 5, 1 on L 3–L 5, 8 on L 4–L 5–S 1 and 1 more on L 3–L 4–L 5–S 1, that is to say that 90% were operated on only one level. 9% on two levels and 1% on three levels. In the later patients, the incision was, of course, larger but a microsurgical technique was always used in the approach to the spaces involved.

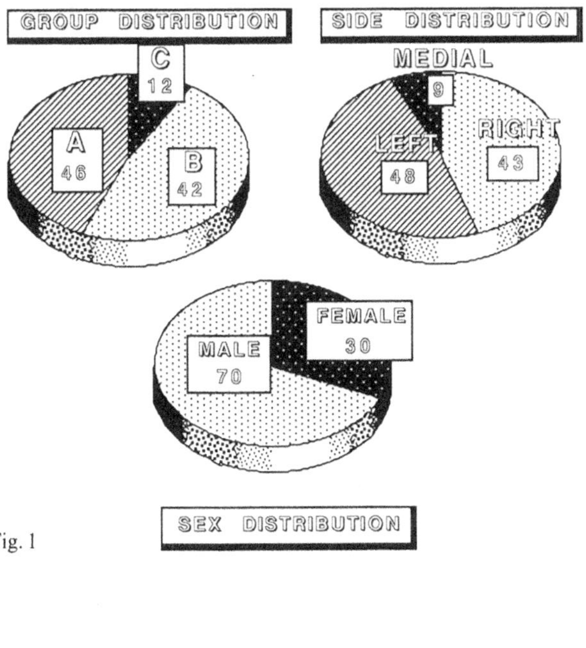

Fig. 1

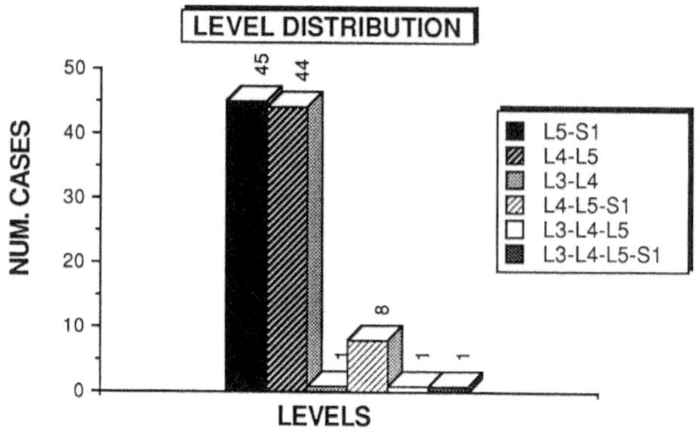

Fig. 2

Patients were studied by plain x-ray, neurophysiological studies, and CT scan that clearly showed the extent of the disc lesion. Radiculography was never used.

50 patients were treated prophylactically with cloxacilline or an intravenous cephalosporine during the first 24 hours. The other 50 cases did not receive prophylactic treatment.

The positioning of the patients on the operating table was always genu-pectoral[1] modified by us with the addition of a support under the gluteal region and tilting the table in anti-Trendelenbourg until achieving a sitting position over the barr supporting the gluteal region. This position allows complete freedom of the abdominal area whilst the patient is supported in the sitting position. The vertical tilting of the table facilitates the access of the operating microscope to the field, leaving the surgeon in a more comfortable working position.

The space involved was located manually and sometimes, in the case of doubt x-ray control was used. A 2.5–3 cm incision was performed along the midline[3, 4, 7, 8, 10, 11] which allowed bilateral access if necessary, to one or two levels. No special instrumentation was required to carry out the above procedure, that is to say, muscular retraction was achieved thanks to the placing of a standard Anderson-Adson retractor and some gauzes rolled up in the shape of a cigarette packing both superiorly and inferiorly. A partial flavectomy is performed and the disc approached following careful retraction of the nerve root. Posterior vertebral common ligamentum is divided following the fiber direction with the help of a fine dissector. Removal of the disc is carried out using a disc rongeur and an angled curette. Finally the lateral recess and the intervertebral foramen are routinely explored. Wound closure is made using reabsorbible stitches.

The patient is mobilized during the first twelve hours following the operation. Discharge has been reduced from six to seven days to one to three days at present.

7 patients have required re-intervention up to now, and in the near future, 3 patients may be added to this group. This second surgical indication was made following the clinical picture observed and the radiological review of the case. A new CT scan together with a functional radiological study of the spine was performed in all cases, and only one patient underwent radiculography. In such cases microsurgical re-operation was indicated depending of the cause of failure. Those who we felt not suitable for microsurgery were operated on conventionally.

Results

As we have already pointed out the results analysed here refer to those obtained medium term, with a fol-

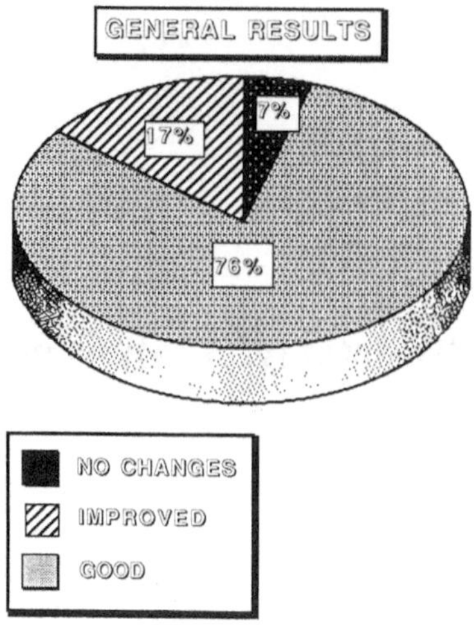

Fig. 3

low-up period oscillating between 6 and 18 months. We are not going to deal with long term progress of the patients over five years, a period of time sufficient to draw definitive conclusions.

We consider as a consequence, those patients able to go back to work and lead a full social life as good results, without disturbances that might restrict their activity. We also include here four patients with foot drop who experienced little or no improvement in the motor loss but didn't have any residual pain. It consists of 76 patients (Fig. 3).

Another group includes 17 cases that were subjectively satisfied with the treatment followed, that have improved and returned to work despite suffering some labour or sporting restriction and occasionally presenting some kind of discomfort.

The third group refers to those who haven't benefitted from the surgical treatment (5 cases) and who have even in some cases worsened (2 patients).

7 patients have been operated on more than once, achieving a greater or lesser degree of improvement in 4 of them (57%) who were then situated in the other groups, and three cases, in spite of new therapy, did not achieve positive results. 3 patients from the group of bad results and who have been operated on only once will be re-operated on in the future.

Group analysis (Fig. 4) is of great interest. In group A 40 patients have attained good results, 4 have improved and 2 have worsened. In group B 31 cases have attained good results, 9 have improved and 2 more have worsened. In group C, 5 patients have attained good results, 4 have improved and 3 have worsened. We should observed that how groups A and B are comparable in terms of bad results, however, group C

GROUP RESULTS

GROUP	GOOD		IMPROVED		NO CHANGES	
A	40	87%	4	8.5%	2	4.5%
B	31	74%	9	21%	2	5%
C	5	41.5%	4	33.5%	3	25%

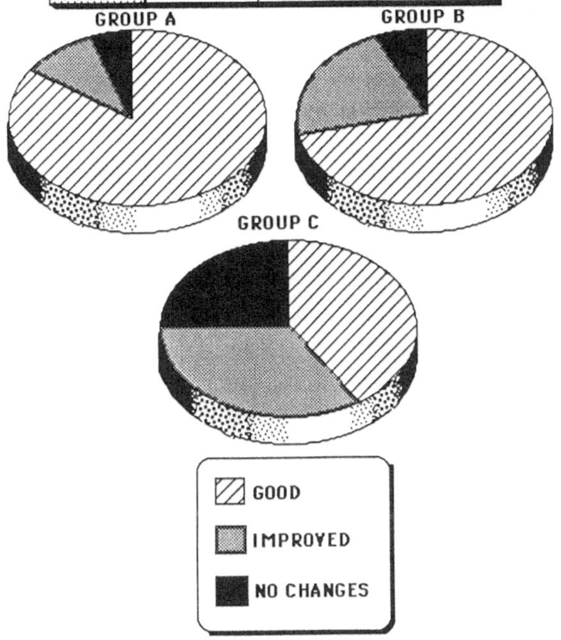

Fig. 4

deviated from them reaching 25% of bad results. Between group A and B we can appreciate subtle differences marked only by the appearance of the greater incidence of minor residual pain disturbances probably in relation to a higher average age.

Intraoperative complications: The most frequent has been the opening of the dural sac with CSF leakage in 8 cases. No complications were derived as a consequence. In 1 case a partial nerve root lesion occurred, fortunately in a patient who previously suffered from foot drop. Operating level was never mistaken.

Early postoperative complications: The most frequent have been the infection of the wound, that has always been solved using topical care and oral antibiotics. 7 cases in all. 3 due to Staphylococcus aureus and 4 due to Staphylococcus epidermididis. No Gram negative infections were reported. Antibiotic prophylaxis was performed in 50% of the cases. Only one patient was infected after following treatment (0.5%), while the remaining 6 (12%) had not received antibiotics.

1 patient had to be re-operated on 48 hours after the first operation on detection of a free fragment disc recurrence. We have not had any case of external CSF fistula.

Later postoperative complications: In 1 case without previous antibiotic prophylaxis the appearance of a serious picture of spondylo-discitis was observed 6 weeks after the surgical procedure. The diagnosis was radiological, without having previous wound infection.

We were not able to identify the germ through a percutaneous punction of the intervertebral space. It was treated for three months as Staphylococcus, being cured with good result after 10 months conservative treatment. Simple discitis was not reported.

3 patients showed signs and symptoms of segmentary instability discovered radiologically with a functional study. One of them who had previously undergone discolysis and later microdiscectomy due to a subligamentous herniated disc presented complete intervertebral space collapse, for which reason an interbody fusion was carried out. The other two cases were not operated on again.

2 patients presented signs of serious epidural fibrosis in the enhanced CT scan[9]. 1 patient presented intense low back pain twelve months after the operation, interpreted as articular that completely resolved after performing a percutaneous facet denervation.

Discussion

There is no doubt that the microsurgical procedure is much more comfortable for the patient and that there is a marked improvement in terms of the length and quality of the postoperative period. However, the patient does not only benefit from the advantages of this technique in relatively simple aspects, but also, in our opinion, the limited view of the operative field at this level is greatly increased thus allowing better exploration of this region.

We should not forget that the contingency of radicular or osseous malformations in the lateral recess[6] are relatively common and that if they go undetected are a frequent cause of surgical failure. For example, in our series we were able to observe at least 4 cases of conjoined root and one of double root, the latter previously operated on by ourselves undetected and therefore without achieving decompression of the second root, which was of course, much more engaged in the lateral recess and required a second operation for us to realise this and carry out effective radicular decompression.

The most common anatomical variation observed by us, is without doubt high exit of the root and less frequently low exit, both usual causes of bad radicular decompression and surgical failure.

In other cases, generally in the elderly, we found in the margin of the inferior intersomatic edge arthritic changes under the crossing point of the root, on its way out through the intervertebral foramen. In these patients, we removed this marginal spur, thus achieving the enlargement of the anterior aspect of the infundibulum of the intervertebral foramen.

We have had a very high index of wound infection that has changed drastically with antibiotic prophylaxis and for this reason we believe this is clearly justified, above all if we take into account the existing possibility of developing a serious spondylo-discitis.

We would like to point out the low index of epidural fibrosis capable of generating clinical symptoms, differing in this respect from the classical series, fitting an easy diagnostic facility available and overused in those patients who didn't go well.

Let's now analyse the group of patients with postoperative pain disturbances and see what kind of pain pattern they follow[5]. Many of them suffer either episodic or chronic simple low back pain. In these cases we have to suspect the existence of musculo-skeletal disturbances. The possibility of acquired segmental instability is something that one should take into account. Other cases developed facet overload which was usually easily controlled.

Other patients were afflicted with disturbances radiating to the legs in a non radicular pattern. In these cases one should also consider the articular or ligamentous origin of the pain. However, the role of a kind of discomfort that often afflicts some patients stands out. It consists of a sensation of pain and heaviness in the legs, related to prolonged standing or walking. In these cases, the hypothesis of a vascular origin of the problem could possibly be considered as due to venous stasis secondary to a vegetative innervation disturbance.

Persistent sciatic pain suggests an involvement of the nerve root, and forces us to think that the nervous structures are entrapped. Only in a small number of cases the real cause of pain is a fibrotic or even less likely arachnoiditic. We must refine our diagnosis as much as possible when reviewing these cases in search of lateral recess stenosis, migrated free disc fragments, intraforaminal or extreme-lateral herniated discs, malformations or even tumoural lesions simulating disc origin and undetected in the CT scan. Notice that those patients initially treated using the microsurgical procedure show a full anatomical integrity and for that reason we mustn't think we are dealing with cases secondary to an extensive and sometimes not indicated surgical procedure.

Conclusions

We believe there are two clear situations in our work, the first one is the utility of re-operation of patients after following a careful study protocol, looking for an explanation for surgical failure, taking into account that the patient damage after re-intervention, is minimal.

The second is the group following chemonucleolysis which clearly went worse in comparison to the other patients. One has to take into account that this group is formed of pre-selected patients who had obtained bad results in the preceding procedure. However, we are able to affirm that at least, those patients developed segmental instability much more frequently, due to failure of the intersomatic structures, aggravated by the operation.

In summary, microdiscectomy is a technique that offers maximum efficiency in terms of solutions to the problem of lumbar disc disease, with a very high margin of safety and with a relatively low index of important complications, which means that in our environment it is today the technique of choice, without completely renouncing other supposedly more conservative kinds of therapy.

References

1a. Caspar W (1977) A new surgical procedure for lumbar disc herniation causing less tissue damage through a microsurgical approach. Advances in Neurosurgery 4: 74–77. Springer, Berlin Heidelberg New York

1b. Caspar W, Loew F (1977) Die mikrochirurgische Operation des lumbalen Bandscheibenvorfalles. Dtsch Ärzteblatt 13: 863–868

2. David MG (1985) Efficacy of chymopapain chemonucleolysis: A long term review of 105 patients. J Neurosurgery 62: 662–666
3. de Divititis E, Spaziante R, Stella L (1979) Some technical modification of surgical treatment of lumbar disc lesions. Neurochirurgia (Stuttg) 22: 95–98
4. Gilsbach J, Eggert HR, Weeger W (1979) Microsurgical operation for herniated lumbar disc. AESCULAP own publication
5. Goald H (1980) Microlumbar discectomy follow-up of 477 patients. J Microsurg 2: 95–100
6. Housner OW, Onofrio BM, Forbes GS, Baker HL (1986) Correlation of radiological features of failure of lumbar intervertebral disc chemonucleolysis. J Neurosurg 64: 736–742
7. Loew F, Caspar W (1978) Surgical approach to lumbar disc herniations. (The micro-approach to the lumbar disc prolapse operation.) In: Krayenbühl H *et al* (eds) Advances and technical

standards in neurosurgery, vol 5. Springer, Wien New York, pp 153–171
8. Maroon JC, Abla AA (1985) Microlumbar discectomy. CNS 33: 407–417. Williams & Wilkins
9. Schubinger O, Valavanis A (1980) CT differentiation between recurrent disc herniation and postoperative scar formation: The value of contrast enhancement. Neuroradiology 22: 251–254
10. Williams RW (1975) Congress of neurological surgeons USA
11. Yaşargil MG (1977) Microsurgical operation of herniated lumbar disc. In: Advances in neurosurgery, vol 4. Springer, Berlin Heidelberg New York

Address for correspondence: Dr. E. Ferrer, Department of Neurosurgery, Hospital de Bellvitge, Barcelona, Spain.

Acta Neurochirurgica, Suppl. 43, 44–47 (1988)

Computed Tomography After Lumbar Disc Surgery: a Comparison Between Symptomatic and Asymptomatic Patients

P. Cervellini[1], **D. Curri**[1], **L. Bernardi**[2], **L. Volpin**[1], and **A. Benedetti**[1]

[1] Department of Neurosurgery and [2] Department of Neuroradiology, City Hospital, Vicenza, Italy

Summary

The evaluation of patients with recurrent symptoms after lumbar disc surgery, is a difficult diagnostic problem. The causes of failure may include recurrent disc herniation, postoperative scarring, arachnoiditis, spinal stenosis, infection and mechanical instability. The most common causes are recurrent herniation and postoperative scarring; the routine x-ray and myelographic differentiation between them is difficult or impossible. High resolution CT has shown some results in the evaluation of the postoperative patients. It requires some knowledge of CT findings of "normal" pictures of the physiologic healing and scarring after disc surgery.

We scanned 30 asymptomatic operated patients and 30 patients with recurrent sciatic nerve pain after disc surgery. From our observations result that it is quite impossible to distinguish "normal" scar from asymptomatic fibrosis. The degree and type of fibrosis are not related to recurrent symptoms.

Keywords: Lumbar disc herniation; computed tomography; postoperative complaints.

Introduction

The evaluation of patients with recurrent symptoms after lumbar disc surgery, is a difficult diagnostic problem both from clinical and radiographical standpoints[1]. The main causes are: recurrent disc herniation, postoperative scarring, arachnoiditis, infections, spinal stenosis and mechanical instability[15]. The most common causes are postoperative scarring and recurrent herniation[1, 2]. High resolution computed tomography has been shown to be effective in giving a picture of anatomical details, but it is not easy to relate them to the operative findings[4]. CT is used extensively in the evaluation of the postoperative patients with recurrent symptoms, but till today it is difficult to differentiate the physiologic healing from symptomatic fibrosis and from recurrent herniation[1, 12]. To provide a "baseline" in the radiological evaluation of these patients we scanned a series of 30 asymptomatic patients at interval ranging from 1 to 3 years after surgery, comparing the findings with those of 30 symptomatic patients operated in the same period.

Material and Methods

We scanned 60 patients at interval from surgery ranging from 1 to 3 years. 30 patients were asymptomatic and 30 were symptomatic. The age of the patients ranged from 25 to 75 years. The levels operated were L 5–L 5 and L 5–S 1. All patients underwent surgery at one level and the surgical procedure consisted in an interlaminar approach and radical disc removal done by a single neurosurgeon (A.B.).

A Somaton DRH equipped with scout view to indentify the scanning planes, was employed in the study. A slice thickness of 5 mm was used. Intravenous contrast material was administered (2 ml/kg) in 3 asymptomatic and in 16 symptomatic patients; in the remaining 14 of this group, the injection of contrast medium was not done because the picture of fibrosis was evident.

The analyzed features were: recurrent disc herniation; type of epidural fibrosis; degree of fibrosis; retraction of the dural tube toward the side of surgery; morphology of the nerve root; vacuum phenomenon; spinal stenosis and advantage of contrast medium administration.

According to Schubiger[18], fibrosis has been divided in 4 types: I-Stripe-like scar limited to the partial hemilaminectomy (Fig. 1 A); II-Stripe-like scar with extension to the lateral recess or neural foramen (Fig. 1 B); III-Knot-like scar (Fig. 1 C); IV-Concentric scar encasing the dural tube. Contrast Medium enhances the image (Fig. 1 D).

The degree of fibrosis has been classified as follows: + low degree: + + middle degree of fibrosis: + + + high degree of fibrosis.

Results

The frequency of the different types of fibrosis is represented in Table 1 A. The most common type was the II one with a regular distribution in both groups of patients. The type I was slightly more frequent in asymptomatic patients and type IV was more present in symptomatic patients. The distribution of the various

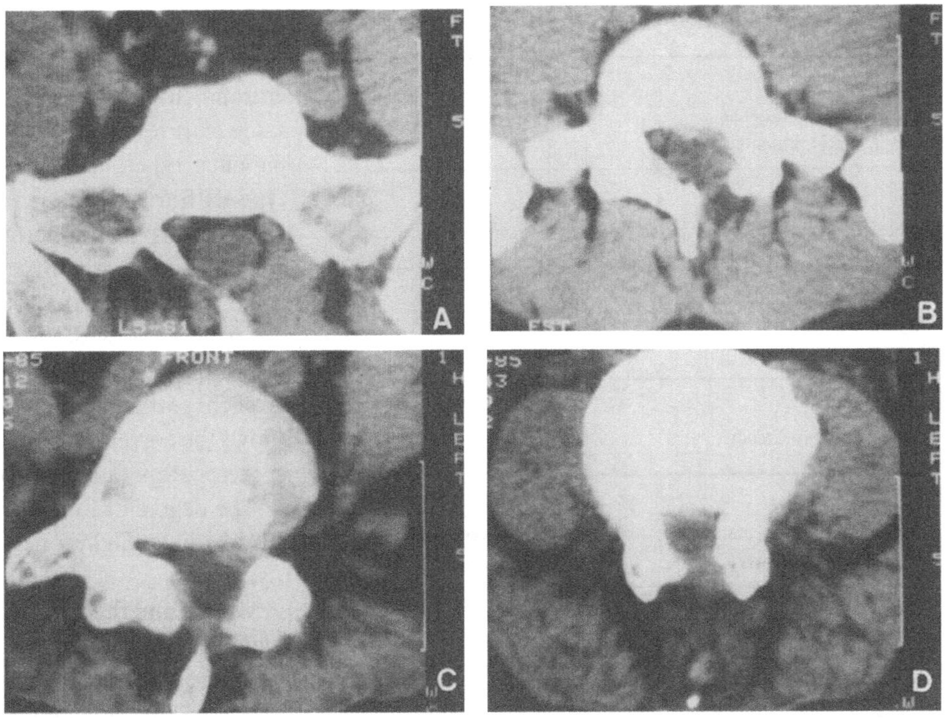

Fig. 1. Types of fibrosis. A) Stripe-like scar limited to the partial hemilaminectomy. B) Stripe-like scar with extension to the lateral recess. C) Knot-like scar. D) Concentric scar encasing the dural tube

Table 1 A. *Epidural Fibrosis*

	N. cases	Asymptomatic	Symptomatic
Type I	15	9	6
Type II	30	16	14
Type III	5	4	1
Type IV	4	1	3
Lack of Fibrosis	6	0	6

Table 1 B. *Degree of Epidural Fibrosis*

	N. Cases	Asymptomatic	Symptomatic
+	26	16	10
+ +	20	7	13
+ + +	8	7	1

Table 1 C. *Relationship Between Type and Degree of Epidural Fibrosis in Symptomatic Patients*

	N. cases	Type I	Type II	Type III	Type IV
+	10	5	4	1	—
+ +	13	1	9	—	3
+ + +	1	—	1	—	—

Table 1 D. *Relationship Between Type and Degree of Epidural Fibrosis in Asymptomatic Patients*

	N. cases	Type I	Type II	Type III	Type IV
+	16	9	3	4	—
+ +	7	—	6	—	1
+ + +	7	—	7	—	—

degrees of fibrosis in the 2 groups is shown in Table 1 B. Low degree is more present in asymptomatic patients; middle degree in symptomatic but without a significant prevalence. Contrary to expectation, high degree fibrosis was found in 7 asymptomatic patients against 1 symptomatic. In Table 1 C, D the relation between the type and degree of fibrosis is analyzed: low fibrosis is mostly represented in type I; middle and high fibrosis are prevalent in type II.

In Table 2, the analysis of 4 features is showed: A – recurrent disc herniation; B – retraction of the dural tube; C – bony stenosis; D – usefulness of intravenous contrast material.

The evidence of the nerve root has been classified as follows: 1) loss of distinction; 2) partial loss of distinction; 3) good distinction; in 25 patients (8 asymptomatic, 6 symptomatic) the nerve root was not visible; in 21 cases (11 asymptomatic, 10 symptomatic) the nerve root was clearly visible; partial loss of distinction was present in 14 cases (8 asymptomatic, 6 sympto-

Table 2 A. *Recurrent Disc Herniation*

N. cases	9
Asymptomatic patients	2
Symptomatic patients	7

Table 2 B. *Dural Tube Retraction*

N. cases	18
Asymptomatic patients	9
Symptomatic patients	9

Table 2 C. *Bony Stenosis*

N. cases	16	
Asymptomatic patients	10	(6 lateral recess stenosis)
Symptomatic patients	6	(5 lateral recess stenosis)

Table 2 D. *Intravenous Contrast Material*

N. cases	19	
Asymptomatic patients	3	not useful
Symptomatic patients	16	8 not useful 8 useful

matic). These data do not show marked differences between the 2 groups.

The vacuum phenomenon was observed in 10 cases, 5 symptomatic and 5 asymptomatic.

Discussion

The treatment of the symptomatic patient after lumbar disc surgery is difficult; reoperation is mostly not helpfull unless in the presence of a recurrent herniation. The most common reoperative finding consists in a epidural fibrosis of the dural tube and nerve root. Surgical scar removal does not relieve the symptomatology. Myelography does not allow one to distinguish a recurrent herniated disc from epidural fibrosis[9, 14, 17]. Meanwhile CT is a noninvasive procedure, widely employed, that with the add of contrast medium should be of value in the differential diagnosis[11, 18]. In our series, contrast enhancement was performed in 19 cases, but in only 8 patients did it help the differential diagnosis. CT-myelography might be helpful in differentiating between the two entities[13].

Fibrosis has been reported to be the aetiology of the "failed back" surgery syndrome in 5% to 12% of cases[3, 6]. However there are no characteristic CT findings which distinguish "normal" postoperative fibrosis from fibrosis that is causing symptoms of nerve root entrapment[1, 11]. In our study, epidural scarring has the same features both in asymptomatic and symptomatic patients: type I and II fibrosis is equally distributed; type III fibrosis is more common in asymptomatic patients and 6 patients, without epidural scar, complained of recurrent symptoms. Thus it means that fibrosis could not be the cause of recurrence of symptoms. In the 6 patients without fibrosis, fat grafts had been placed on the dura and around the nerve root. The use of such grafts reduces epidural fibrosis[7, 8, 10, 19] and allows the correct interpretation of the CT examination in the postoperative patient.

The degree of fibrosis does not distinguish between the 2 groups: highest degrees are more represented in asymptomatic subjects. In our series, the aspect of the nerve root and the dural tube retraction, that are tightly related to the scar, do not demonstrate differences between the 2 groups.

The vacuum phenomenon is thought to result from liberation of nitrogen in the degenerated disc[5] or in that injured by mechanical trauma[16]. This feature has no importance in the postoperative period[1].

Postoperative bony stenosis was observed more frequently in pain free patients; thus we believe that this anatomical condition is not an important factor in "failed back" surgery syndrome. Nevertheless other authors found that spinal stenosis, occurring in 50–60% of patients, is a main cause of recurrent symptoms[3].

CT scan in "failed back" surgery syndrome does not provide enough information for reoperation. From our observations it is impossible to distinguish the "normal" physiologic healing scar from "symptomatic" fibrosis. It is also difficult to make a differential diagnosis between "symptomatic" scar and recurrent disc herniation in spite of a metrizamide enhanced CT examination.

References

1. Braun IF, Lin JP, Nenjamin HV, Kricheff II (1984) Computed tomography of the asymptomatic postsurgical lumbar spine: analysis of the physiologic scar. AJR 142: 149–152
2. Burton CV, Kirkaldy-Willis WH, Yong-Hing K (1981) Causes of failure of surgery on the lumbar spine. Clin Orthop 157: 191–199
3. Burton CV (1981) Symposium: the role of spine fusion. Spine 6: 291–298
4. Fankhauser H, Mousouri B, de Tribolet N (1986) Ré-intervention après cure de hernie discale. Confrontation entre radiologie et constatation opératoire dans 18 cas. Neurochirurgie 32: 52–55

5. Ford LT, Gibula LA, Murphy WA (1977) Analysis of gas in vacuum lumbar disc. AJR 128: 1056–1057
6. Frymayuer JW (1981) Symposium: the role of spine fusion. Spine 6: 284–290
7. Jacobs RR, McClain O, Naff J (1980) Control of postlaminectomy scar formation, an experimental and clinical study. Spine 5: 223–229
8. Keller JT, Dunsker SB, McWhorter JM, Ongkiko CM, Saunders MC, Mayfield FH (1978) The fat of autogenous grafts of the spinal dura. An experimental study. J Neurosurg 49: 412–418
9. Knutsson F (1949) The myelogram following operation for herniated disc. Acta Radiol 32: 60–65
10. Langenskiöld A, Kiviluoto O (1976) Prevention of epidural scar formation after operations on the lumbar spine by means of free fat transplants. Clin Orthop 115: 92–95
11. Mall JC, Kaiser JA, Heithoff KB (1983) Postoperative spine. In: Newton TH, Potts DG (eds) Computed tomography of the spine and spinal cord. Claveladel Press, San Anselm, pp 187–204
12. Mall JC, Kaiser JA (1982) Computed tomography of the postoperative spine. In: Genant HK *et al* (ed) CT of the lumbar spine. University of California, San Francisco, pp 245–252
13. Meyer JD, Latchaw RE, Rappolo HM (1981) CT myelography of the postoperative spine presented at the annual meeting of the Radiological Society of North America, Chicago
14. Moseley I (1977) The oil myelogram after operation for lumbar disc lesions. Clin Radiol 28: 267–276
15. Quencer RM, Murtagh RF, Post MJD, Rosomoff HL, Stokes NA (1978) Postoperative bony stenosis of the lumbar spinal canal: evaluation of 164 symptomatic patients with axial radiography. AJR 131: 1059–1064
16. Resnick D, Niwayama C, Guerra JJr, Vint V, Usselman J (1981) Spinal vaccum phenomena: anatomical study and review. Radiology 139: 341–348
17. Rothman RH, Campbell RE, Henkowitz E (1974) Myelographic patterns in lumbar disc degeneration. Clin Orthop 99: 18–29
18. Schubiger O, Valvanis A (1986) Diagnostic entre epidurite et contrainte d'origine herniaire. Neurochirurgie 32: 48–81
19. Yong-Hing K, Reilly J, Korompay V, Kirkaldy-Willis WH (1980) Prevention of nerve root adhesions after laminectomy. Spine 5: 59–64

Address for correspondence: Dr. D. Curri, Divisione di Neurochirurgia, Ospedale Civile, Viale Rodolfi, I-36100 Vicenza, Italy.

Acta Neurochirurgica, Suppl. 43, 48–54 (1988)

Transfacet Decompression with Dowel Fixation: a New Technique for Lumbar Lateral Spinal Stenosis

Ch. D. Ray

Institute for Low Back Care Minneapolis, Minnesota, U.S.A.

Summary

Auto-stabilization consequent to spinal segment instability involves osteophyte formation. The most common lumbar spinal stenoses are due to uncinate process spurs at L 5. These spurs inaccessibly lie ventral to the facet joints. Most surgical methods for lateral stenotic lesions do not address the incipient instability. Facet destructive approaches further destabilize the segment. The new decompressive technique here preserves and stabilizes posterior supporting structures: ligaments are left intact; laminas (and facets) are distracted; 11 mm transfacet holes are bored, exactly dorsal to both the spurs and entrapped ganglia; the entrapment (even a lateral herniated disc) is decompressed through the drilled hole. 12 mm bone dowels are then driven into the holes, immediately stabilizing the segment. The dura is not exposed. Reviewed are fifty transfacet cases. In four, a posterior interbody fusion was performed via the transfacet holes. The procedure presents a new window to spinal lesions.

Keywords: Lateral spinal stenosis; uncinate spurs; segmental instability; facet dowel fixation; lumbar spinal fusion.

Introduction

The evolution of lateral spinal or foraminal stenosis relative to osteophytosis and instability has been studied and reported[2, 3, 6, 8, 12, 20]. The term "segmental instability" is not used here to refer to gross mechanical instability, but to a syndrome having pain on micromotions, usually torsional, associated with degenerative loss of normal mechanical integrity, the tearing of outer anular fibers and osteophytic proliferation[20]. In a real sense this is segmental microinstability. Most clinical cases have a mechanical low-back pain syndrome, rather similar to that arising from facet arthralgia[12]. More important, however, is the stenosis where the spur typically compresses the ganglion cephalad against the pedicle of the same segment[6]. Swelling of the ganglion, as a result of minor trauma or circulatory changes, entraps the nerve structure leading to neurogenic claudication. Generally, hard neurological signs are absent, probably due to the intermittency of the compression. In almost all cases of lateral spinal stenosis there is a mixture of low back and leg pain varying widely, independent of the findings as shown in imaging techniques[6, 8, 12].

In a series of 200 CT scan cases in whom the primary diagnosis was lumbar lateral stenosis, Heithoff and Ray found that 92% occurred at the lumbosacral level and were due to uncinate spurs[6]. Since symptoms and signs of microinstability are also present in most cases having lateral stenosis, simple decompression by removal of the offending spurs is insufficient treatment of the overall problem; the segment must either be stable or become stabilized in order to stop both the back pain and the osteophytogenic process[11, 17]. The importance of and methods for decompressive foraminotomies in lateral lumbar spinal stenosis have been well documented[2, 6–9, 11, 14, 20].

The ganglion and nerve lie immediately ventral to the facet capsule; both nerve and spur are thus quite inaccessible[2, 3, 6, 8, 11, 14]. Surgical approaches to decompression are clearly easier if the facet joint and lateral lamina are amputated, a technique that will likely worsen the instability; segmental fusion often therefore being required. Facet undercutting, lateral pars cutting or a wide paralateral approach will not improve on but may instead permit further weakening of the segment and the pars[7, 11, 20]. The new technique presented here takes direct access to the spur and entrapped ganglion by boring through the distracted facet joint. Following the decompression, the remaining facet is then plugged with a bone graft dowel, producing instant stability. Since distraction and stabilization result, the method may be used instead of pedicle screws and plates in selected cases. For convenience, this com-

Table 1

CASES=50: MALE=38 FEMALE=12; AGE RANGE=20-80, AVERAGE=41.6 YEARS

PRIMARY DIAGNOSES:	NUMBER OF CASES
LATERAL SPINAL STENOSIS: L5-S1	39 CASES
" " " : L4-L5	4
DISCOGENIC PAIN (SEGMENTAL INSTABILITY) ASSOCIATED WITH LATERAL HERNIATED DISC	7
SECONDARY DIAGNOSES (IN ADDITION TO LATERAL STENOSIS)	
HERNIATED DISC: L5-S1	14 CASES
" " : L4-L5	6
" " : L3-L4	2
" " : L2-L3	1
DISCOGENIC PAIN AT L4-5, STENOSIS AT L5-S1	17 CASES
LATERAL SPINAL STENOSIS AT FUSION TRANSITION	3 "
" " " WITH SPONDYLOLISTHESIS	4 "

Table 2

TRAFAC PROCEDURES (50 CASES) PERFORMED AT:	NUMBER OF CASES
ONE LEVEL	26
(L5-S1=19 CASES, L4-5=4 CASES, L3-4=3 CASES)	
TWO LEVELS	18
THREE LEVELS	6

FACET DOWEL STABILIZATION ONLY	14 CASES
TRAFAC PLUS PLIT FUSIONS	32 CASES
(ONE LEVEL=10 CASES; TWO LEVELS=18 CASES; THREE LEVELS=4 CASES)	
TRAFAC PLUS TRANSFACET PLIF DOWEL FUSION	4 CASES
TRANSFACET DISCECTOMIES AS PART OF A TRAFAC PROCEDURE	12 CASES

FOLLOW-UP: RANGE (TIME SINCE PROCEDURE) = 24 TO 8 MONTHS
AVERAGE = 11.5 MONTHS

RESULTS:	DOWEL FAILURES	PLIT FAILURES	<50% CLINICAL IMPROVEMENT
FIRST 25 CASES:	7	9	6
SECOND 25 CASES*:	0	1	2

(*Significant procedural changes in the PLIT technique accounts for the marked reduction in failures between the two groups; PLIT failure promotes facet dowel failure, in cases with high instability.)

bined transfacet decompression with stabilization is referred to as a TRAFAC procedure*.

* Instrumentation (patents pending) used in this procedure is available from: CeDaR Surgical Inc. 15265 Minnetonka Blvd., Minnetonka, MN 55345, U.S.A.

There appears to be no reported technique similar to the one presented here, although bone dowels have been widely used for interbody fusions and for facet arthrodeses[1, 5, 7, 18, 19]. When the segment is highly mobile, a posterolateral intertransverse process (PLIT) fusion is also applied. In stenotic cases where the segment has already become highly autostabilized, transfacet dowels alone have been used with good success (14 of the cases in the present series).

This technique has evolved from a series of surgical methods reported by the author attempting to address lateral stenosis without creating, or at least without worsening, segmental instability[9, 11, 14, 17]. In many cases so operated previously, it became obvious that the remaining, unaddressed instability led to persistent or recurrent symptoms; in several of these, osteophytes recurred and a fusion was subsequently required. In some cases, the discs reherniated.

Patient Selection

As given in Tables 1 and 2, the 50 patients in this series were those having lateral spinal stenosis together with segmental microinstability. CT and MR studies were performed in all cases. Uncinate spurs or other osteophytes were almost invariably present on one or both sides causing a lateral stenosis. A few cases also had far lateral herniated discs. Three-dimensional reconstructions of CT scan images were employed in some patients to demonstrate the pathoanatomy with related structures and were felt valuable to demonstrate the surgical results. Magnetic resonance scans generally showed severe dehydration in the involved discs. Some patients had dynamic saline acceptance and contrast discography to confirm the correlation between structural and symptomatic details, especially when deciding which segments required stabilization without decompression. Some cases had repeatedly herniated discs until it became apparent that the underlying cause was segmental instability, accompanied by uncinate spurs and other changes.

Many patients were selected for TRAFAC procedures at one level; a few earlier ones on only one side. Due to the early uncertainty regarding long-term stability of the dowels, PLIT fusions were performed routinely; later only on the more grossly unstable cases where there were preoperative (and sometimes intraoperative) indications.

More recently, it has become obvious that the dowels alone may have insufficient long-term strength, unless a posterolatent fusion is also produced.

Surgical Technique

The size of the hole, its location and angulation of the drilling are clearly important in the TRAFAC procedure; they are planned using appropriate CT scan images and then marked on them. Fig. 1 illustrates the TRAFAC technique. The patient is placed on a surgical frame promoting lordosis and paraspinous muscle relaxation[10]. A standard midline incision is made; gentle traction is applied to one side, then the other[13, 16]. Ligaments (interspinous and ligamentum flavum) are carefully preserved. The facets to be drilled are dorsally denuded of tissue using a cautery hot-loop at high cutting current (the "poor man's laser")[14]. Laminas and facets are maximally distracted (about 4–6 mm) using a lamina spreader, stretching the ligaments, opening the foramena and foward flexing the segment.

A plunge-cutting tip drill bit, *e.g.*, Anspach (Anspach Effort, Inc. Lake Park Florida, 33403 U.S.A.) type 18X (11 mm diameter), has proven reliable for making the holes. The Midas Rex M-18 drill tip may also be used but this bit is smaller and the resulting hole has more uneven walls. A drill guide (CeDaR Surgical Inc. Minnetonka, Minnesota 55345 U.S.A.) is placed over the bearing shaft and adjusted to the depth as measured on the CT scan images. The drill shaft is carefully stabilized with a long surgical clamp when first starting the hole; a constant stream of sterile saline keeps the drill bit cool. The small shelf of pedicle seen in the depth of the hole will later act as a limit stop for the driven dowel. Through the drill hole, the decompression is performed; this may include a discectomy. The spur is removed by impaction with suitable instruments (CeDaR Surgical, Inc.) then the residual, fragmented tissue is pulled out[14, 15]. The toe of the footed impactor is finally passed ventral to the ganglion and the spur is hammered into the vertebra. Small curettes and rongeurs are also used until the ganglion is free. Small ball-tipped probes explore and evaluate the decompression.

The remarkable part of the method is that all of the important elements, decompression, discectomy and impaction of osteophytes, are all performed through the 11 mm transfacet hole. A more complete description of the technique is to be published elsewhere[17]. Where the stenosis is caused by a ventral hypertrophy of the facet (or a facet chondroma), boring the hole completes the decompression. Once the ganglion is freely movable, it is covered by a small fat-fascial graft[14]. A 12 mm homologous bone graft dowel, preserved by a freeze-dried method, is then driven into each hole. Autologous dowels may be cut from the patient but they are not quite as strong as the homologous ones. Small facets may require smaller dowels; here, 10 mm dowels are driven into 9 mm holes. After seating the dowels, the lamina spreader is removed; the elastic recoil of the ligaments, the stretch of the hole by oversized dowels and the rough walls of the holes, hold the dowels immovable.

When properly drilled, a 11 mm transfacet hole may be continued beyond the appropriately protected ganglion and into the vertebral bodies to perform a transfacet PLIF fusion; this has been done successfully in 4 cases of this series.

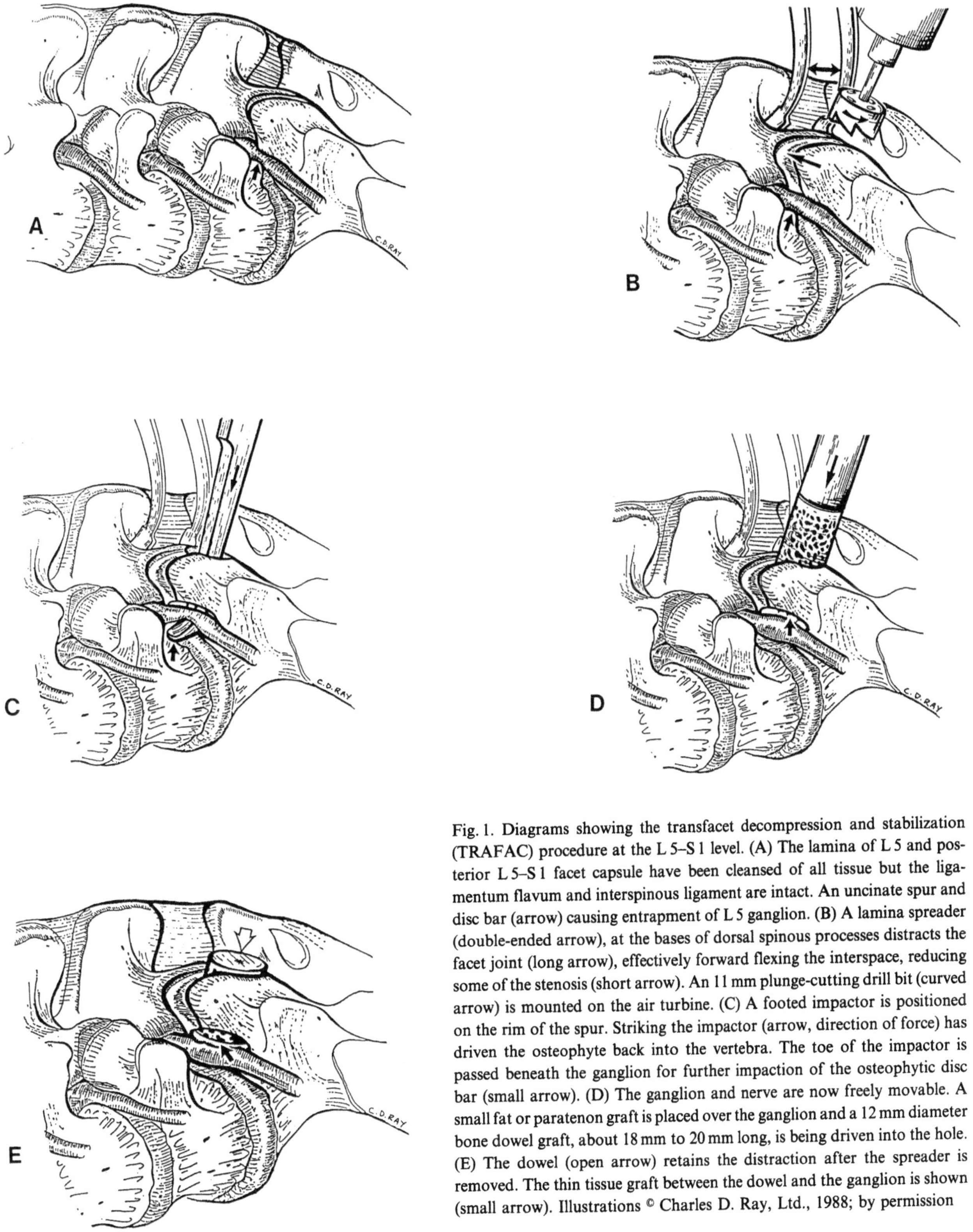

Fig. 1. Diagrams showing the transfacet decompression and stabilization (TRAFAC) procedure at the L 5–S 1 level. (A) The lamina of L 5 and posterior L 5–S 1 facet capsule have been cleansed of all tissue but the ligamentum flavum and interspinous ligament are intact. An uncinate spur and disc bar (arrow) causing entrapment of L 5 ganglion. (B) A lamina spreader (double-ended arrow), at the bases of dorsal spinous processes distracts the facet joint (long arrow), effectively forward flexing the interspace, reducing some of the stenosis (short arrow). An 11 mm plunge-cutting drill bit (curved arrow) is mounted on the air turbine. (C) A footed impactor is positioned on the rim of the spur. Striking the impactor (arrow, direction of force) has driven the osteophyte back into the vertebra. The toe of the impactor is passed beneath the ganglion for further impaction of the osteophytic disc bar (small arrow). (D) The ganglion and nerve are now freely movable. A small fat or paratenon graft is placed over the ganglion and a 12 mm diameter bone dowel graft, about 18 mm to 20 mm long, is being driven into the hole. (E) The dowel (open arrow) retains the distraction after the spreader is removed. The thin tissue graft between the dowel and the ganglion is shown (small arrow). Illustrations © Charles D. Ray, Ltd., 1988; by permission

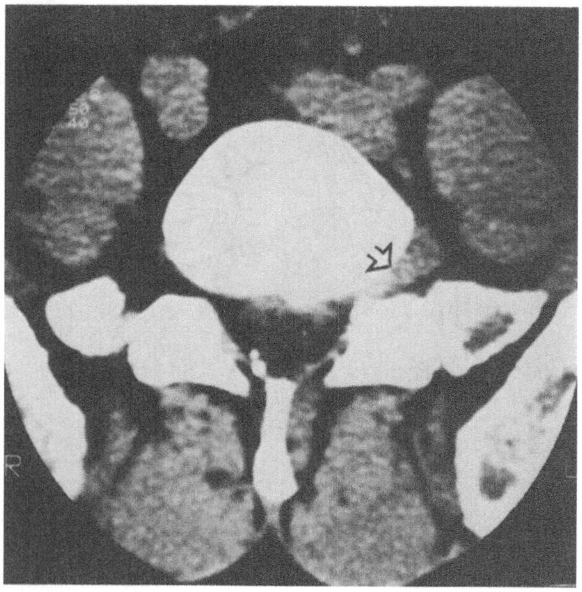

A

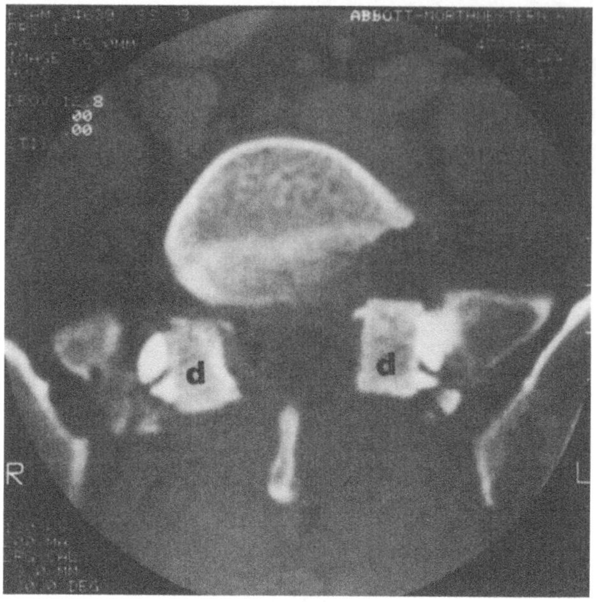

C

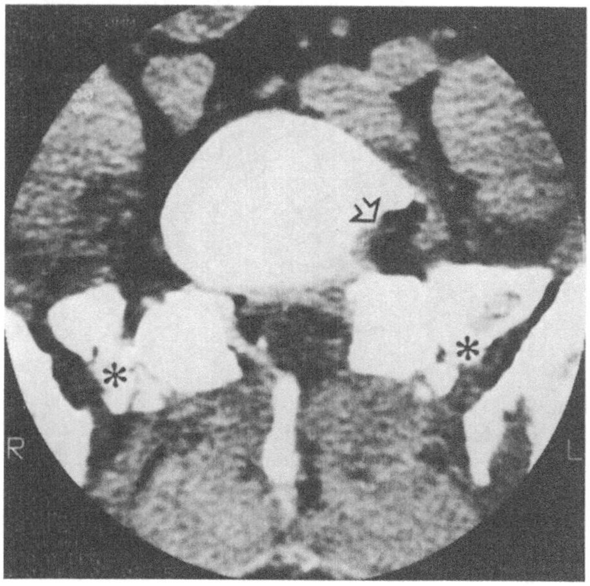

B

Fig. 2. A 43 year-old male with low back and leg pain. (A) Preoperative CT scan showing the herniated disc and uncinate spur (open arrow), left L 5–S 1. (B) One month postoperative CT scan at L 5–S 1 showing the decompression with dorsally placed fat graft (open arrow). The nerve lies lateral to the fat. The immature PLIT fusion is indicated by the asterisks. Contrast window setting, 2,000. (C) Same CT image as (B), window 250. Dowels indicated by letters *"d"*. Illustrations © Charles D. Ray, Ltd., 1988; by permission

Results

Results of the TRAFAC procedure in the 50 cases are given in Table 2. The Table shows that the cases were divided into one group of 25 earlier cases and then a second, more recent group of 25 cases. Several important technical details were changed and improvements in both dowel and PLIT fusion results are seen in the Table. 6 dowels of the earlier group became simply or multiply fractured, but none displaced. Most of the dowels fuse to the facets. Even when facet fusion does not occur, the dowels apparently become part of a stabilizing, fibrous complex mass. In relatively autostabilized cases facet dowels alone provided the only stabilizing elements needed; 14 cases were so operated.

One patient sustained a ganglionic injury, not from the hole drilling but as a sharp cut while dissecting into the tough fibrous tissue of a previous discectomy that encapsulated the ganglion; sympathetic dystrophy and a partial foot drop followed. Another early case had persistent nerve compression due to a facet fragment pushed against the ganglion by the dowel; a subsequent, limited decompression was required. There have been no other complications. All patients but six in the first

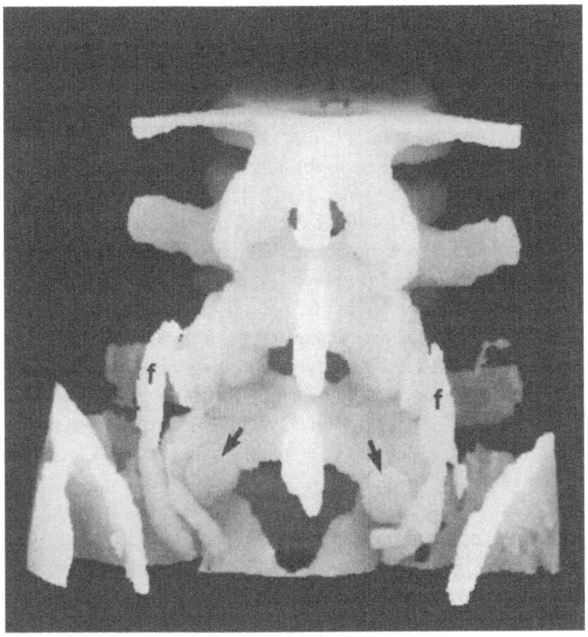

A

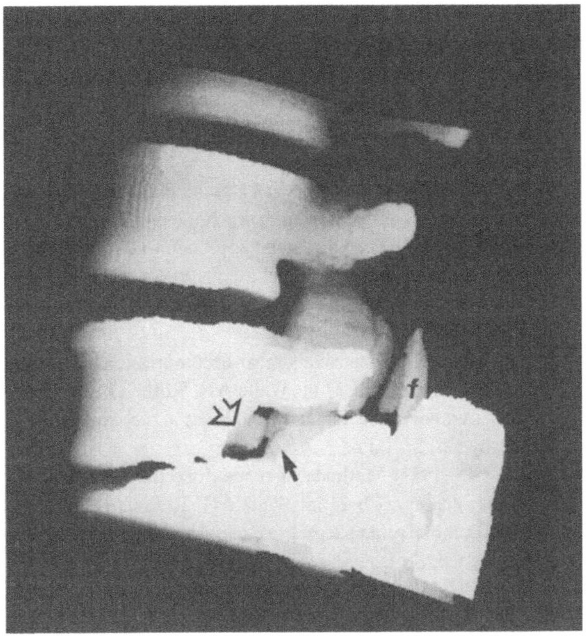

B

Fig. 3. Three-dimensional reconstructions from scans of Fig. 2. (A) Direct posterior view. Dowels indicated by arrows. Fusion, posterior cortical bone struts indicated by letters *"f"*. (Cancellous bone material not visualized.) (B) Lateral view. Dowel location at short arrow; fusion strut at letter *f*. Open arrow indicates location of resected osteophytic spur and bar. The right side of the neural arch is visible through the decompression defect, beneath the open arrow. Illustrations © Charles D. Ray, Ltd., 1988; by permission

group and 2 in the later group have had good to excellent, *e.g.*, better than 50%, clinical improvement as judged by patient comments, drug usage, independent examinations and return to work. Routine checks of fusion status involved plain x-ray films and standard tomography; follow-up CT scans were performed three months post-procedure.

Figs. 2 and 3 show pre- and postoperative CT scans, sagittal and 3-dimensional reconstructions on selected cases in this series.

Conclusion

The TRAFAC (transfacet decompression and stabilization) technique provides a new window of access to lumbar spinal lesions. It is felt that this technique may prove to be technically useful for selected cases having lateral spinal stenosis and segmental instability. The combination of decompression and stabilization in distraction makes the method a worthwhile alternative to the use of metal instrumentation, such as pedicle screws and plates, in selected cases. In the author's practice (limited to conservative and surgical care of the lumbar spine) the TRAFAC procedure, often accompanied by one of the fusion methods, is now the most frequently utilized surgical approach in cases having lateral spinal stenosis. A semi-stereotaxic drilling technique for combined PLIF and TRAFAC procedures is currently under development by the author.

On longer follow-up it has been noted that the dowels were in several cases too soft (cancellous); a good posterolateral fusion was therefore important. A stronger substitute for such cancellous dowels is being sought.

References

1. Blume HG (1985) Unilateral posterior lumbar interbody fusion: simplified dowel technique. Clin Orthop 193: 75–84
2. Burton CV (1987) Lumbar nerve entrapment: Central, foraminal and extra-foraminal zones. In: Genant HK *et al* (eds) Spine update 1987. Radiol Res & Educational Foundation, San Francisco
3. Chafetz N, Genant HK (1983) Computed tomography of the lumbar spine. Orthop Clin North Am 14: 147–169
4. Cloward RB (1963) Lesions of the intervertebral discs and their treatment by interbody fusion methods. Clin Orthop 27: 51–77
5. Harrington PR, Dickson JH (1973) An eleven year clinical investigation of Harrington instrumentation: a preliminary report of 578 cases. Clin Orthop 93: 110–112
6. Heithoff KB, Ray CD (1987) Computed tomographic assessment of lateral spinal stenosis. In: Genant HK *et al* (eds) Spine update 1987. Radiol Res & Educational Foundation, San Francisco

7. Hutter CG (1985) Spinal stenosis and posterior lumbar interbody fusion. Clin Orthop 193: 103–114
8. Kirkaldy-Willis WH, McIvor GWD (1976) Symposium: spinal stenosis. Clin Orthop 115: 1–144
9. Ray CD (1982) New techniques for decompression of lumbar spinal stenosis. Neurosurg 10: 587–592
10. Ray CD (1987) A new kneeling attachment and cushioned face rest for spinal surgery. Neurosurg 20: 266–296
11. Ray CD (1987) The paralateral approach to decompression for lateral stenosis and far lateral lesions of the lumbar spine. Ch 12 in: Watkins RG (ed) Lumbar discectomy. Aspen Press, Rockville, MD
12. Ray CD (1987) Extensive lumbar decompression: patient selection and results, Ch 17 in: White AH, Rothman RH, Ray CD (eds) Lumbar spine surgery: techniques and complications. CV Mosby Co, St. Louis
13. Ray CD (1987) Methods of tissue dissection and resection in lumbar surgery, Ch 21 in: White AH, Rothman RH, Ray CD (eds) Lumbar spine surgery: techniques and complications. CV Mosby Co, St. Louis
14. Ray CD (1987) Methods of resection of bone spurs, osteophytes and bony encroachment, Ch 23 in: White AH, Rothman RH, Ray CD (eds) Lumbar spine surgery: techniques and complications. CV Mosby Co, St. Louis
15. Ray CD (1986) Bone impactors: New instruments for spinal decompression. Spine 11: 1051–1053
16. Ray CD (1987) New malleable force-fulcrum retractors for lumbar spinal surgery. Spine 12: 418–420, 1987
17. Ray CD (1988) Transfacet decompression and dowel fixation (TRAFAC): a new surgical technique for lateral stenosis with instability. Spine 13 (in press)
18. Wiltberger BR (1953) The prefit dowel intervertebral body fusion as used in lumbar disc therapy. Am J Surg 86: 723–731
19. Wiltberger BR (1964) Intervertebral body fusion by the use of posterior bone dowel. Clin Orthop 35: 69–79
20. Yong-Hing K, Kirkaldy-Willis WH (1983) The pathophysiology of degenerative disease of the lumbar spine. Orthop Clin North Am 14: 491–504

Address for correspondence: Ch. D. Ray, M.D., FACS, Associate Director Institute for Low Back Care, 2737 Chicago Ave. So., Minneapolis, MN 55407, U.S.A.

Acta Neurochirurgica, Suppl. 43, 55–57 (1988)

Automated Percutaneous Nucleotomy – Initial Experience in Twenty-Five Cases of Contained Lumbar Disc Herniation

F. Gomes

Comprehensive Spine Center, Tampa, Florida, U.S.A.

Summary

The posterolateral percutaneous approach using the automated nucleotome system has been used to treat twenty-five patients presenting with lumbar radiculopathy refractory to conservative care. The procedures were performed on an outpatient basis and under local anaesthesia and sedation. The patients were either worked up with the MRI, (Magnetic Resonance Imaging) or intra-thecal enhanced CT scan and/or lumbar myelography. All the patients were shown to have a contained central or lateral disc herniation that correlated well with the clinical findings.

The overall results within a short-term follow-up period (mean of six months) were comparable to those obtained with standard lumbar discectomy. A low postoperative morbidity was associated with this procedure and no significant intraoperative complications have occurred.

The indications as well as the pitfalls and potential complications of this surgical technique are emphasized in this report.

This preliminary study suggests that in carefully selected patients and in experienced hands this procedure offers a viable alternative to conventional discectomy.

Introduction

In the United States alone close to eight million patients per year seek medical and surgical treatment for relief of low back and leg pain. It is estimated that every year a quarter of a million patients in the United States undergo a surgical procedure for lumbar disc herniation. The economic impact is estimated to be in the range of billions of dollars.

Even though the majority of patients treated by means of surgical intervention for disc herniation have shown significant improvement, 20–25% continue to have fair-to-poor results. In recent years there has been increasing interest in developing new methodology for the treatment of these patients. This included microsurgical techniques from the 70s up to the chemonucleolysis in the 80s.

In 1984 Dr. Gary Onik leading a team of orthopaedic surgeons, neurosurgeons and radiologists began the clinical evaluation of a new suction-cutting probe called the nucleotome. The preliminary clinical studies were carried out in 126 cases during a period of thirty months; from June 1984 to December 1986 by fifteen investigators at sixteen centres across the United States. They concluded that the nucleotome system when used as indicated for the treatment of uncomplicated herniated lumbar discs, was a safe and effective mode of treatment with no significant complications and with an overall success rate of 70%. They emphasized a careful preoperative planning, selection of patients along with meticulous intraoperative technique and fluoroscopic guidance, and the low cost of hospitalization.

In this report we are presenting our initial experience with twenty-five cases of contained lumbar disc herniations treated by the posterolateral percutenaous approach using the automated nucleotome system.

Materials and Methods

Twenty-five consecutive patients with a mean age of 39 years were treated by the percutaneous posterolateral approach using the nucleotome system. The selection criteria included the primary complaint being radicular pain, greater than back pain, at least six weeks conservative therapy, a positive neurological finding, positive straight leg raising, test, a loss of sensation following a specific dermatomal distribution and a positive radiological workup, either by intra-thecal enhanced CT and/or MRI or lumbar myelography.

Patients who have had previous low back surgery, previous intradiscal therapy, or have been diagnosed to have lateral lumbar spinal canal stenosis, free fragment of disc, degenerative facet disease, or severe hypertrophy of the ligamentum flavum, were not considered for suction discectomy.

All the patients were treated on an outpatient basis, with an average mean stay in the hospital of five hours. The operations were done under local anaesthesia and sedation. The procedure was performed with the patient in full lateral position with the symptomatic side up on a fluoroscopy operative room table. A C-arm fluoroscope system was used with video disc storage capacity and AP and lateral views of the patient's L 3 to S 1 level were obtained. Once the disc space was identified the skin was marked with a marking pen and the patient was subsequently prepped and draped and a local anaesthetic was used.

We used a posterolateral approach, approximately 10 cm from the midline as the insertion point for a 25 cm long 18 gauge stainless steel trocar with a trocar point. A 2 mm skin incision was made and the trocar was then, under C-arm control, guided towards the centre of the posterior margin of he intervertebral disc as seen on the lateral view. The trocar was then advanced to the centre of the disc space and the patient was continuously monitored for any signs of radicular pain. If the patient experienced radicular pain, the trocar was withdrawn fully and redirected. At all times we used AP and lateral images to confirm the position of the tip of the trocar in the centre of the disc space. A 2.8 mm cannula, with a dilator, was then placed over the trocar and advanced to the annulus of the disc. The dilator was removed leaving the trocar and the cannula in place. A 2 mm trephine was then placed over the trocar and advanced down with the cannula to the annulus. An approximately 2 mm hole, was then cut into the annulus and the trephine and trocar were removed, leaving the cannula in place. The nucleotome probe with a 2 mm tip was then inserted through the cannula into the disc space.

The nucleotome probe has a blunt closed tip and a single sided cutting port. The cutting blade which is pneumatically driven, was activated with a foot pedal generating an average aspiration pressure of 600 mmHg and a 180 cycle per minute cutting rate.

Results

The technique of percutaneous posterolateral approach, has now been used in twenty-five patients all of whom had contained disc herniations, either central or focal, confirmed by intra-thecally enhanced CT and/or myelography – MRI. All the patients had at least six weeks of failed conservative therapy including physical therapy modalities, anti-inflammatory agents, muscle relaxants, analgesics and an exercise programme. Except for mild paraspinal muscle spasms in the immediate postop period, no complications have occurred in the group of twenty-five patients. Two out of twenty-five patients have subsequently required open laminectomy and discectomy. In retrospect, one had a free fragment of disc and another one moderate degree of lumbar spinal stenosis.

The patients have been followed for a relatively short period of time with a mean of six months. The criteria used to determine success, included the need for no further intervention, significant or total improvement in the radicular pain, and the patient was no longer taking narcotic analgesics for leg or back pain. Additional criteria included: – that the patient's preoperative functional status must have been significantly improved postoperatively, and that both the patient and the surgeon were satisfied with the initial results of the procedure.

All patients but two, returned to work within six weeks and 50% of all successful cases returned within three weeks to gainful employment.

Discussion

In the United States low back pain is the most common source of disability to people under the age of 45 and it is estimated that approximately eight million people have a permanent disability as a result of chronic low back and leg pain. It has been estimated in the United States that some 60–80% of the population will be effected with low back pain at some point during their lives. These facts have led to the development of less invasive procedures to treat patients afflicted with low back pain.

The primary advantage of the procedure used in this group of twenty-five patients, is that there is no interference with the normal anatomical structures, and the fact that the procedure can be usually done on an outpatient basis and under local anaesthesia thus reducing the surgical trauma and the risk of intra- and postoperative complications.

Our initial experience with twenty-five cases has shown that the procedure can provide immediate relief from the symptoms offering a more economic mode of treatment with relative short recovery time. We found it to be most useful to engage these patients immediately after operation in an intensive rehabilitation programme including physical therapy. We believe that extremely careful selection of the patients is the most important feature in the final outcome, along with precise intraoperative monitoring and an aggressive approach to the rehabilitation of these low back patients. Patients who have had previous low back surgery, have central or lateral lumbar spinal stenosis, free fragments of disc or congenital anomalies of the lumbosacral spine leading to spinal stenosis, are not candidates for percutaneous discectomy. The ideal candidate for this procedure, is a young patient, with a recent history of back and leg pain, on whom adequate radiological workup has shown a central or focal disc herniation on a contained disc, with an intact annulus fibrosis and with sciatic pain greater than low back pain. This preliminary study of twenty-five consecutive patients with contained disc herniations treated by the percutaneous posterolateral approach suggests that in carefully selected patients and in experienced hands, this procedure is a viable alternative to conventional open laminectomy and discectomy.

In comparison with lumbar discectomy the results were similar. There was a low morbidity and there was a significantly reduced postoperative hospitalization.

Even though the initial experience with these twenty-five cases in encouraging, the need for a multicentre study in order to obtain further data on long-term follow-up results is emphasized.

References

1. Friedman WA (1983) Percutaneous discectomy: an alternative to chemonucleolysis? Neurosurgery 13: 542–547
2. Hijikata S, Yamagishi M, Nakayama T *et al* (1975) Percutaneous discectomy: a new treatment method for lumbar disc herniation. J Toden Hosp 5: 5–13 (Jpn)
3. Hijikata S, Nakayama T, Yamagishi M, Ichihana M (1985) Percutaneous nucleotomy for low back pain, presented at the Sicot XIV World Congress, Kyoto, Japan, October 15–20
4. Houser OW, Onofrio BM, Forbes GS *et al* (1986) Correlation of radiological features to failure of lumbar intervertebral disc chemonucleolysis. J Neurosurg 64: 736–742
5. Kambin P, Gellman H (1983) Percutaneous lateral discectomy of the lumbar spine. A preliminary report. Clin Orthop 174: 127–132
6. Maroon JC, Alba AA (1985) Microdiscectomy versus chemonucleolysis. Neurosurgery 16: 644–649
7. Onik G, Helms CA, Ginsberg L *et al* (1985) Percutaneous lumbar discectomy using a new aspiration probe. AJNR 6: 290–293; AJR 144: 1137–1142
8. Onik G, Helms CA, Ginsberg L *et al* (1985) Percutaneous lumbar discectomy using a new aspiration probe: porcine and cadaver model. Radiology 155: 251–252
9. Suezawa Y, Jacob HAC (1986) Percutaneous nucleotomy – an alternative to spinal surgery. Arch Orthop Trauma Surg (in press)

Address for correspondence: F. Gomes, M.D., Comprehensive Spine Center, 3104 W. Waters Ave., Suite 206, Tampa, FL 33614, U.S.A.

Acta Neurochirurgica, Suppl. 43, 58–62 (1988)

Automated Percutaneous Discectomy: Preliminary Experience

G. Onik, J. Maroon, A. Day, and **C. Helms**

Allegheny General Hospital, Department of Neurological Surgery, Pittsburgh, PA, U.S.A.

Summary

In 1985, Onik, described a new automated lumbar disc aspiration technique that utilizes a probe with the guillotine cutting technology used in vitrectomy and arthroscopic instrumentation. The probe's small size (2 mm in diameter) minimizes the risk of nerve injury, while its automated action permits rapid, safe removal of disc material. The technique and our preliminary results in patients treated with this method are described.

Keywords: Herniated lumbar disc; disc prolapse; percutaneous aspiration technique; indication; results.

Introduction

The treatment for herniated lumbar discs, surgical removal through laminectomy, benefits most patients, however, still entails the risk of injury to soft tissues, joints and neural structures and can lead to prolonged disability. To eliminate the complications associated with open surgical removal of lumbar disc herniations, percutaneous methods such as chemonucleolysis have been introduced. Chymopapain, however, has been associated with complications, including anaphalyxis, subarachnoid haemorrhage, and transverse myelilitis with associated paraplegia which have greatly curtailed the use of this procedure. Percutaneous discectomy by mechanically decompressing the disc, hopes to gain the beneficial effect of chymopapain without its associated complications.

As previously practiced[1–7], percutenaous discectomy methods have not become popular or widely used due to the need for relatively large cannulas to gain access to the disc space and the use of pituitary forceps to remove the disc material in a piecemeal fashion.

In 1985, Onik[8–11] introduced a new percutaneous technique that utilizes an aspiration probe called a Nucleotome* to remove the disc material. The use of this automated instrument has allowed the down-sizing of the instrumentation, the probe itself is 2 mm in diameter, while still allowing relatively rapid removal of the disc material. We will present here the preliminary results of a multi-institutional study to evaluate percutaneous automated discectomy with this instrumentation.

* Surgical Dynamics.

Materials and Methods

Patient Selection

The following criteria were used in assessing candidates for automated percutaneous discectomy: a) a major complaint of sciatica (leg pain must be unilateral, leg pain greater than the back pain); b) a history of paraesthetic discomfort in a specific dermatomal distribution; c) positive findings on a straight leg raising test, cross-over pain, or a positive bow string sign; and d) the presence of two of four possible neurological findings (wasting, weakness, sensory alteration, and reflex alteration). To be included in this prospective study, the patient had to satisfy at least two of the four above criteria. In addition, all patients underwent a computed tomography scan, or MRI scan which showed a herniated nucleus pulposus in an area consistent with the clinical findings. Patients with a positive myelogram still underwent a CT examination to confirm the diagnosis and to exclude other lumbar abnormality. Patients must have had at least 6 weeks of conservative therapy without success and must otherwise been candidates for a laminectomy. Patients were excluded from the study if they had any of the following: a) history of previous lumbar surgery; b) previous chymopapain injection; c) Workmen's compensation claim; d) any other cause of back pain as revealed on the CT scan, such as severe degenerative facet disease, lateral recess stenosis, or evidence of a free fragment; and e) any contraindication to local anaesthesia.

Procedure

The procedure itself is conducted in either the Operating Room or Radiology Department under C-ARM fluoroscopic control. A strict, sterile procedure was followed, and anaesthesia staff was on stand-by. The procedure was carried out under local anaesthesia; it is contraindicated to perform the procedure under general anaesthesia due to the risk of neural injury. The procedure was performed in the either the lateral dicubitus position or in the prone position.

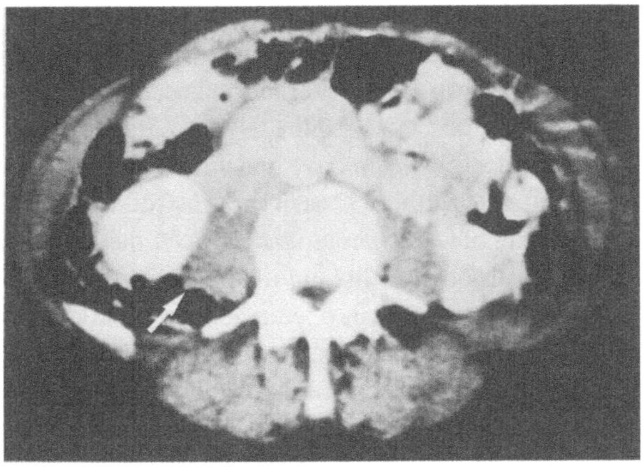

Fig. 1. CT scan through the lower abdomen at the level of the L_4 vertebral body. The arrow indicates air filled colon displaced posteriorly behind the psoas muscle

When performed in the prone position the patient is placed over a bolster on a Collis table in order to open up the disc spaces posteriorly. When performed in the lateral dicubitous position, attention again was paid to opening up the disc spaces posteriorly as well as making sure the patient did not rotate out of a straight lateral position. The entry point for the procedure is chosen from a preoperative localization CT scan obtained through the disc slice of interest on the side of the patients symptoms. A side from choosing an exact entry point for the posterolateral approach to the disc space, this scan, which is done in the prone position, rules out the presence of any retroperitoneal structures, such as bowel that may be in the path of the trocar. Recent radiologic literature indicates that marked posterior displacement of the colon, which could interfere even with this posterolateral approach, can occur in a relatively high percentage of patients (Fig. 1). By means of this posterolateral approach, first a 22 gauge Green needle (Cook Incorporated) is used to localize the exact trajectory to the disc. An 18 gauge needle with a removal hub is then placed beside this needle in a tandem fashion down to the disc space. The trocar placement is monitored in a lateral x-ray view. From the lateral view, the trocar should be parallel to and midway between the vertebral body end plates and the tip of the trocar should be directed toward the center of the disc. The point of the trocar should be touching the posterior vertebral body line when the gritty sensation of touching the annulus is felt. If the trocar is anterior to this line and the annulus is felt, the trocar is directed too anteriorly and is withdrawn and redirected. The patient is monitored continually for any sign of radicular pain. If radicular pain is experienced by the patient, the trocar is withdrawn its full length and redirected. When the annulus is felt and the tip of the trocar (on the lateral view) is at the posterior vertebral body line, the AP view is then checked to confirm that the tip of the trocar is not medial to a line connecting the medial aspect of the pedicles. This view insures that the trocar is not traversing the thecal sac. If the trocar tip is lateral to this line the trocar is then advanced into the disc till the tip is in the center of the disc on the AP view. The lateral view is then checked to confirm the central placement of the trocar (Fig. 2). The cannula, with a tapered dilator in place, is inserted over the trocar. The cannula and dilator are inserted down to the wall of the annulus, and their position is confirmed radiographically in two views. Once confirmed to be in place, the tapered dilator is removed from the cannula leaving the trocar and cannula in place. The trephine is placed over the trocar and through the cannula. The fluoroscopic unit at this point is perpendicular to the cannula and trephine confirming that they are actually against the annulus before the trephine is used. After the disc is incised, the trephine and trocar are removed leaving the cannula in place and the Nucleotom aspiration probe is placed into the disc.

After the position of the nucleotome is confirmed the instrument is turned on and the disc material is aspirated (Fig. 3). The port of the instrument is first placed in the direction of the herniation, and after no more disc material can be removed, the direction of the port is changed to remove disc material from other areas in the disc.

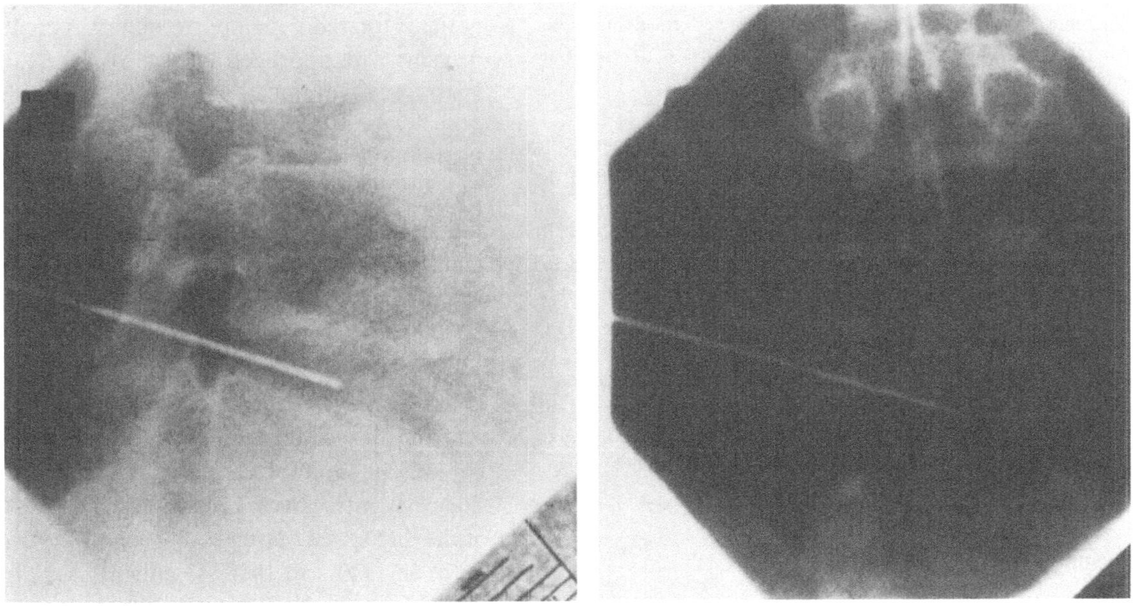

Fig. 2. AP and lateral x-ray showing the trocar in the center of the disc

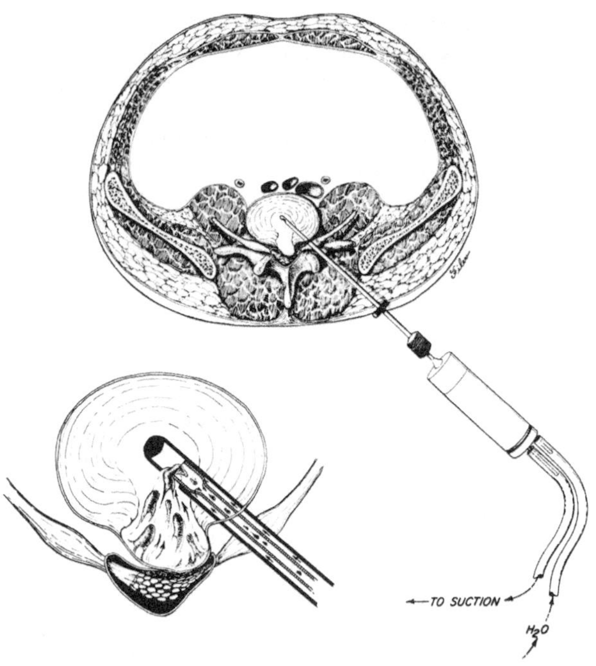

Fig. 3. Nucleotome inserted into lumbar disc showing aspiration and cutting elements

After no more disc material can be removed from these areas, the rigid cannula can then be angled to change the position of the Nucleotome within the disc. As much disc material as possible is removed through a single puncture site. The disc material, however, is aspirated for at least 40 minutes in each case. While the instrument is actively removing disc material, firm pressure, directed toward the annulus is applied to the cannula as a precaution to keep it against the annulus. When no further disc material can be removed, the nucleotome is turned off and pulled back into the cannula and both are removed. The puncture site is then covered with a bandage. In most cases, the patient is held for three hours and then discharged the same day of surgery. When the surgery is performed later in the afternoon or the patient does not have a relative or friend to accompany him home, the patient is held overnight.

Patient Evaluation

The patients were evaluated by the physician who performed the operation or his associate immediately after the procedure, at 6 weeks, 6 months, and 6 month intervals thereafter. The success of the procedure was judged on criteria that considered the need for further intervention and the patient's pain and functional status. Therapy was judged successful if the radicular pain was moderately to totally improved; if the patient was no longer taking narcotic analgesics; if the patient's functional status was impaired before the procedure, the patient's functional status was improved; and if the patient and physician were both satisfied with the results. If any of these criteria were not met, the treatment was considered unsuccessful.

Results

At the time of this writing, 120 patients have met the prospective criteria for the study and have had the procedure and have at least 6 weeks follow-up. Ninety-three of these patients have a follow-up of 6 months or greater. Using the criteria for success previously outlined for this study, 74% of the patients with at least 6 weeks follow-up have had successful results. Of those patients with at least 6 months follow-up 71% have met all of the criteria for success. Of the 16 failures in whom surgical results are available, 8 had demonstrated free fragments at operation. Seven patients had herniations under the posterior longitudinal ligament or bulging discs. One patient had lateral recess stenosis. Fifteen patients had no surgical results at this time.

In the 120 patients, no serious complications were reported. There was no nerve damage or great vessel injury associated with the procedure. There was one instance of vaso-vagal syncopy associated with the procedure and the one presumably psoas haematoma which resulted in 4 days of paresthesias in the distribution of the lateral femoral cutaneous nerve on the side of the procedure. This haematoma was of no haemodynamic significance and the patient was still discharged the same day of the procedure. In November of 1986, the instrumentation for the procedure was released for sale and since that time over 2,000 procedures have been completed outside of this study protocol, again, with no serious complications reported[12].

Discussion

In 1975, Hijikata *et al.* presented their percutaneous technique for herniated lumbar discs[1]. In 1978, they reported a 68% good to excellent results in the 80 patients treated with a posterolateral percutaneous approach[2]. This approach consisted of inserting a cannula against the annulus and then, removing the disc material with a long grasping forceps after a hole was made in the annulus. They subsequently reported their experience with this technique in over 100 patients with the same positive results[3]. In 1983, Kambin and Gellman, using a similar approach, reported their results in 9 patients who were all relieved of their leg pain with no reported serious complications. Suezawa and Jacobson in Switzerland in 1979, carried out percutaneous discectomies by means of a forceps introduced into the disc space similar to the procedure used by Hijikata in Japan. In addition, from the contralateral side, they introduced a discoscope to view the interior of the disc space during extraction of the disc material. Between 1979 and 1985, 40 patients were treated with this technique; they reported 70% success rate. Again, no complications were reported. In 1981, Jacobson (written communication)[6] developed a percutaneous discectomy technique that involved a straight lateral approach in which a 40-French chest tube was inserted against the lateral annulus. Again, a grasping forceps were used to remove the nucleus pulposus. In 1983, Freedman[7], using Jacobson's technique confirmed that good results were possible but emphasized the potentials risks of bowel perforation, neural injury and vascular injury. Subsequently, he abandoned the procedure. All of these techniques removed the disc material by hand with a grasping forceps. Consequently, the procedure needed large cannulas to gain access to the disc space increasing the potential for nerve injury upon introduction, while still being tedious and time consuming.

Percutaneous automated discectomy, as this report describes, takes advantage of a reciprocating suction cutter than can cut separate pieces of discs up to 180 times per minute. The procedure can therefore be completed in a reasonable period of time while minimizing the size of the introduction cannula to 2.8 mm. This, then, reduces the possibility of nerve injury. An added advantage of this technique is that once the instrument is placed within the disc, the operation can usually be completed without removing it thereby theoretically decreasing the risk of infection.

The advantages of the procedure over the traditional laminectomy are that no general anaesthesia is needed, and because trauma to the patient is minimized, early ambulation is possible. The procedure is now being carried out routinely on an outpatient basis. Perhaps the major benefit of the procedure is that the route taken to the disc does not violate the spinal canal and the procedure, therefore, is not associated with epidural fibrosis. If a patient fails with a percutaneous discectomy, they are, therefore, not precluded from having a good result from a laminectomy that follows. Lastly, only a small portion of the disc material can be removed through this procedure which has resulted in only minimal disc space narrowing at the level of the procedure. Therefore, this procedure may only have minimal biomechanical affects.

This procedure appears to have minimal morbidity as indicated by the study data and data that is now being accumulated by application of the technique in the community. It is remarkable that at this point in the learning process that no serious complications have occurred in over 2,000 cases performed[12].

Although there have been no significant complications, the potential for injury to nerves and great vessels still exists. Several precautions have been incorporated into the procedure to minimize this risk of injury. Perhaps the most important precaution for the procedure is that it is carried out under local anaesthesia. Therefore, if there is a the nerve in the path of the trocar, the patient reacts and the placement of the instrument can be readjusted. The instrument and procedure have been carefully designed to insure that the instrument stays within the safe confines of the disc space, isolated from surrounding nerves and vessels. The instrument is blunt-ended so it cannot be pushed through the opposite annulus inadvertantly, as has been shown in cadaver experiments[8]. The nature of the annulus is such that it cannot be deformed into the port of the instrument. This fact, coupled with the fact that the port of the instrument is recessed from the end of the probe, insures that the instrument cannot cut its way out of the disc space. With careful attention to technique, complications of this kind should not occur. Lastly, CT planning allows an exact entry point to be chosen for the procedure making sure that the intended path to the disc does not traverse abnormally placed bowel or vessels. It is clear from the radiologic literature that this bowel displacement occurs in the prone position in a significant number of patients. This fact probably accounts for some of the infections that were seen with chymopapaine and discography.

To balance against the advantage of this procedure is its major disadvantage, a lower success rate than in open laminectomy. This decreased sucess rate is secondary to the unrecognized free fragments of discs that have broken out of the disc space and extruded through the posterior longitudinal ligament. The selection of patients, by ruling out those patients with extruded fragments may improve in the future with more experience with MRI scanning of the lumbar spine. In addition, the role of discography and CT discography are being examined to see if they will help in this regard.

Only about 10% of the cases in the series were performed at the L 5–S 1 level. The approach to this disc space is much more difficult due to the obstruction to the disc space by the iliac crest. New instrumentation is now available that consists of a curved cannula through which the flexible Neucleotom can be passed which allows approach to this space in a consistent manner. We are now doing the L 5–S 1 disc on a routine basis without significant problems using this new instrumentation. The success of doing the L 5–S 1 disc is largely dependent on careful attention to the anatomy, particularly an assessment of the space between the L 5 transverse process and the sacrum.

In conclusion, automated percutaneous discectomy,

as indicated by the preliminary data, appears to be an efficacious means of treating uncomplicated lumbar disc herniations. Clearly, larger patient numbers with longer follow-up is needed to determine what future role this procedure will play in the management of these patients. Certainly, however, this study along with the experience now being gained in the community indicates that the procedure has a very low morbidity rate.

References

1. Hijikata S, Yamagishi M, Nakayama T, Oomori K (1975) Percutaneous discectomy: a new treatment method for lumbar disc herniation. J Toden Hosp 5: 5–13
2. Hijikata S, Nakayama T, Yamagishi M, Ichihara M (1978) Percutaneous nuclectomy for low back pain. Presented at the 14th World Congress of the Societé Internationale de Chirurgie Orthopédique et de Traumatologie, Kyoto, Japan, October 15–20, 1978
3. Hijikata S (1981) Percutaneous nuclectomy for low back pain: the second report. Presented at the 15th World Congress of the Societé International de Chirurgie Orthopédique et de Traumatologie, Rio de Janeiro, August 30–September 4, 1980
4. Kambin P, Gellman H (1983) Percutaneous lateral discectomy of the lumbar spine. Clin Orthop 174: 127–132
5. Suezawa Y, Jacob HAC. Percutaneous nucleotomy: an alternative to spinal surgery. Arch Orthop Trauma Surg (in press)
6. Personal communication, 1986
7. Friedman WA (1983) Percutaneous discectomy: an alternative to chemonucleolysis. Neurosurgery 13: 542–547
8. Onik G, Helms C, Ginsberg L *et al* (1985) Percutaneous lumbar discectomy using a new aspiration probe: procine and cadaver model. Radiology 155: 251–252
9. Onik G, Helms C, Ginsburg L, Hoaglund F, Morris J (1985) Percutaneous lumbar discectomy using a new aspiration probe. AJNR 6: 290–293
10. Onik G, Maroon J, Helms C *et al* (1987) Automated percutaneous discectomy initial patient experience. Radiology 162: 129–132
11. Maroon J, Onik G (1987) Percutaneous automated discectomy: a new method for lumbar disc removal. Neurosurg 66: 143–146
12. Written communication, Ron Allan Surgical Dynamics, Inc. May 87

Address for correspondence: J. Maroon, M.D., Director of Neurosurgery, Allegheny General Hospital, 320 East North Avenue, Pittsburgh, PA 15212-9986, U.S.A.

Acta Neurochirurgica, Suppl. 43, 63–68 (1988)

Atlas Fractures: Treatment and Long-term Results

V. K. H. Sonntag, M. N. Hadley, C. A. Dickman, and **C. M. Browner**

Barrow Neurological Institute, Phoenix, Arizona, U.S.A.

Summary

Fractures of the first cervical vertebra (C 1) represent approximately 7% of all acute cervical spine fractures. Frequently, the atlas will be fractured in combination with the axis (44% of cases). Treatment of isolated C 1 fractures should be governed by the rules of Spence. The treatment of combination C 1–C 2 fractures is dictated by the type and severity of the C 2 fracture. Experience with 57 cases of acute traumatic atlas fractures are reviewed. The long-term outcome from an atlas fracture is good (median follow-up, 40 months).

Keywords: Atlas fractures; cervical spine; spine fractures; combination atlas-axis fractures; spine trauma.

Introduction

The atlas, uniquely positioned between the skull and the remainder of the vertebral column, is predisposed to a variety of acute traumatic injuries. The incidence of acute cervical one vertebral (C 1) fractures ranges between 3% and 13%; despite this relative frequency, little attention has been give to atlas fractures in the literature[1, 10, 13, 15, 16]. This review describes the management of 57 patients with acute atlas fractures and outlines their diagnostic work-up, treatment, and long-term outcome. The management principles that govern the treatment of atlas fractures and combination atlas-axis injuries will be presented.

Material and Methods

Fifty-seven patients who sustained acute traumatic atlas fractures were identified among 860 patients treated for acute cervical vertebral column injuries between 1976 and 1986. Individuals with acute occipito-atlantal or atlanto-axial dislocations, congenital abnormalities of C 1, atlas pathology due to neoplasm or rheumatoid disease, or gunshot injuries to the atlas were excluded.

The medical records and radiographs of each patient were reviewed, and each case was characterized according to type of fracture, aetiology of injury, degree of neurologic impairment, treatment administered, and outcome. Follow-up was achieved in 52 of 57 patients (91%) with a median follow-up of 40 months (range, 2 to 101 months).

Results

The 57 patients with acute atlas fractures represented 6.6% of the total population of acute cervical spine injured patients treated over a period of 10 years. Thirty-four patients were male with a median age of 40 years (range, 14 to 82 years), and 23 patients were female with a median age of 42 years (range, 17 to 86 years). Thirty-two patients had isolated C 1 fractures, and 25 sustained combination C 1–C 2 fractures. Motor vehicle accidents were the most common cause of injury (68% of fractures) followed by falls (13%) and motorcycle accidents (7%). Twelve percent of the patients sustained an atlas fracture from miscellaneous causes. The diagnostic radiographic studies performed on these patients included plain cervical radiographs in all cases (anteroposterior, lateral, and open-mouth odontoid views), tomography (18 patients), and computed tomography (CT) in 30 patients.

Not one of the 32 patients with an isolated C 1 fracture sustained neurological injury (0%). Several complained of dysesthesias at the base of the skull and neck at presentation. Three patients with acute combination C 1–C 2 fractures had neurological deficits upon admission (3 of 25 = 12%). Two patients had central-cord findings, the third had a complete neurological injury at the C 2 level. The latter patient died on day 4 postadmission and represents the only death in the series.

Isolated Atlas Fractures (N = 32)

The majority of isolated C 1 fractures (18 of 32 = 56%) were acute bilateral or multiple fractures of the ring of the atlas (Fig. 1). There were 10 unilateral ring fractures (31%) (Fig. 2 A and B) and 4 isolated lateral mass fractures (13%). Twelve of the isolated atlas fractures were without C 1–C 2 dislocation as determined by antero-

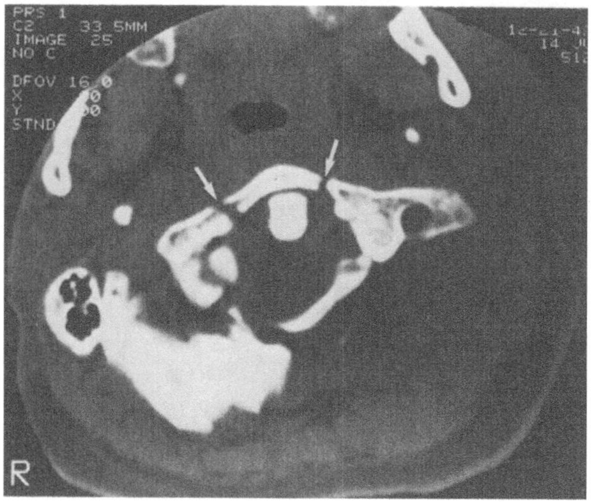

Fig. 1. Multiple ring fractures of the arch of C 1 (arrows)

posterior (AP), open-mouth view radiographs. Fifteen of the fractures had less than 6.9 mm of displacement of C 1 and C 2 (determined by measuring and summing the spread of the lateral masses on the AP x-ray), and 5 fractures were displaced by greater than 6.9 mm.

Treatment of the 32 isolated C 1 fractures depended on the degree of C 1–C 2 lateral mass dislocation (Table 1). Not one of 32 patients required surgical stabilization or fusion. The median length of treatment for the group was 12 weeks with a range of 8 to 16 weeks. Three of 32 patients were lost to follow-up. The remaining 29 patients had no evidence of nonunion or instability on follow-up radiographs, including dynamic studies (median follow-up, 40 months).

Combination Atlas-Axis Fractures (N = 25)

The most common type of C 1–C 2 combination fracture was a C 1-odontoid Type II fracture (10 of 25 = 40%) (Fig. 3 A, B and C). There were 5 C 1-odontoid Type III injuries (20%), 3 C 1-Hangman's fractures

Table 1. *Treatment*

Isolated Atlas Fx		
* Nondisplaced	(N = 12)	1 Soft Collar 10 Rigid Collar 1 SOMI Brace
* Displaced < 7.0 mm	(N = 15)	8 Rigid Collar 3 SOMI Brace 4 Halo Vest
* Displaced ≥ 7.0 mm	(N = 5)	5 Halo Vest (Median TX = 12 weeks) (Range 8 to 16 weeks)

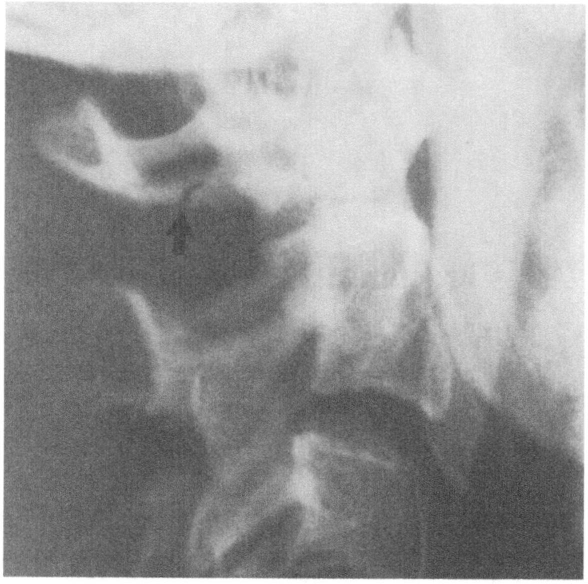

A

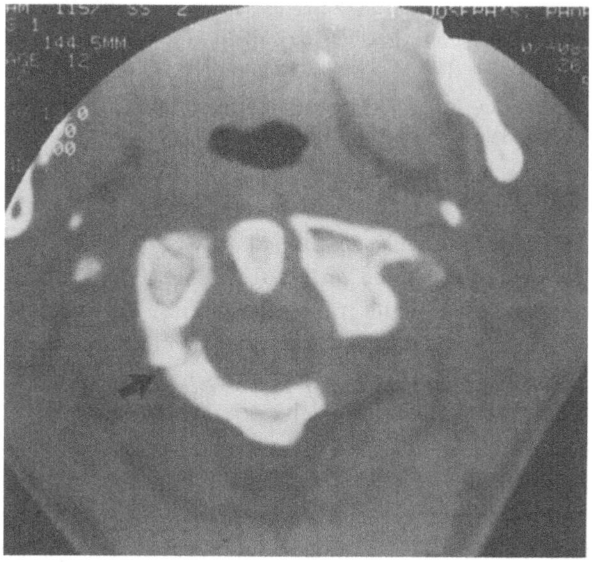

B

Fig. 2. (A) Lateral x-ray reveals C 1 arch fracture (arrows). (B) CT of atlas depicts unilateral fracture (arrow)

(12%), and 7 C 1-miscellaneous C 2 fractures (nonodontoid, nonHangman's fractures) representing 28% of the combination-injury population. The distribution of atlas fractures in these combination injuries was similar to that identified for the isolated C 1-fracture group.

Three patients in this group sustained neurological deficits from their combination injuries. The one pa-

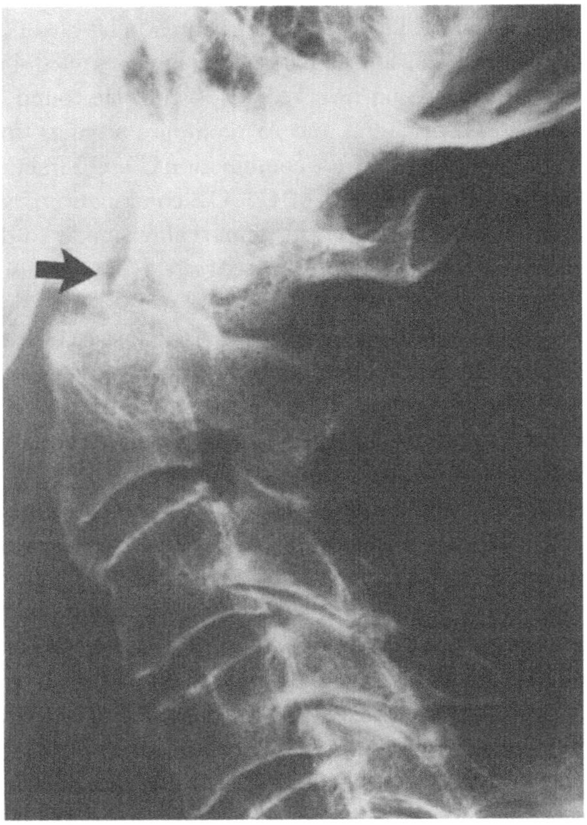

A

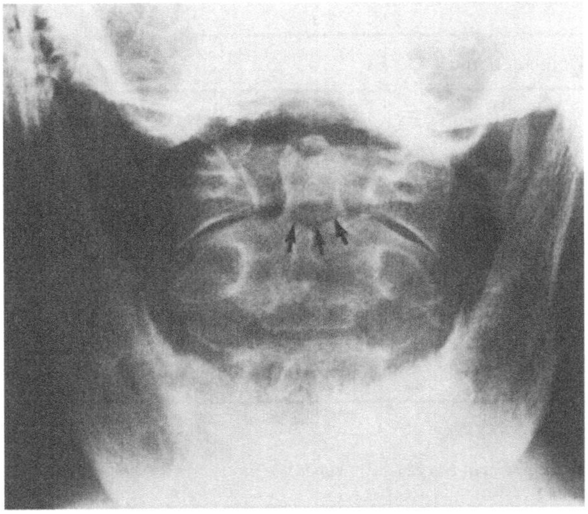

B

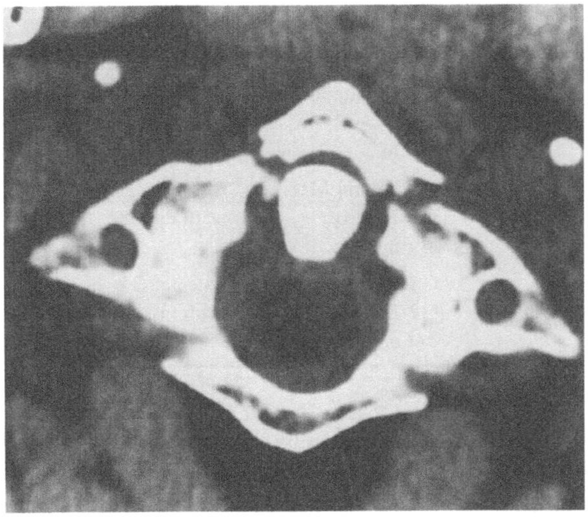

C

Fig. 3. (A) Lateral radiograph depicts Type II odontoid fracture (arrow). (B) AP x-ray documents fracture at base of the odontoid (arrows). (C) CT of atlas reveals multiple anterior ring fractures (arrows)

tient with the complete C 2 level injury had a combination C 1-odontoid Type II fracture. He sustained severe thoracic and abdominal injuries and died from cardiopulmonary complications on his 4th hospital day. The two other patients had central cord injuries from combination C 1-miscellaneous C 2 fractures. (Both had C 2 body fractures.) These two patients demonstrated improvement in neurological function over time; however, both had persistent neurological sequellae at last follow-up (36 and 54 months respectively).

The treatment of the 24 surviving combination C 1–C 2 fracture patients is presented in Table 2. Treatment was determined by the type of C 2 fracture present. The median duration of external immobilization for these patients was 12 weeks with a range of 10 to 22 weeks. Four patients with C 1-odontoid Type II fractures with dens dislocation greater than 6.0 mm were treated with early surgical reduction and stabilization (Table 3).

Two of 24 patients were lost to follow-up (1 C 1-Hangman's fracture and 1 C 1-miscellaneous C 2 fracture). None of the remaining 13 patients with C 1-Hangman's, C 1-odontoid Type III, or C 1-miscellaneous C 2 fractures had evidence of nonunion or instability on follow-up posttreatment radiographic studies. All 9 C 1-odontoid Type II fracture patients were identified for follow-up. The 4 who underwent early surgery had evidence of fusion without instability at last evaluation. Four of 5 patients with C 1-odontoid Type II fractures treated with 12 weeks of Halo vest immobi-

Table 2. *Treatment*

Combination C 1-C 2 Fx		
* C 1-Hangman's Fx	(N = 3)	3 Halo Vest
* C 1-Odontoid II Fx	(N = 9)	5 Halo Vest 4 ORIF
* C 1-Odontoid III Fx	(N = 5)	1 SOMI Brace 4 Halo Vest
* C 1-Miscellaneous Fx	(N = 7)	1 SOMI Brace 6 Halo Vest (Median TX 12 weeks) (Range 10–22 weeks)

Table 3. *Early Surgical Treatment*

Patient No.	Injury Type	Procedure Performed
(1)	C 1-Unilateral Arch C 2 Type II > 6 mm	C 1–C 2 wiring with fusion
(2)	C 1-Unilateral Arch C 2-Type II > 6 mm	C 1–C 2 wiring with fusion
(3)	C 1-Lateral Mass C 2 Type II > 6 mm	C 1–C 2 wiring with fusion
(4)	C 1-Bilateral Arch C 2-Type II > 6 mm	Occiput-C 2 wiring with fusion

lization had good bony union without evidence of subluxation on follow-up flexion and extension radiographs. The fifth patient, a 34-year-old with an odontoid Type II dens dislocation of 5 mm had persistent nonunion with recurrent dens dislocation despite good alignment in the Halo immobilization vest. He underwent operative wiring of C 1–C 2 and iliac crest bone graft fusion of C 1–C 2 at 12 weeks postinjury. He had a good bony fusion without evidence of subluxation 18 months later. At last follow-up (median 41 months), no patient with an acute atlas fracture had developed new neurological symptoms.

Discussion

The atlas, interposed between the skull and the remainder of the cervical spine, has a morphology distinctly different from other vertebral segments[7, 11, 14]. Axial loading injuries are the most common forces which lead to atlas fractures, although a variety of forces translated to the atlas can result in fracture-dislocation[7, 11, 14].

The Jefferson fracture or bursting atlas fracture is the hallmark C 1 axial loading fracture. In his historical report of 1920, Sir Geoffrey Jefferson presented 4 of his own cases and reviewed 42 others[7]. He found 21 isolated C 1 fractures and 25 "complicated atlas fractures", 19 of which were combination C 1–C 2 fracture injuries. The incidence of C 1–C 2 combination fractures in his report (41%) is remarkably similar to the incidence discovered in the present series (25 of 57 cases, 44%).

The treatment of atlas fractures is almost exclusively nonsurgical (particularly for isolated C 1 fractures) and depends on the type of atlas fracture and the presence of associated axis or other cervical vertebral body injuries. Thin-section computerized tomography (CT) is the radiographic diagnostic study of choice to accurately define the extent of a C 1 (and/or C 2) fracture-injury[2, 9, 12, 15]. Correct radiographic diagnoses can be made with plain radiographs; however, only polytomography or CT can unequivocally delineate the precise nature of the fracture(s) in difficult cases[2, 9, 12].

Spence *et al.*[16] have provided criteria that help determine appropriate therapy for isolated C 1 fractures, uncomplicated by an axis fracture or other associated cervical spine fracture (for which more rigid external immobilization or even surgery may be required). They documented that if the sum of the spread of the lateral masses of C 1 over C 2 as determined on the AP C 1–C 2 radiograph exceeded 6.9 mm, then the likelihood of transverse ligament disruption was great. In cadaver studies this degree of dislocation was associated with transverse ligament disruption and C 1–C 2 instability. The authors concluded that injuries of this magnitude required a more aggressive approach to treatment and advocated surgical stabilization of C 1 and C 2[16].

Experience has demonstrated that isolated acute C 1 fractures rarely require early surgical intervention (in the absence of bone compromising the spinal cord)[1, 7, 11, 13, 15]. Spence's recommendations are nonetheless applicable. Isolated atlas fractures with spread of the lateral masses less than 6.9 mm on the AP x-ray can be effectively treated with less rigid cervical support (typically a Philadelphia collar), for a duration of 8 to 12 weeks (Fig. 4 A and B). More rigid external immobilization is advocated for atlas fractures with 6.9 mm (or greater) dislocation of the lateral masses. This is most effectively accomplished with the Halo vest immobilization device for a duration of 10 to 14 weeks[15]. Periodic clinical and radiological follow-up (including flexion and extension x-rays) of these patients is essential to document patient compliance and adequate

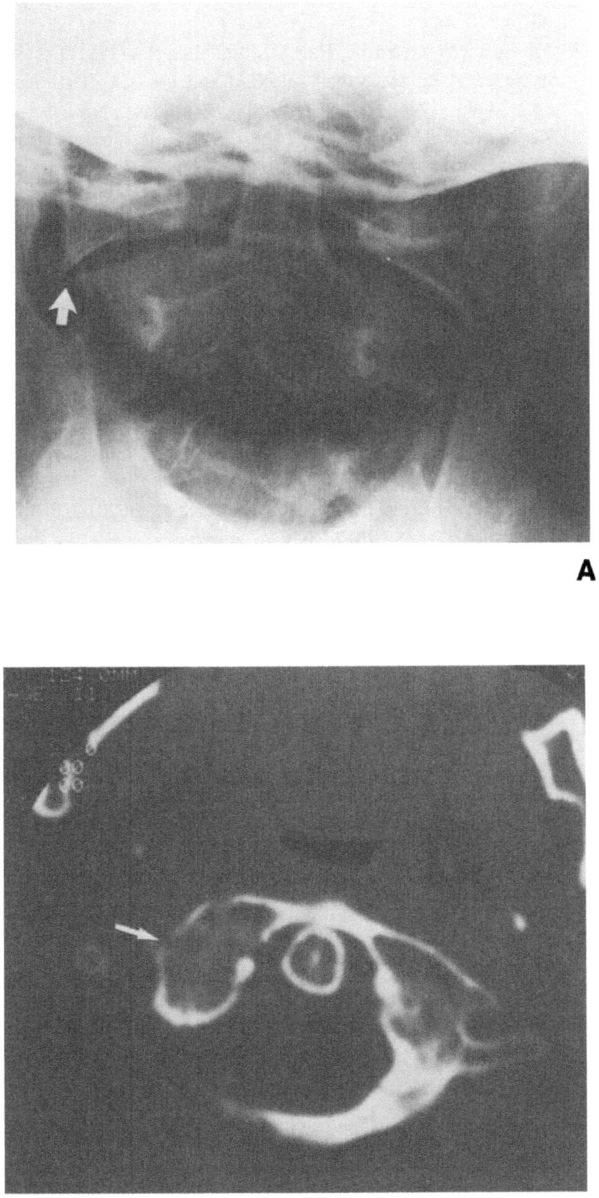

A

B

Fig. 4. (A) AP x-ray reveals minimal offset at C 1–C 2 facet interface on right (arrow) and none on left, implying a stable fracture. (B) CT study reveals right lateral mass fracture without disruption of the ring of C 1

alignment and fusion. A small percentage of isolated C 1 fractures will have persistent C 1–C 2 instability and will require delayed surgical stabilization[1, 11, 13, 16].

The treatment of combination C 1–C 2 fractures is more complicated. Lipson has advocated external immobilization in the Halo vest for 6 to 8 weeks to allow the *atlas* fracture to heal, followed by surgical wiring and fusion as definitive treatment of the *axis* fracture-dislocation[10]. Recent experience with the management of patients with axis fractures (combined with the knowledge that not all atlas fractures are unstable) has led us to conclude that a more refined approach to combination C 1–C 2 fractures is necessary[3–6, 11, 15]. A unilateral arch fracture of the atlas or an isolated C 1 lateral mass fracture does not preclude early surgical wiring of C 1 to C 2 in the treatment of combination C 1–C 2 fracture injuries. If the atlas fracture is of a bursting type (bilateral or multiple arch fractures), then C 1 cannot be incorporated in the wiring, and an occiput to C 2 wiring and fusion procedure should be considered. The additional immobilization of patients for 6 to 8 weeks prior to definitive surgical therapy of their combination C 1 to C 2 injury as advocated by Lipson is thus obviated.

The decision to perform early surgery (within 3 to 10 days postinjury) on patients with combination C 1–C 2 fractures is based upon the type of axis fracture present. Several large series with acute C 2 fractures reveal that Hangman's fractures, odontoid Type III fractures, and virtually all miscellaneous C 2 fractures heal with external immobilization alone when treated for 8 to 14 weeks[1, 3, 4, 6, 11, 15]. Odontoid Type II fractures with odontoid dislocations less than 5.0 mm to 6.0 mm heal well with rigid external immobilization (90%) if adequate alignment and immobilization can be maintained for the duration of treatment[3, 6, 15]. Type II odontoid fractures with dens dislocations of 6.0 mm or greater have a high incidence of nonunion (70%) despite rigid external immobilization regardless of patient's age, the direction of den's dislocation, or the degree of neurological impairment[6, 15]. These patients should be considered for early surgical stabilization.

Conclusions

We recommend that patients with combination C 1–C 2 fractures be evaluated carefully with conventional radiographs and CT. Based on these studies, the identification of the features of both the atlas and axis fractures should be possible. If the axis fracture component of the C 1–C 2 combination injury is an odontoid Type II fracture with 6.0 mm dens dislocation (or greater), then early operative intervention should be considered. The type of atlas fracture present will dictate whether C 1 and C 2 will be the only levels included in the surgical wiring and fusion (unilateral arch fractures or lateral mass fracture), or whether the atlas injury is unstable (bilateral or multiple ring fractures) and requires an occiput to C 2 wiring and fusion procedure. All other axis fractures (Hangman's, Odontoid Type III and miscellaneous fractures) encountered in combination with a C1 fracture should be initially managed with rigid external immobilization. We favor the Halo vest for its superior immobilization characteristics[8, 15, 17]. All patients require compulsive periodic follow-up to assess the effectiveness of therapy. Nonunion can occur (as in one of the patients presented) and late surgical therapy may be required[6, 10, 15].

References

1. Bohlman HH (1979) Acute fractures and dislocations of the cervical spine: An analysis of three hundred hospitalized patients and review of the literature. J Bone Joint Surg 61-A(8): 1119–1142
2. Brant-Zawadski M, Miller EM, Federle MP (1981) CT in the evaluation of spine trauma. AJR 136: 369–375
3. Clark CR, White AA III, Cooper P (1985) Fractures of the dens: A multicenter study. Presented at the 35th Annual Meeting of the Congress of Neurological Surgeons, Honolulu, Hawai, September, 1985
4. Hadley MN, Browner C, Sonntag VKH (1985) Axis fractures: A comprehensive review of management and treatment in 107 cases. Neurosurgery 17(2): 281–290
5. Hadley MN, Browner C, Sonntag VKH (1985) Miscellaneous fractures of the second cervical vertebra. BNI Quarterly 1(4): 34–39
6. Hadley MN, Sonntag VKH (1986) Acute axis fractures. Contemporary Neurosurgery 9(2): 1–6
7. Jefferson G (1920) Fracture of the atlas vertebra. Report of four cases, and a review of those previously recorded. Br J Surg 7: 407–422
8. Johnson RM, Owen JR, Hart DL, Callahan RA (1981) Cervical orthoses: A guide to their selection and use. Clin Orthopaed 154: 34–45
9. Keene GCR, Hone MR, Sage MR (1978) Atlas fracture: Demonstration using computerized tomography. J Bone Joint Surg 60-A: 1106–1107
10. Lipson SJ (1977) Fractures of the atlas associated with fractures of the odontoid process and transverse ligament ruptures. J Bone Joint Surg 59-A(7): 940–943
11. Pierce DS, Barr JS (1983) Fractures and dislocations at the base of the skull and upper cervical spine. In: The cervical spine research society (ed) The cervical spine. JB Lippincott, Philadelphia, PA, pp 196–206
12. Prager PJ (1983) Differential diagnosis and radiological workup in bilateral lateral atlantoaxial offset. Eur J Radiol 3(4): 309–313
13. Sherk HH, Nicholson JT (1970) Fractures of the atlas. J Bone Joint Surg 52-A(5): 1017–1024
14. Sköld G (1983) Fractures of the arches of the atlas: A study of their causation. Z Rechtsmed 90(4): 247–258
15. Sonntag VKH, Hadley MN (1988) Nonoperative management of cervical spine injuries. Clin Neurosurg 34: 630–649
16. Spence KF, Decker S, Sell KW (1970) Bursting atlantal fracture associated with rupture of the transverse ligament. J Bone Joint Surg 52-A(3): 543–549
17. Wolf JW, Johnson RM (1983) Cervical orthoses. In: The cervical spine research society (ed) The cervical spine. JB Lippincott, Philadelphia, PA, pp 54–61

Address for correspondence: V. K. H. Sonntag, M.D., F.A.C.S., Editorial Office, Barrow Neurological Institute, 350 West Thomas Road, Phoenix, AZ 85013, U.S.A.

Acta Neurochirurgica, Suppl. 43, 69–74 (1988)

The Transoral Approach to the Anterior Superior Cervical Spine. A Review of 29 Cases

R. F. Spetzler, M. N. Hadley, and V. K. H. Sonntag

Division of Neurological Surgery, Barrow Neurological Institute, Phoenix, Arizona, U.S.A.

Summary

The transoral surgical approach is the most direct operative approach to pathology ventral to the brain stem and superior spinal cord. This approach is useful in the treatment of extradural compressive cervicomedullary lesions in selected patients.

Twenty-nine patients with ventral extradural brain stem or cervical cord compressive lesions who were treated via the transoral surgical approach are presented. Their evaluation, treatment, and long-term outcome will be reviewed (median follow-up, 24 months). Specific features of the transoral procedure, including intraoperative monitoring, methods of retraction, and microsurgical techniques that contributed to minimal operative morbidity (3%) and zero operative mortality, will be discussed.

Keywords: Transoral operative approach; cervicomedullary junction; brain stem compression; cervical spine; odontoid process; treatment; results.

Introduction

Several pathological processes involving the craniovertebral junction can lead to compromise of the brain stem and superior cervical spine. When the offending lesion is ventral to the spinal cord at these levels, the transoral surgical approach represents an efficacious, direct operative option. The most common application of the procedure has been for resection of the odontoid process secondary to chronic traumatic dens dislocation or for basilar invagination from rheumatoid disease or congenital deformities of the skull base. Tumours may be treated via this approach in selected cases. Extradural tumour masses can be resected by this route provided they are not too broad-based. Intradural pathology has been treated via the transoral-transclival approach; however, there is a high morbidity and mortality from these procedures due to cerebrospinal fluid (CSF) fistula and meningitis[2, 5, 7, 19].

In the following discussion, specific operative techniques including methods of retraction and intraoperative monitoring, which contributed to the effectiveness of this procedure in 29 consecutive cases of extradural cervicomedullary junction pathology, will be reviewed (Table 1).

Indications

Patients with high cervical cord compression usually present with signs and symptoms of a progressive spastic quadriparesis. The initial complaints may be those of mild arm or leg weakness or minimal bladder or bowel dysfunction. Patients may have vague sensory complaints including numbness, pain, dysesthesias, and/or mild incoordination. As the disease progresses, hyperreflexia with Babinski and Hoffman signs can be present. Patients with brain stem compromise from basilar invagination may have lower cranial nerve dysfunction in addition to their myelopathic findings. The presenting symptoms and physical findings of the 29 patients treated via the transoral approach for ventral extradural cervicomedullary junction pathology are outlined in Tables 2 and 3.

Table 1. *Outline of Pathology at the Craniovertebral Junction Treated via Transoral Approach*

Craniocervical Junction Pathology		
Odontoid Pathology:	23	
Congenital		7
Rheumatoid arthritis		13
Traumatic C 1–2 sublux		3
Other Pathology:	6	
Extradural met. neoplasm		2
Extradural tumor		4
Total	29	

Table 2. *Neurological Symptoms (29 Patients)*

Dysesthesias	29
Incoordination/clumsiness	29
Numbness	26
Pain	19
Voice change	12
Urinary incontinence/frequency	6
Difficulty swallowing	4

Table 3. *Neurological Signs (29 Patients)*

	Upper Ext.		Lower Ext.
Decrease range of head/ neck motion		29	
Nasal voice		9	
Decreased gag/palatal lift		6	
Uncoordinated swallow		3	
Reduced rectal sphincter tone		8	
Sensory dysfunction			
Pin/touch	5		3
Position	0		9
Motor weakness	24		27
Muscle atrophy	10		3
Spasticity	15		29
Cerebellar dysfunction	0		0
Ataxia	12		29

It is essential to define the precise nature and location of the craniocervical pathology and the mechanism(s) of brain stem or spinal cord compression[7, 10, 15, 18]. In some patients with rheumatoid arthritis and other forms of C 1–C 2 instability, the displacement and resultant cervicomedullary compression can be reduced. If the subluxation is reducible, then a posterior fusion from C 1 to C 2 or C 3 can often stop the progression of neurologic deficits and effect some improvement. In the patient with a subluxation that is not reducible, a posterior fusion may be hazardous[4, 16, 17].

In certain patients, particularly those with basilar invagination, a combination of surgical procedures may be required. When ventral pathology is present it should be treated first, and if indicated, a posterior procedure should follow to provide stabilization (and to remove dorsal compressive pathology if present).

The decision of which surgical approach to employ for an individual patient is based on the findings on preoperative radiographic studies, which include plain skull base and spine roentgenograms (including dynamic flexion and extension views), computerized tomography (CT), water-soluble contrast myelography with CT, and magnetic resonance imaging (MRI) with flexion and extension views (Fig. 1). Three-dimensional computerized tomography (3 D CT) may help better delineate the precise nature of the bony anatomy in patients with complex skull base deformities[6].

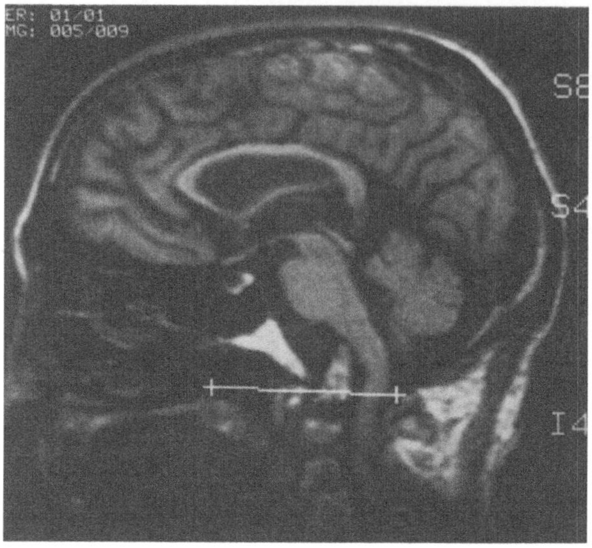

Fig. 1. Preoperative MRI documenting extension of odontoid above the foramen magnum and marked compression of the brain stem

Surgical Techniques

The patient is intubated with a flexible oral endotracheal tube and positioned in the supine position on the operating room table. The patient's head is secured in a Mayfield fixation device with the head and neck slightly extended, unless the patient was placed in a Halo vest immobilization device prior to surgery. The surgeon works while sitting from above the patient's head. A self-retaining McGarver three-ring retractor is inserted allowing the lips and gums to be spread apart and the tongue and endotracheal tube to be displaced caudally. Two small, red rubber catheters are inserted through the nostrils into the oropharynx and sutured to the uvula with a 2-0 silk suture. The uvula and soft palate are retracted into the nasopharynx by placing traction on the two rubber catheters and tying them to the McGarver retraction apparatus (Fig. 2). This combination of retraction allows a wide operative exposure of the posterior pharyngeal wall and eliminates the need for a tracheostomy[1, 7, 9, 12–14].

The patient's mouth and oropharynx are prepared with iodine solution, the patients is draped, and the

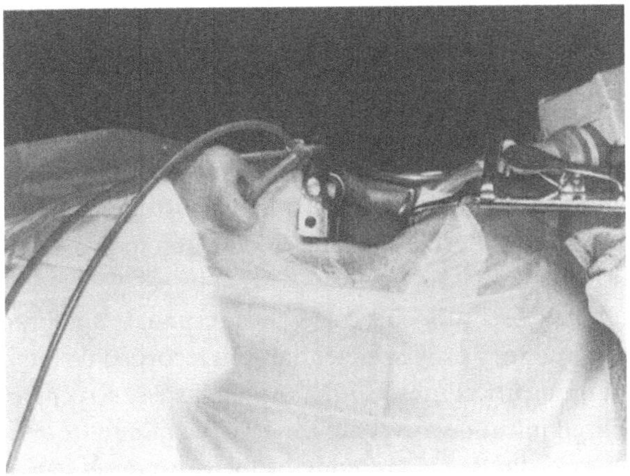

Fig. 2. Lateral view of McGarver 3-ring retractor and use of red rubber catheters through the nostrils

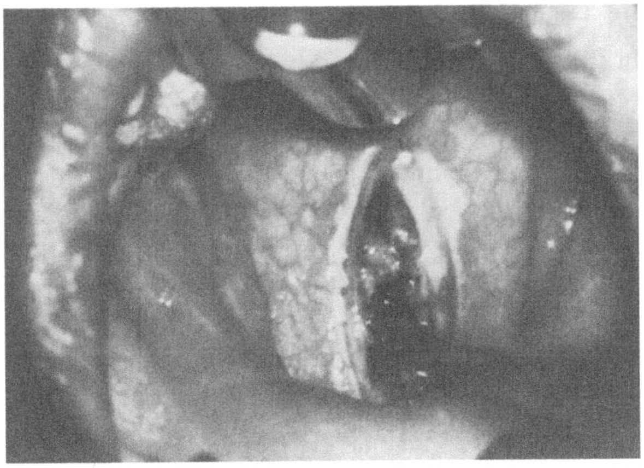

Fig. 3. Intraoperative photograph of the initial midline incision. The surgeon works from above the patient's head. The endotracheal tube is depressed caudally by the McGarver retractor (superior in this photo)

operating microscope with a 300 mm lens is positioned. The tubercle of the atlas can be palpated through the posterior pharyngeal wall and is used as a landmark. A lateral radiograph can be used to confirm localization in difficult cases.

A longitudinal midline incision is made in the posterior pharyngeal tissues using the Shaw hemostatic scalpel (Oximetrix Inc., Mountain View, CA), which allows a bloodless opening (Fig. 3). For odontoid pathology the incision is made directly over the tubercle and extended approximately 2.5 cm inferiorly. For pathology at the lower clivus, a more cephalad incision is required, extending upward through the posterior pharyngeal wall of the nasopharynx. The muscles and soft tissues of the posterior pharynx are incised along the midline down to the periosteum overlying the atlas

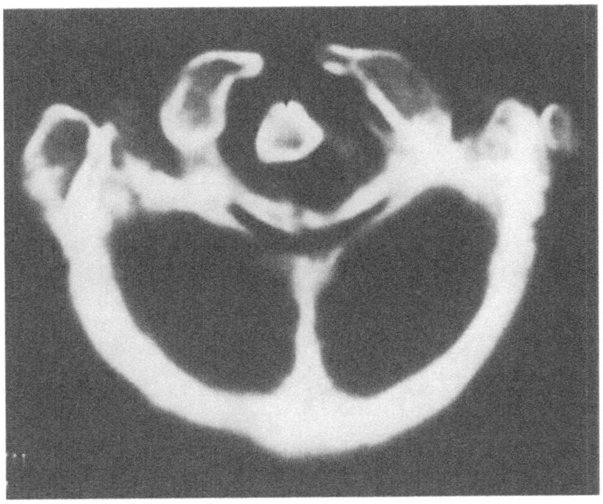

Fig. 4. Metrizamide CT demonstrates marked cervicomedullary compression by the odontoid process

and axis. A fine periosteal elevator is used to dissect these tissues laterally from the anterior surface of the vertebrae, and traction sutures or a large eyelid retractor are used to retract them laterally. The C 1–C 2 and C 2–C 3 interspaces may be identified utilizing intraoperative radiography.

A Midas Rex high speed drill (Midas Rex, Ft. Worth, TX) with a small cutting burr is used to perform the bone dissection, either superiorly through the tubercle of the atlas and the inferior clivus or inferiorly at the bodies of C 2 or C 3. The depth of bone dissection can be assessed by several ways: 1) by direct inspection, dissecting down to the posterior longitudinal ligament, 2) by intraoperative radiography or fluoroscopy placing a dissection tool in the operative site, or 3) by placing contrast material into the resection cavity and obtaining lateral (and occasionally A–P) cervical radiographs. The latter technique allows the most accurate estimates of the depth and width of bone removal. A fine diamond tip burr is used to resect the remaining cortical bone layer to expose the posterior longitudinal ligament (dura at the clivus level). Occasionally, the tip of the odontoid can be grasped with a rongeur, elevated away from the brain stem and spinal cord, and dissected free from its ligamentous attachments with straight and angled microcurettes.

When performing this procedure for odontoid resection (Figs. 4, 5 and 6), the posterior longitudinal ligament is opened if it is thick and calcified or if it does not bulge into the resection cavity following odontoid removal. In patients with basilar invagination from rheumatoid arthritis, there may be nonosseous reactive granulation tissue (rheumatoid pannus) around the

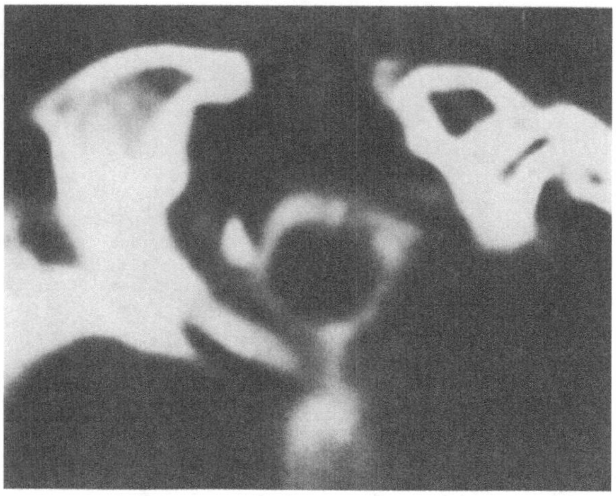

Fig. 5. Postoperative metrizamide CT documents odontoid resection and decompression of cervicomedullary structures

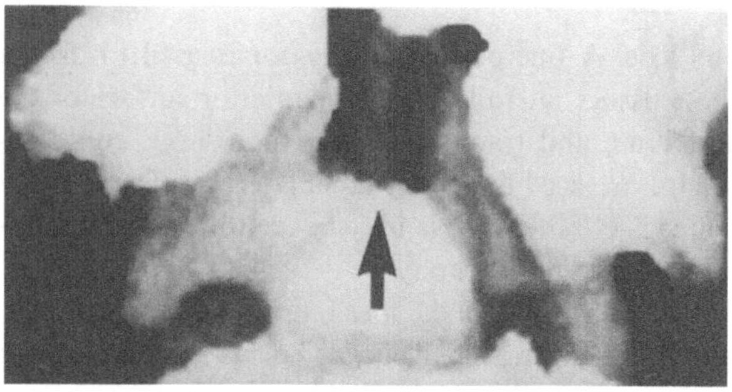

Fig. 6. Three-dimensional CT depicts region of odontoid resection (arrow)

odontoid process that may contribute to ventral cervicomedullary compression[2, 13, 15]. This tissue must be excised. The dura mater then assumes a more normal position, a change which can be documented radiographically when contrast material is placed into the resection cavity.

For tumour removal between C 3 and the clivus, this approach allows an extensive anterior exposure for excision or biopsy. Using microneurosurgical techniques, we have resected six extradural tumours (2 metastatic neoplasms and 4 chordomas) without operative morbidity or mortality.

Prior to closure the entire wound is irrigated with antibiotic solution, and a multilayer closure of the posterior pharyngeal tissues is performed. The deep muscular layers are approximated with interrupted absorbable sutures, the superficial muscular layers are approximated with a continuous absorbable suture, and finally the posterior pharyngeal mucosa is approximated with a continuous absorbable suture. The posterior pharyngeal wall of the nasopharynx is thin and fragile making a good closure at the superior extent of the incision difficult[7, 8]. A lumbar drain is inserted intraoperatively in those cases in which the dura mater has been violated. CSF hypotension is maintained with continuous drainage for 3 to 4 days postoperatively to facilitate a watertight wound closure and to avoid CSF fistula.

Somatosensory cortical evoked responses and brain stem auditory evoked potentials are recorded throughout these procedures. Computerized display screens are utilized intraoperatively to observe the minute-to-minute recordings. A hard copy of their responses can be generated for comparison to pre- and/or postoperative studies.

Discussion

In the past, significant morbidity and mortality have been reported after the transoral-transclival surgical procedure-factors that have limited its general acceptance and application. Refinements in microneurosurgical techniques combined with effective retraction, intraoperative neurophysiologic monitoring, and intraoperative radiography result in a safe and efficacious surgical procedure[3, 7, 9, 12–14]. Of the 29 patients in this review, only 1 had a postoperative complication (3% morbidity) (Table 4). The patient re-presented 2 weeks after discharge with a brain stem stroke. Follow-up radiographic studies revealed vertical C 1 and C 2 subluxation with vertebral artery thrombosis. He was improving neurologically on the rehabilitation ward when he died from severe pneumonitis 9 weeks after surgery. While the operative mortality in this series was zero, two patients, including the patient described above, eventually died (median follow-up 24 months). The second patient succumbed to a massive pulmonary embolus 12 weeks after the transoral procedure to decompress the spinal cord ventral to C 2 and C 3. Six weeks postoperatively she underwent a posterior decompression procedure (for metastatic neoplasm) with fusion. She died six weeks following the second procedure.

The routine use of a flexible oral endotracheal tube, secured caudally by the McGarver 3-ring retractor in combination with red rubber catheters to retract the uvula and soft palate superiorly, provides a wide exposure of the posterior pharyngeal wall and obviates the need to perform a tracheostomy or to split the mandible, tongue, and palate. The endotracheal tube is left in place for a least 24 hours postoperatively to

Table 4. *Outcome of Twenty-Nine Patients Treated Surgically via the Transoral Approach*

Follow-up transoral approach	Morbidity	Mortality
Odontoid pathology:		
Congenital	0	0
Rheumatoid	1 Vertical sublux. with brainstem stroke	Severe pneumonitis 9 weeks
Traumatic C 1–2 sublux	0	0
Other pathology:		
Extradural met. neoplasm	0	Pulmonary embolus 12 weeks
Extradural tumor	0	0
(median follow-up 24 months)		

avoid potential respiratory compromise from glottic and tongue swelling. Patients are monitored in the intensive care unit for 24 hours after extubation, and a tracheostomy tray is maintained at the bedside following removal of the endotracheal tube as a precautionary measure should respiratory compromise occur.

Somatosensory cortical and brain stem auditory potential recordings are sensitive intraoperative neurophysiological monitoring techniques when performing neurosurgical procedures of high risk[3, 7, 7, 9, 14]. Experience has demonstrated their predictive value in instances of potential spinal cord or brain stem injury and their correlation between observed intraoperative responses and clinical outcome.

The use of intraoperative radiography (both AP and lateral planes) is an important surgical adjunct[7, 9, 11, 13, 14]. The depth and width of bone dissection and decompression can be accurately assessed, particularly in combination with contrast material that is placed into the resection cavity. Intraoperative fluoroscopy can document the adequacy of spinal cord decompression by the degree of dural elevation following odontoid resection or tumour excision.

In summary, the transoral approach is a viable alternative in the treatment of ventral compressive pathology at the lower clivus or superior cervical spine. Best results can be obtained by using the surgical adjuncts and monitoring techniques outlined above with close attention to sterile microneurosurgical techniques.

References

1. Blazier CJ, Hadley MN, Spetzler RF (1986) The transoral surgical approach to craniovertebral pathology. J Neurosci Nurs 18(2): 57–62
2. Crockard HA (1985) The transoral approach to the base of the brain and upper cervical cord. Ann R Coll Surg Engl 67: 321–325
3. Croft TJ, Brodkey JS, Nulsen FE (1972) Reversible spinal cord trauma: A model for electrical monitoring of spinal cord function. J Neurosurg 36: 402–406
4. Dastur KD, Wadia NH, Desai AD *et al* (1965) Medullospinal compression due to atlanto-axial dislocation and sudden haematomyelia during decompression. Brain 88: 897
5. Drake CG (1973) Management of aneurysms of posterior circulation. In: Youmans JR (ed) Neurological surgery, vol 2. WB Saunders, Philadelphia London Toronto, pp 787–806
6. Hadley MN, Sonntag VKH, Amos MR, Hodak JA, Lopez LJ (1987) Three-dimensional computed tomography in the diagnosis of vertebral column pathological conditions. Neurosurgery 21: 186–192
7. Hadley NM, Spetzler RF (1986) The transoral surgical approach to the craniocervical junction. In: Samii M (ed) Surgery in and around the brain stem and third ventricle. Springer, Heidelberg, pp 467–475
8. Hadley MN, Martin NA, Spetzler RF, Sonntag VKH, Johnson PC (1988) Comparative transoral dural closure techniques: A canine model. Neurosurgery 22: 392–397
9. Masferrer R, Hadley MN, Bloomfield S, Spetzler RF, Sonntag VKH (1985) Transoral microsurgical resection of the odontoid process. BNI Quarterly 1(3): 34–40
10. Menezes AH, VanGilder JC, Graf CJ, McDonnell DE (1980) Craniocervical abnormalities: A comprehensive surgical approach. J Neurosurg 53: 444–445
11. Owen MP, Brown RH, Spetzler RF, Nash CL Jr, Brodkey JS, Nulsen FE (1979) Excision of intramedullary arteriovenous malformation using intraoperative spinal cord monitoring. Surg Neurol 12: 271–276
12. Selman WR, Spetzler RF, Brown R (1981) The use of intraoperative fluoroscopy and spinal cord monitoring for transoral microsurgical odontoid resection. Clin Orthop 154: 51–56
13. Spetzler RF (1983) Transoral approach to the upper cervical spine. In: Evarts CM (ed) Surgery of the musculoskeletal system. Churchill Livingstone, New York, pp 4: 19–4: 24
14. Spetzler RF, Selman WR, Nash CL, Brown RH (1979) Transoral microsurgical odontoid resection and spinal cord monitoring. Spine 4 (6): 506–510
15. Stevens JM, Kendall BE, Crockard HA (1986) The spinal cord in rheumatoid arthritis with clinical myelopathy: A computed myelographic study. J Neurol Neurosurg Psychiatry 49: 140–151

16. Stratford J (1957) Myelopathy caused by atlanto-axial dislocation. J Neurosurg 14: 97
17. Thompson H (1970) Transpharyngeal fusion of the upper cervical spine. Proc R Soc Med 63: 893
18. VanGilder JC, Menezes AH (1982) Craniovertebral abnormalities and their treatment. In: Schmidek HH, Sweet WH (eds) Operative neurosurgical techniques. Grune and Stratton, New York, pp 1221–1235
19. Yamaura A, Makino H, Isobe K, Takashima T, Nakamura T, Takemiya S (1979) Repair of cerebrospinal fluid fistula following transoral transclival approach to a basilar aneurysm. J Neurosurg 50: 834–836

Address for correspondence: Prof. R. F. Spetzler, MD, F.A.C.S., Division of Neurological Surgery, Barrow Neurological Institute, 350 West Thomas Road, Phoenix, AZ 85013, U.S.A.

Acta Neurochirurgica, Suppl. 43, 75–78 (1988)

Posterior Acrylic Surgical Fixation of Odontoid Fractures

D. Gambacorta and **F. Reale**

Department of Neurosurgery, Ospedale "Le Scotte", Siena, Italy.

Summary

Fourteen cases of odontoid fractures had, as primary treatment, a posterior surgical fixation (C 1 + C 2 + C 3 and occiput in some cases) by means of acrylate and stainless steel wires, without waiting for the unpredictable final results of external systems of stabilization. The youngest patient was 20 years old, the oldest 84 years old. No immediate or late neurological complications occurred (follow-up: one year to ten years). All the patients were mobilized early without external fixation. Union of the dens took place in eleven cases, and good stabilization in the remainder. Thirteen patients recovered, one improved. Posterior acrylic fixation is the treatment of choice in cases of odontoid fracture, particularly in elderly patients in whom a solid union rarely is obtained.

Keywords: Odontoid fractures; vertebral fractures; posterior cervical fixation.

Introduction

If marked displacement occurs odontoid fractures need, like many other vertebral fractures, to be reduced and stabilized as soon as possible because of the high risk of early or delayed neurological complications or deterioration. Moreover the system of stabilization ought to be reliable, without risk, easy to perform, as comfortable as possible for the patients, and should allow prompt and easy mobilization in safety.

On the basis of several reports collected from the literature and from our experience, we believe that an immediate surgical fixation (posterior approach and use of acrylate and wires) is a rational method of treatment.

Material

We have operated on 14 patients affected by odontoid fractures. Their main clinical and radiological findings are summarized in Table 1.

The youngest patient was 20 and the oldest 84 years old. In the case 2 the fracture was discovered 4 years after the trauma; case 11 had been imprisoned in a plaster cast for more than one year.

Surgical fixation was performed in all the cases not later than 24 hours after admission. Transcranial traction was applied before the operation and left in place during it in presence of gross displacement.

Technique

The technique of posterior acrylic fixation employed by us is that described by Kelly *et al.*[25] in 1972 and later by others[19, 20, 31, 41].

It consists in applying a single block of acrylate on the posterior aspect of C 1, C 2 and C 3 and between the spinous processes. The vertebrae are sutured to the acrylate by means of stainless steel wires passed beneath the posterior arch of the atlas and through or beneath the base of the spinous processes of C 2 and C 3. It is advisable to include in the fixation also the inferior part of the occiput in case of marked displacement or instability of C 1/C 2 or occiput/C 1/C 2. A layer of Gelfoam has to be interposed between acrylate and bone, and the acrylate is continuously irrigated by cold water during the curing process: both manoeuvres are aimed at reducing the effects of the exothermic reaction on neural structures[21].

Results

Every patient, except the tetraparetic one, left his bed on the second postoperative day without external immobilization. Neither immediate nor late neurological complications were observed in a follow-up ranging from six months to nine years.

Wound infections caused by Staphylococcus epidermidis occurred in two cases: in the first, which came on two weeks after the operation, antibiotics solved the problem, but in the second one (late infection about three months after the operation) we had to remove, with some difficulty, the acrylic mass, but by that time, there was already a good union of the fracture.

Late clinical controls showed full recovery in thirteen cases and very evident improvement in the severely tetraparetic one (case 1), but probably, from this point of view, we were particularly lucky with our cases.

A good union was seen in eleven cases, but in three (3, 8 and 14) the alignment was not perfect. In the cases 2 and 11 operated on 4 years and 14 months after the

Table 1

No. Case	Sex	Age	Clinical findings	Displacement	Complications	Follow-up	X Rays	Clinical
1) B.V.	m	63	Tetraparesis Traction	Post. > 5 mm		9 years	Union	IMP.*
2) C.B.	m	53	Trauma 4 years before Hypertonic attacks in head flexion			8 years	Non union	REC.**
3) B.C.	m	53	Neck pain	Ant. > 5 mm		8 years	Union. Not perfect alignement	REC.
4) P.T.	m	20	Upper limbs weakness			5 years	Union	REC.
5) M.E.	f	84	Neck pain			4 years	Non union	REC.
6) D.L.	f	53	Lower limbs paraesthesiae		Wound infect. 2 months postop.	3 year	Union	REC.
7) F.G.	m	76	Neck pain		Wound infect. 3 months postop.	3 year	Union	REC.
8) M.L.	m	57	Paraparesis	Ant. slight		3 year	Union. Not perfect alignement	REC.
9) O.A.	m	20	Paresis left arm	Ant. slight		2 years	Union	REC.
10) R.A.	m	22	Neck pain. Vertigo. Left hand paraesthesiae			2-1/2 years	Union	REC.
11) B.Q.	m	72	Trauma 14 months before.			1 year	Non union	REC.
			Neck pain. Head flexed on left shoulder			1 year	Non union	REC.
12) S.P.	m	21	Neck pain			1 year	Union	REC.
13) F.I.	m	60	Neck pain	Post. > 5 mm		1 year	Union	REC.
14) B.P.	m	27	No one	Ant. > 5 mm		1 year	Union. Not perfect alignement	REC.

* IMP. = improved, ** REC. = recovered.

trauma respectively, in the case 5 union did not occur, but, owing to the persistently good stabilization, the shifting of the dens during the movements of the head and neck was minimal and so the final outcome was very good for this patient too.

Significant reduction of flexion, extension, and rotation movements was complained of only when the occiput was also included in the acrylic fixation (two cases of our serie).

Discussion

Odontoid fractures are not really rare lesions. Hadley *et al.*[23] found 107 (17%) fractures of the axis among 625 cervical fractures with 39 (= 6%) of these 107 involving only the odontoid.

They show various anatomopathological findings – fractures of the apex or type 1, fractures through the base or type 2, fractures involving the body or type 3, displaced of not fractures[3, 16] – and different clinical aspects: there are cases which are suddenly fatal, cases with more or less severe neurological deficits, cases with minimal signs like neck pain, and even unrecognized cases[30, 37]. In general, neurological deficits, when present, appear soon after the trauma but a delayed onset is possible[6, 30] as in our case 2.

However, once an odontoid fracture has been diagnosed it is mandatory to try a reduction in case of marked displacement and, above all, to perform an early stabilization to prevent neurological deterioration and to facilitate union of the fractured segments.

Stabilization can be tried or by means of external, non surgical, devices, or by surgical procedures, or by both in succession. All the external devices – cervical collar[3, 8, 22] Minerva jackets, plaster and halo cast[5, 12, 13, 14, 16, 18, 22, 29, 34, 35, 39, 40] and so on – are more or less uncomfortable, and often difficult to fit or to maintain, they reduce physical activities and, as underlined by Wisoff[42], expose the patient to the risk of numerous complications. In particular, the use of a snug fitting vest or cast and the protracted traction in bed carry an "unacceptable incidence of respiratory insufficiency, pneumonia or both", which is sometimes lethal especially in old patients or in patients with posttraumatic respiratory problems. Moreover external systems have in general an unpredictable final result, with an high rate of non-union, related to the patient's age and to the degree of displacement. So it is not surprising that after the long period – 3–6 months[10, 13] of external immobilization usually required to knit together the fractured bones, the union has not taken place and that one is then obliged to perform a surgical fixation. Even in patients treated by halo cast (which is probably at present the most effective external system of immobilization[4, 16, 24, 33, 38]), the rate of non-union is high, ranging from 41%[16] to 64%[38] to 86% in the aged[3]. That is why, in line with many others[1, 2, 24, 31, 37, 41, 42] we prefer to treat the fractures of the odontoid directly by operation, thus avoiding the long and uncomfortable wait for an uncertain final result; operative treatment also reduces the risk of orthopaedic (pseudoarthrosis) and neurological complications, which are mainly associated with non-surgical systems of treatment. Surgical fixation can be done by an anterior[8, 17, 35] or by a posterior approach. The anterior approach allows a direct synthesis of the fractured segments, usually by means of a long screw. Recently this approach has been emphasized again[26, 27] and the pros and cons of different surgical anterior approaches (transoral, antero-lateral, transcervical) discussed[27].

Direct osteosynthesis is a valid technical solution when posterior fixation cannot be performed (*e.g.* in case of agenesia or fracture of the posterior of C 1), and it does not impair the articular function at C 1/C 2, in particular the rotation movement.

However, a part from the risks related to a specific anterior approach, every anterior approach presents some general limitations, risks and contraindications:

1) Surgical difficulty is greater than by posterior approach.

2) Only fractures of type 2 and without concomitant rupture of the transverse ligament can be treated[27].

3) Fractures due to massive bony destruction like those from rheumatoid arthritis cannot be treated.

4) The exact placement of a rather big screw in a rather small dens, in particular if there is a marked displacement, can be difficult[26].

5) Lesions of the mucuous membrane of the pharyngeal cavity can lead to septic complications

6) Retropharyngeal haematomas are a spontaneous complication of the odontoid fractures[42]. An anterior operation and the screwing of the fractured bones can increase the danger of this complication.

7) A long postoperative external immobilization is required after direct synthesis.

By the anterior approach a bilateral arthrodesis C 1–C 2 is also possible[27]: but this block does not offer any definite advantage over the posterior one. Moreover we were not able to find any reliable long-term follow-up of the cases treated by anterior approach.

Every posterior surgical fixation consists in fixing together the laminae and spinous process of C 1, C 2 and C 3, and sometimes, but rarely, the occiput as well.

Posterior fixation can be performed in different ways, that is by:

– Simple stainless wires[28, 37]. This technique is not reliable enough because the wires can break (unless very thick, but very difficult to manoeuvre, wires are employed) and because it is almost impossible to twist the wires tightly enough to prevent any movement.

– Metallic plates, screw and wires[9, 11]. Experience with these devices and on their long-term reliability is too limited for proper judgement.

– Autologous bone grafts placed on the sacrified spinal grooves and secured with wires[1, 2]. Possibility of graft breakage, and failure of fusion[30], pain and trouble at the donor site, are the best known drawbacks of this technique.

– Fusion by acrylic block as we have done. The main advantages of this technique can be summarised as follows:

a) Fast and simple execution.

b) Reliable immobilization/stabilization because the acrylate fits on and between bone structures like a cement so that even if breakage of the wires occurs it does not fail as the time goes by, in addition non-union of the dens can be accepted without untoward consequence.

c) Possibility of very early active and/or passive mobilization and, if required, standing. This operation has no specific contraindications apart, of course, from those related to the patient's general and neurological condition. Very old patients have been operated in our department without problems and we have not found in the literature any complications definitely related to the exothermic reaction.

As in every heterologous grafting the risk of infection is slightly increased in the above mentioned technique but this only means that in such cases we have to be more careful without aseptic measures to reduce the risk to zero.

References

1. Alexander E, Davis CH (1969) Reduction and fusion of fracture of the odontoid process. J Neurosurg 31: 580–582

2. Alexander E jr, Forsyth HF, Davis CH jr, Nashold BS (1958)

Dislocation of the atlas on the axis: the value of early fusion of C 1, C 2 and C 3. J Neurosurg 11: 353–371

3. Althoff B, Bardhom P (1979) Fractures of the odontoid process. A clinical and radiographic study. Acta Orthop Scand [Suppl] 177: 61–95
4. Anderson LD, D'Alonzo RT (1974) Fractures of the odontoid process of the axis. J Bone Joint Surg (Am) 56 A: 1663–1674
5. Apuzzo MLJ, Heiden JS, Weiss Mh, Ackerson TT, Harvey JP, Kurze T (1978) Acute fracture of the odontoid process. An analysis of 45 cases. J Neurosurg 48: 85–91
6. Askenasy HM, Braham J, Kosary IZ (1960) Delayed spinal myelopathy following atlanto-axial fracture-dislocation. J Neurosurg 17: 1100–1104
7. Aymes EW, Anderson FM (1956) Fracture of the odontoid process. Arch Surg 72: 377–393
8. Böhler J (1982) Anterior stabilization for acute fractures and non union of the dens. J Bone Joint Surg 64 A: 18–28
9. Bret Ph, Lecuire J, Carlioz P, Deruty R (1980) Atlanto-axial dislocation due to ossiculum terminale persistens. Report of two cases treated by a new technique of posterior fusion. Acta Neurochir (Wien) 52: 105–112
10. Blockey NJ, Purser DW (1956) Fractures of the odontoid process of the axis. J Bone Joint Surg 38 B: 794–817
11. Cantore GP, Chiappetta P, Delfini R (1984) New steel device for occipito-cervical fixation. J Neurosurg 60: 1104–1106
12. Chan RC, Scheigel JF, Thomson GB (1983) Halo-thoracic brace immobilization in 188 patients with acute cervical spine injuries. J Neurosurg 58: 508–515
13. Cooper PR (1981) Comment to the article of Six E, Kelly DC (23): Technique for C 1, C 2 and C 3 fixation in cases of odontoid fracture. Neurosurgery 8: 374–377
14. Cooper PR, Maravilla Kr, Sklar FH, Moody SF, Clark WK (1979) Halo immobilization of cervical spine fractures. J Neurosurg 50: 603–610
15. Cornish BC (1968) Traumatic spondylolisthesis of the axis. J Bone Joint Surg 50 B: 31–43
16. Ekong CEU, Schwartz ML, Tator CH, Rowed D, Edmonds VE (1981) Odontoid fracture: management with early mobilization using the halo-device. Neurosurgery 9: 631–637
17. Fang HSY, Ong GB (1972) Direct anterior approach to the upper cervical spine. J Bone Joint Surg (Am) 44 A: 1588–1604
18. Fried LC (1973) Atlanto-axial fracture-dislocation. Failure of posterior C 1-C 2 fusion. J Bone Joint Surg 55 B: 490–496
19. Gambacorta D (1978) Early posterior acrylic fixation of traumatic odontoid fractures. Neurochirurgia (Stuttg) 30: 16–18
20. Gambacorta D, Scarfo' GB, Mascolo G (1979) Stabilizzazione delle fratture del dente dell'epistrofeo mediante innesti posteriori di resina acrilica. Chir Org Movim 65: 61–65
21. Goldner JL (1981) Comment to the article of Six and Kelly: Technique for C 1, C 2 and C 3 fixation in cases of odontoid fracture (41). Neurosurgery 8: 374–377
22. Grogono BJS (1954) Injuries of the atlas and axis. J Bone Joint Surg 36 B: 397–410
23. Hadley MH, Sonntag VKH, Grahm W, Masferrer R, Browner C (1986) Fractures resulting from motor vehicle accidents. The need for occupant restraints. Spine 11: 861–864
24. James JIP (1960) Fracture dislocation of cervical spine. JR Coll Surg Edinburgh 5: 232–233
25. Kelly DR jr, Alexander E jr, Courtland HD jr, Smith JM (1972) Acrylic fixation of atlanto-axial dislocation. Technical note. J Neurosurg 36: 360–371
26. Knoringer P (1984) Zur Behandlung frischer Fracturen des Dens axis durch Kompressionsschraubenosteosynthese. Neurochirurgia (Stuttg) 27: 68–72
27. Lesoin F, Autricque A, Franz K, Villette L, Jomin M (1987) Transcervical approach and screw fixation for upper cervical spine pathology. Surg Neurol 27: 459–465
28. Lipscomb PR (1957) Cervico-occiput fusion of the congenital and posttraumatic anomalies of the atlas and axis. J Bone Joint Surg (Am) 39 A: 1289–1301
29. Marar BC, Tay CK (1976) Fracture of the odontoid process. Aust NZ J Surg 46: 231–236
30. McLaurin RL, Vernal R, Salmon JH (1972) Treatment of fractures of the atlas and axis by wiring without fusion. J Neurosurg 36: 773–780
31. Oh S, Bosiger C (1981) Acrylic fixation in displaced dens fractures. Acta Neurochir (Wien) 56: 95–98
32. Paradis GR, James JM (1973) Posttraumatic atlantoaxial instability: the fate of the odontoid process fracture in 46 cases. J Trauma 13: 359–367
33. Prolo DJ, Runnels JB, Jameson RM (1973) The injured cervical spine: immediate and long term immobilization with the halo. JAMA 224: 591–594
34. Roberts A, Wickstrom J (1973) Prognosis of odontoid fractures. Acta Orthop Scand 44: 21–30
35. Roy Camille R, Saillant G, Lapresie Ph, Leonard P (1983) Fractures recentes de l'odontoide. Facteur de prognostic. Memoire Presse Med 36: 2233–2236
36. Ryan MD, Taylor TK (1982) Odontoid fracture: a rational approach to treatment. J Bone Joint Surg 64 B: 416–421
37. Salmon GH (1977) Fractures of the second cervical vertebra: internal fixation by interlaminar wiring. Neurosurgery 2: 125–127
38. Schtzker J, Rorabeck CH, Waddel JP (1971) Fractures of the dens (odontoid process). An analysis of thirty-seven cases. J Bone Joint Surg (Br) 53 B: 392–405
39. Schweigel JB (1979) Halo thoracic brace management of odontoid fractures. Spine 4: 192–194
40. Saljeskong EL (1978) Nonoperative management of acute upper cervical injuries. Acta Neurochir (Wien) 41: 87–100
41. Six E, Kelly DL (1981) Technique for C 1/C 2 and C 3 fixation in cases of odontoid fracture. Neurosurgery 8: 374–376
42. Wisoff HS (1984) Fracture of the Dens in the Aged. Surg Neurol 22: 547–555

Address for correspondence: D. Gambacorta, M.D., Department of Neurosurgery, Osp. "Le Scotte", I-53100 Siena, Italy.

Acta Neurochirurgica, Suppl. 43, 79–84 (1988)

Realignment and Surgical Fixation of Atlanto-axial and Subaxial Dislocations in Rheumatoid Arthritis (RA) Patients

S. C. Zygmunt[1], **B. Ljunggren**[1], **M. Ålund**[1], **H. Brattström**[2], **H. G. Säveland**[1], **S. Holtås**[3], **E. M. Larsson**[3], and **I. Redlund-Johnell**[4]

Departments of [1] Neurosurgery, [2] Orthopaedic Surgery, and [3] Neuroradiology, University Hospital, Lund, Sweden, and [4] Department of Radiology, Malmö, Sweden

Summary

One hundred consecutive patients with atlanto-axial dislocation(s) were subjected to posterior occipito-cervical fusion and followed up to 16 ½ years postoperatively (mean: 4 ½ years). The mean age at surgery was 60.3 years. At the time of follow-up, 45 patients had died, mostly due to cardiopulmonary disease. Out of the 100 patients, 67% showed a major improvement and in an additional 14% there was a slight or moderate improvement. Some patients with a stable fusion later developed subaxial dislocation(s) necessitating an anterior fusion which in patients with RA is difficult due to poor bone quality. MRI revealed pannus formation around the odontoid peg in many patients, in several to such a degree that cord compression was evident. Following the posterior fixation, the pannus disappeared or was reduced which may at least partly explain the generally favourable operative outcome following the fixation procedure. Early occipito-cervical fusion appears to prevent further vertical dislocation or upwards migration of the odontoid which is a more severe condition, and does not appear to result in increased incidence of subsequent subaxial dislocation(s). Such luxation(s) pose a technically difficult problem since the quality of the patients' own bone usually does not permit a common anterior fusion procedure with autologous graft. In such cases with cord compression, acrylic cement may be used to obtain a normalized realignment.

Keywords: Rheumatoid arthritis; cervical spine; atlanto-axial dislocation; subaxial dislocation; odontoid migration; surgical fixation.

Introduction

Garrod in his monograph "A Treatise on Rheumatism and Rheumatoid Arthritis" presented evidence of frequent involvement of the cervical spine in chronic rheumatoid arthritis (RA) based on clinical observations five years before Wilhelm Conrad Röntgen took the first "x-ray film"[1]. Garrod found that out of 500 patients with chronic RA, one hundred and seventy-eight (36%) patients showed evidence of cervical spine involvement[1]. Many subsequent investigators, with more sophisticated radiological methods such as conventional radiographic modalities, myelography, and more recently computed tomography (CT) and magnetic resonance imaging (MRI) have verified this frequent involvement of the cervical spine in chronic RA[2–7]. Such findings represent a continuum from slight degrees of horizontal instability or subluxation at the C 1–C 2 level to severe luxation and odontoid upward migration with pronounced cranial settling and subaxial luxation(s) (staircase formation) with secondary myelopathy and in the most advanced cases cord compression causing total physical incapacitation of the patient[8–9] (see Fig. 1 a–e).

It has been emphasized that surgical fixation should be carried out as soon as instability is recognized in order to avoid the development of a cervical myelopathy since the results in patients who already show evidence of cord compression have been unsatisfactory. Brito and coworkers thus stressed that in cases with signs of cord involvement or "the beginning of even slight neurological signs of a cervical myelopathy an operation is necessary"[10]. In this context, it should also be emphasized that spinal cord damage secondary to a previously proven atlanto-axial instability in RA patients is not an infrequent cause of sudden death in afflicted patients[11].

The present study is focused on long-term follow-up results after posterior occipito-cervical fusion performed over a period of 15 years (1970–1984) with special regard to later mortality and the causes of death in the deceased individuals. Furthermore, the additional information that may be obtained by investigations pre and postoperatively with MRI is discussed.

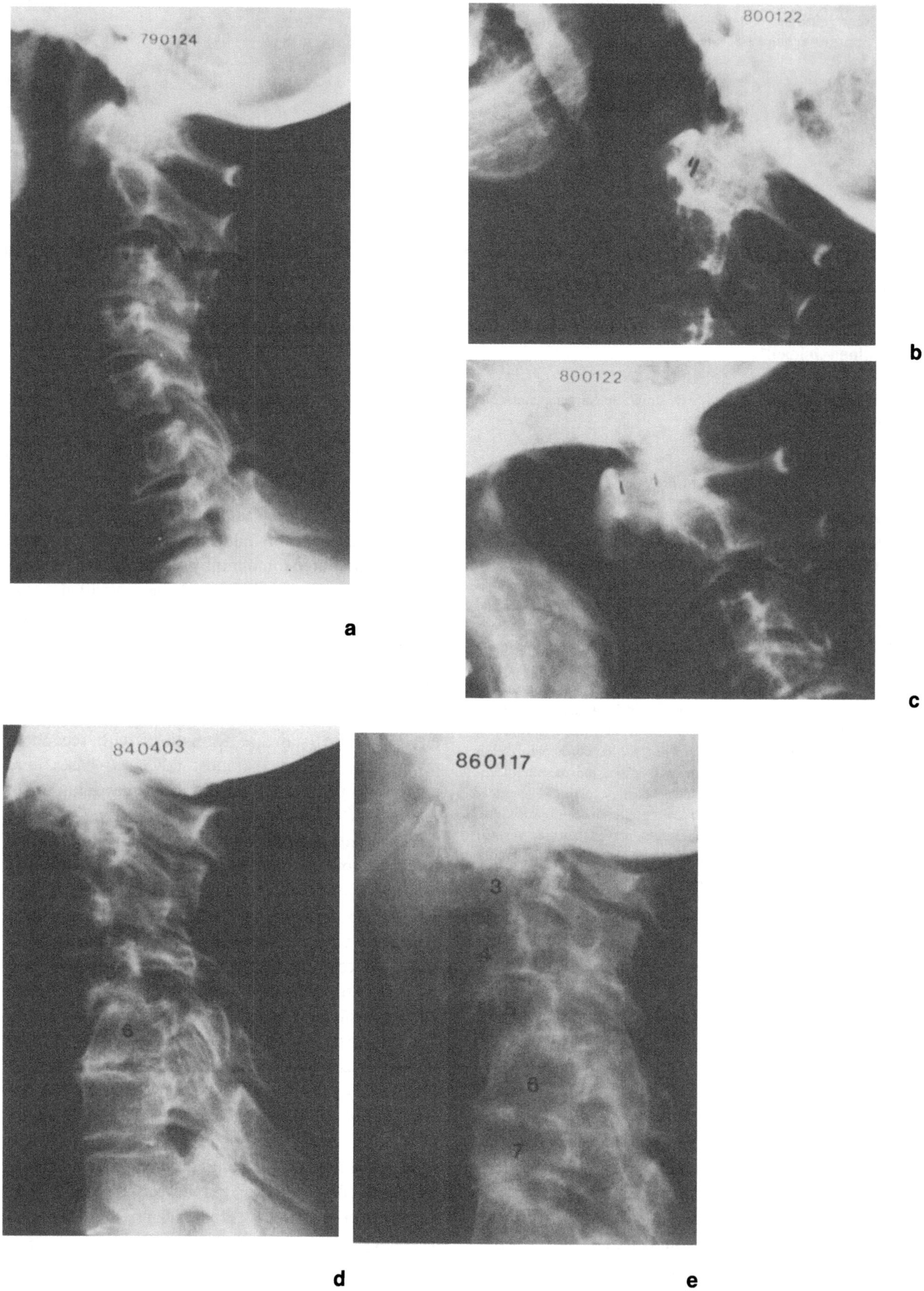

Fig. 1 a–e. Chronic RA patient with fairly normal x-ray of the cervical spine in 1979 (a). Two years later a marked horizontal atlanto-axial instability is observed (b–c). Later x-ray investigations reveal severe odontoid upwards migration and subaxial dislocations (d–e)

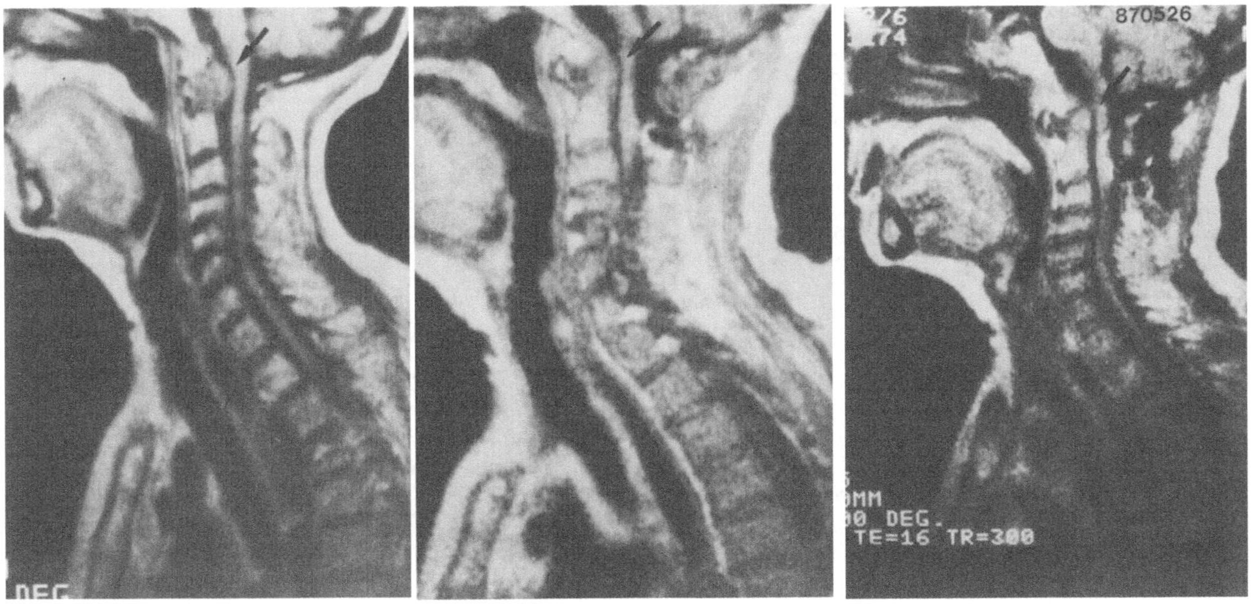

a b c

Fig. 2a–c. MRI reveals extensive pannus formation in synovial joints around the odontoid peg in RA patient with marked atlanto-axial instability prior to posterior fusion (a). Early postoperative MRI shows pannus reduction (b) and late MRI reveals further significant reduction of the synovitis

Patient Material

One hundred consecutive patients, who presented with different types of rheumatoid instability at the C 1–C 2 level and who were operated upon with a posterior occipito-cervical fusion, were followed up (mean time: $4\frac{1}{2}$ years; longest follow-up: $16\frac{1}{2}$ years). Seventy-eight were female, and 22 male; age at time of surgery ranged from 26 to 80 (mean: 60.3 years).

Symptomatology

A full neurological assessment of patients with advanced RA involving the cervical spine is difficult. Limb stiffness, walking disturbances, and general muscle weakness may all be attributed to active RA disease with limb joint affections as well as involvement of the peripheral or the central nervous system and furthermore, in the presence of painful or ankylosed joints it may not be possible to assess the tendon reflexes[12]. A wide spectrum of signs and symptoms including brain stem and cranial nerve root dysfunction, cervical myelopathy and cervical nerve root involvement with poorly localized motor abnormalities ranging from fatiguability to tetraparesis may thus be encountered. Sensory dysfunction manifested as severe suboccipital headaches in the sensory C 2 root distribution has often been described as a nearly constant feature. Vascular disturbance may also occur with symptoms including syncopé and vertigo.

Radiological Assessment

Conventional Radiology

All patients were examined with a plain view of the lateral cervical spine in maximal flexion and extension. Horizontal anterior atlanto-axial dislocation was considered when the distance between the lower posterior border of the anterior arch of the atlas and the nearest part of the odontoid peg exceeded 3 mm on provoked radiographies in maximal flexion and extension. Vertical atlanto-axial dislocation was considered when the centre of the lower end plate of C 2 was located at a distance of less than 34 mm in males and 29 mm in females from the palato-occipital or McGregor's line[13]. Subaxial dislocation was said to exist when the distance between the posterior borders of two adjacent vertebral bodies was more than 2 mm.

Computerized Tomography (CT)

A preoperative CT was performed in some patients in the latter part of the series and in a few cases offered additional information about bony and soft tissue abnormalities.

Magnetic Resonance Imaging (MRI)

MRI proved to be a superior tool to establish a full anatomical picture of both hard and soft tissues. In many cases, an extensive pannus formation in the synovial bursae (atlanto-dental bursa, apico-dental bursa, cruciate tectorial bursa, etc.) immediately adjacent to the odontoid peg[14] could be visualized. This synovitis may offer another possible explanation for the severe suboccipital headaches often encountered in afflicted patients (see Fig. 2a–c) as well as explain other signs of medullary compression.

Operative Technique

All patients except the first five were operated upon under general anaesthesia with the patient on a Stryker bed[15–19]. Preoperative traction with a Gardner-Wells tong is applied after which the patient is turned to the prone position. Traction prevents hazardous movements of the cervical spine when the patient is positioned and may also allow the fusion to be performed in the most favourable position. The head is placed on a cerebellar headrest, which ensures that no

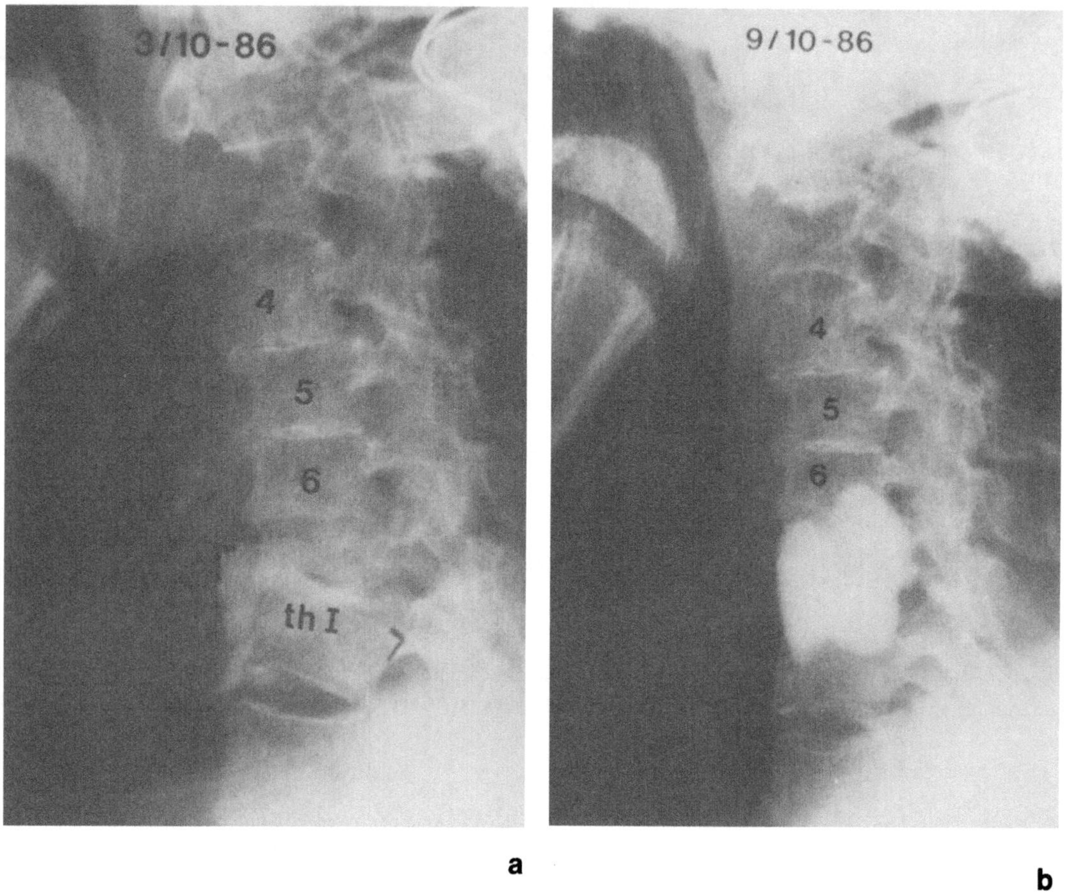

Fig. 3 a–b. RA patient with successful posterior occipito-cervical fusion who later developed subaxial dislocations with clinical signs of cord compression at C 4–C 5 level. Anterior fusion with Kiel bone graft was primarily successful. However, one week after surgery the patient agian experienced signs of cord compression and x-ray revealed collapsed graft (a). In this situation, acrylic bone cement was used to obtain realignment of the spine at the collapsed level (b)

inadvertent pressure is placed around the eyes. A midline incision is then performed from the inion to the spinous process of CIV and the paracervical muscles are dissected off the spinous processes and laminae. The suboccipital muscles are dissected from the occipital bone subperiostally. An occipito-cervical fusion is then performed. Four burr holes are made with the trephine in the occipital bone, two on each side of the midline and with the lower holes approximately 7–8 mm above the posterior rim of the foramen magnum. The posterior arch of the atlas is dissected free and a nylon thread is passed extradurally and ventral to the posterior arch. By means of this nylon thread a loop of stainless steel wire is pulled under the arch. The loop is then cut and the two free wires are crossed and passed in and out through the burr holes on one side and the other wire is similarly passed through the burr holes on the other occipital side. The upper wire ends are then twisted in the midline below the occipital protuberance. A pin, originally designed to fix a fractured medial malleolus, is then inserted through the spinous process of C 2, its way through the process having been prepared by introducing both ends of a heavy towel clip. One end of the pin is bent slightly to improve later incasement in bone cement. One caudal steel wire end is now passed dorsal to the pin and twisted together with the other end. When the arch of the atlas is very thin or when it is displaced rostrally and ventrally thereby reducing the anteroposterior diameter of the spinal canal and by no means can be retracted caudally and dorsally, a laminectomy of the C 1 arch is preferred. In such cases, the steel wires are instead passed ventrally to the pin in the C 2 spinous process.

On one side, the two occipital burr holes, the proximal and distal wire twistings, the C 2 spinous process and the bent end of the pin in the process are encassed in acrylic bone cement. The exothermic heat production during polymerisation is counteracted by heavy rinsing with normal saline solution. The uncovered occipital bone on the other side and laminae on that side are partly decorticated with a chisel and covered with bone, chips obtained from the iliac crest and bone meal from the burr holes, so as to result in a future bony fusion. After closure of the wound in layers, the traction tongs and the endotracheal tube are removed. After a postoperative radiographic control, the patients are mobilized without external fixation.

Some patients later developed subaxial dislocations which in a few instances led to signs of lower cervical cord compression. These patients pose a major problem to the surgeon since the quality of these patients' bone usually is very poor and does not permit the use of autologous bone grafts in performing an anterior fusion. Furthermore, the use of heterogeneous bone graft (Kiel bone graft) in our hands has been discouraging because of the tendency of these

Table 1. *Atlanto-axial Dislocation.* Results of posterior occipital-cervical fusions (100 cases)

Major improvement	(painfree, neurological signs reduced, stable)	67%
Slight improvement	(less pain, neurological signs partly reduced, stable)	14%
Unchanged		9%
Worsened		5%
Unknown		5%

Mean observation time: $4^1/_2$ years (Longest $16^1/_2$ years)

grafts to collapse (Fig. 3a). In such cases, an anterior fusion using acrylic cement may be a possible alternative in a desperate situation (Fig. 3b).

Results

Table 1 shows the overall results in all 100 patients in the series. An improvement was achieved in 81% of the patients, whereas 9% were unchanged, 5% were worsened, and 5% were lost to follow-up.

At the time of follow-up, forty-five patients had died. The causes of death are listed in Table 2.

Table 3 shows the incidence of new or aggravated subaxial dislocation(s) following posterior fusion in the present series as compared to a series of conservatively treated patients.

Table 2. *Atlanto-axial Dislocation.* Causes of death (45 out of 100 cases)

1. Cardiopulmonary disease	22
2. Carcinoma	5
3. Gastrointestinal bleed	4
4. Nephropathy	4
5. Cerebrovascular stroke	3
6. Peritonitis	1
7. Generalized infection	1
8. Unknown	5
Total deaths	45
operative mortality	0

Table 3. *Relation Between New Subaxial Dislocation(s) in Patients with Atlanto-Axial Dislocation Treated Conservatively** (follow-up: 1–9 years) *or with Occipito-Cervical Fusion* (present series; follow-up: 1–15 years)

	New subaxial dislocation(s)
Conservatively treated patients	42%
Fusion operated patients	40%

* From Inga Redlund-Johnell: Dislocations of the cervical spine in Rheumatoid Arthritis. Academical Thesis, Malmö 1984.

Fifteen patients had postoperative wound infections. Five of these were deep and required reoperation with extirpation of all foreign material followed by a renewed fusion at a later stage.

Fracture of implanted wires occurred in thirteen patients and led to reoperation in two patients.

Discussion

A previous study from the University of Lund revealed an unexpectedly high and not previously reported incidence (11%) of fatal cord compression in RA patients with atlanto-axial instability[11] and recently Crockard and coworkers stated that since "the outcome of conservative treatment of progressive cervical myelopathy has been uniformly fatal, surgical intervention alone may offer hope to these patients"[12].

Our results offer further evidence for the beneficial effects of posterior occipito-cervical fusion in RA patients with atlantoaxial dislocation(s). The disappearance or at least amelioration of severe suboccipital pain in more than 80% of the patients and the neurological improvement in many patients with signs of cord compression may at least partly be explained by the frequent MRI findings of a markedly reduced synovitis in the bursae adjacent to the odontoid peg.

Crockard and coworkers have emphasized that computerized tomography with three-dimensional reconstruction may show that rheumatoid pannus, together with the odontoid peg, can contribute significantly to anterior cervico-medullary compression. Consequently, these authors have recommended a combined surgical procedure consisting of a transoral anterior decompression followed by a posterior occipito-cervical fusion in one session[20, 21]. In view of the present MRI findings, it appears that a posterior fusion only may lead to sufficient decompression by resolution of the pannus formation around the peg.

Some authors have preferred a posterior C1–C2 fusion not including the occipital bone suggesting that occipito-cervical "fusion may contribute to the rate of destruction in the subaxial region by increasing the load on the lower segments and by compensatory increased movement in these segments"[22, 23]. We investigated the incidence of new or increased subaxial dislocation(s) in the present series and found no evidence for this assumption as shown in Table 3. In our opinion, incorporation of the occipital bone offers a more stable fusion and may result in a better prevention against later odontoid peg upwards migration.

Schürmann found a dorsal suboccipito-cervical decompression unavoidable in his series of 43 cases of atlanto-axial dislocation with cord compression and also proposed the use of a "Baycast"-jacket for 6 to 8 months[24]. In our experience the major advantage with our method is that the patients, who often do not tolerate semi-soft or even soft collars, can be mobilized immediately following surgery without any external fixation.

The late mortality rate in our series reflects the relatively high age of the patients at time of surgery and the long follow-up periods. The most common cause of death was cardiopulmonary disease and there was no known case with a sudden death.

Acknowledgements

This study was supported by grants from Konung Gustav V:s 80-årsfond, Greta & Johan Kocks Stiftelser, Riksförbundet mot Reumatism, Alfred Österlunds Stifelse and Gyllenstiernska Krapperup Stiftelsen.

References

1. Garrod AE (1890) A treatise on rheumatism and rheumatoid arthritis. C Griffin, London, pp 1–342
2. Ball J, Sharp J (1971) Rheumatoid arthritis of the cervical spine. In: Hill AGS (ed) Modern trends in rheumatology, vol 2. Butterworth, London
3. Bland JH (1974) Rheumatoid arthritis of the cervical spine. J Rheumatol 1: 319–342
4. Conlon PW, Isdale IC, Rose BS (1966) Rheumatoid arthritis of the cervical spine. Ann Rheum Dis 25: 120–126
5. Dirheimer Y (1977) The craniovertebral region in chronic inflammatory diseases. Springer, Berlin Heidelberg New York
6. Martel W, Page JW (1960) Cervical vertebral erosions and subluxations in rheumatoid arthritis and ankylosing spondylitis. Arthritis Rheum 3: 546–556
7. Smith PH, Benn RT, Sharp J (1972) Natural history of rheumatoid cervical luxations. Ann Rheum Dis 31: 431–439
8. Menezes AH, van Gilder JC, Graf CJ *et al* (1980) Craniocervical abnormalities. A comprehensive surgical approach. J Neurosurg 53: 444–455
9. van Gilder JC, Menezes AH (1982) Craniovertebral abnormalities and their treatment. In: Schmidek HH, Sweet WH (eds) Operative neurosurgical techniques, indications, methods, and results. Grune & Stratton, Orlamdo FL, London, pp 1221–1235
10. Brito N, Klawki P, Schwarz M, *et al* (1987) Late results in cases of atlanto-axial dislocations after dorsal fusion. In: Voth D, Glees P (eds) Diseases in the cranio-cervical junction. Walter de Gruyter, Berlin New York, pp 279–286
11. Mikulowski P, Wollheim FA, Rotmil P *et al* (1975) Sudden death in rheumatoid arthritis with atlanto-axial dislocation. Acta Med Scand 198: 445–451
12. Crockard HA, Essigman WK, Stevens JM *et al* (1985) Surgical treatment of cervical cord compression in rheumatoid arthritis. Ann Rheum Dis 44: 809–816
13. Redlund-Johnell I (1984) Dislocations of the cervical spine in rheumatoid arthritis. Doctoral Dissertation Lund LUMEDW/MEXM-1001/1-140/1984
14. McAfee PC, Bohlman HH, Han JS *et al* (1986) Comparison of nuclear magnetic resonance imaging and computed tomography in the diagnosis of upper cervical spinal cord compression. Spine 11: 295–304
15. Brattström H, Granholm L (1972) Operative realignment and fixation of atlanto-axial dislocation in rheumatoid arthritis. Orthop Surg Traumatol, Excerpta Medica Int Congr Series 291: 407
16. Brattström H, Granholm L (1973) Chirurgie der Halswirbelsäule bei Patienten mit rheumatoider Arthritis. Orthopädie 2. Springer, Berlin Heidelberg New York, pp 118–120
17. Brattström H, Granholm L (1976) Atlanto-axial fusion in rheumatoid arthritis. Acta Orthop Scand 47: 619–628
18. Brattström H (1981) Surgery in atlanto-axial dislocation in rheumatoid arthritis. Reconstr Surg Traumat 18: 16–29. Karger, Basel
19. Brattström H, Brandt L, Ljunggren B (1987) Atlanto-axial dislocation in rheumatoid arthritis – signs and symptoms, radiographic pathology, operative techniques, and results. In: Voth D, Glees P (eds) Diseases in the cranio-cervical junction. de Gruyter Verlag, Berlin New York, pp 261–268
20. Stevens JM, Kendall BE, Crockard HA (1986) The spinal cord in rheumatoid arthritis with clinical myelopathy: a computed myelographic study. J Neurol Neurosurg Psychiatry 49: 140–151
21. Crockard HA, Pozo JL, Ransford AO *et al* (1986) Transoral decompression and posterior fusion for rheumatoid atlanto-axial subluxation. J Bone Joint Surg (Br) 68-B: 350–356
22. Toolanen G (1986) Rheumatoid atlanto-axial instability. Umeå University Medical Dissertation No 166
23. Meijers KAE, van Beusekom GT, Luyendijk W *et al* (1974) Dislocation of the cervical spine with cord compression in rheumatoid arthritis. J Bone Joint Surg 56-B: 668
24. Schürmann K (1987) Operative stabilization of atlanto-axial dislocation combined with cervical cord compression (myelopathy) in rheumatoid arthritis. In: Voth D, Glees P (eds) Diseases in the cranio-cervical junction. de Gruyter Verlag, Berlin New York, pp 249–259

Address for correspondence: S. Zygmunt, Department of Neurosurgery, University Hospital, Lund, S-221 85, Sweden.

Acta Neurochirurgica, Suppl. 43, 85–90 (1988)

Prognosis and Treatment of Cervical Spine Injuries with Associated Head Trauma

W. I. Steudel[1], **D. Rosenthal**[1], **R. Lorenz**[1], and **W. Merdes**[2]

Departments of [1] Neurosurgery and [2] Neuroradiology, Johann-Wolfgang-Goethe-University, Frankfurt am Main, Federal Republic of Germany

Summary

Pathological and biomechanical studies indicate that many injuries of the head and neck are to be considered as an entity, because cervical spinal fractures and head injuries are frequently combined. Depending on the localization and severity of the injury to the cervical spine, the incidence of an associated head trauma is reported in the literature as being between 6% and 63%. In 59 patients with cervical spinal fractures hospitalized in our department since 1980, we have examined the influence of head trauma on the prognosis and the management. Forty-three patients had fractures and/or dislocations in the lower cervical spine, and 16 had injuries in the upper region; head injuries were present in 33 (56 percent): minor injuries in 15, moderate and severe injuries in 10. Lesions requiring operation were a depressed skull fracture, extracerebral haematomas and scalp lacerations. Six out of 59 patients died; of these, four had suffered a severe head trauma. As a rule, surgery of the head and immobilization of the spine fracture were carried out as the first step followed by realignment of the spine with possible decompression and stabilization of the affected segment as a second step.

Conclusions:

– Fractures and/or dislocations of the cervical spine are frequently accompanied by head injuries.

– The severity of associated brain lesions has a marked influence on the prognosis.

– The correct identification of cervical lesions was delayed in about every second patient with minor head injury but rarely in comatose patients.

– In patients with altered consciousness an early spinal stabilization may be useful.

Keywords: Head injuries; spinal injuries; spinal fusion.

Consideration of cervical spine injuries in head trauma is of interest for various reasons. Pathological and biomechanical studies indicate that many injuries of the head and neck are to be regarded as an entity[9, 10, 12, 32, 57, 74]. In order to emphasize this close relationship, the term "cervico-cranium" was introduced by Schneider *et al.* in 1965[64, 65].

Combinations of cervical spinal fractures with head injuries are therefore frequent. The percentage reported in the literature varies greatly. It ranges from 63% in autopsy cases and 45% in fractures and dislocations to a mere 6% of all cervical spinal whiplash traumas[1, 2, 10, 66]. On the other hand, it is known that the simultaneous presence of cervical lesions can be readily overlooked in a head injured patient. This applies both to the atlanto-occipital region and to the lower cervical spine[10, 29, 49, 60, 73, 74]. The prognosis of cervical injuries is appraised very divergently in the literature. The mortality extends from 1 to 30 percent[1, 10, 22, 26, 40, 43, 48, 56, 65, 74]. One reason for this may be the different frequency of associated head injuries.

Therefore, we have examined the influence of head trauma on the prognosis and management in 59 patients with cervical spinal fractures hospitalized in our department since 1980.

Methods and Patients

In a retrospective study, we have followed up 59 patients with cervical spine fractures and/or dislocations who were treated from April 1980 to April 1987 in the Department of Neurosurgery, University of Frankfurt am Main. These patients comprise 14 women and 45 men (mean age: 36 years). The youngest patient was 13 years old and the oldest 79 years old.

The head injuries were classified according to Tönnis and Loew[72] and the level of consciousness was scored in accordance with the Glasgow-coma-Scale[31].

The neurological status and outcome with regard to the cord lesion were classified in the categories established by Frankel *et al*[24]. The roentgenographic evaluation comprised plain films and as a rule computer tomography (CT). Magnetic resonance imaging was carried out in four cases.

Each patient's outcome was determined between 6 months and 5 years after injury. In 12 patients, who were lost to follow-up, the outcome was considered in accordance with the condition of discharge.

Results

Incidence

Cervical spinal injury was combined 33 times with head injury (56%). In addition, abdominal injury was present six times and extremity fractures 12 times (Table 1). The most frequent cervical spine injuries were combinations of dislocations and fractures (39 times), followed by dislocations without fractures (15 times). Pure fractures were present in only five cases. With regard to the location of the cervical injury, a head trauma occurred in lesions of the upper cervical spine in three quarters of the cases (12 out of 16), and in the region of the lower cervical spine (22 out of 43) in about half of the patients (Table 2).

Accident Causes

The most frequent causes were traffic accidents. Motorcycle riders suffer the most severe injuries (Table 3).

Table 1. *Associated Injuries in 59 Patients with Cervical Spine Injury*

Associated injuries	Number of lesions	Frequency (%)
None	19/59	32
Head	33/59	56
Facial injuries	5/59	9
Thoracic	6/59	10
Abdominal	3/59	5
Limb fracture-dislocation	12/59	20
Thoracic spine fracture-dislocation	3/59	5

Table 2. *Incidence of Head and Neck Injuries Related to Localization of Cervical Spine Fractures and/or Dislocations*

Localization of spine injury	Head injury			Total (dead)
	No	Yes		
		GCS* 3–8	GCS 9–15	
Upper cervical spine (including) spondylolisthesis of C 2/C 3	4	6 (3)	6	16 (3)
Lower cervical spine	22 (1)	4 (1)	17	43 (3)
Total	26 (1)	10 (4)	23	59 (6)

* Glasgow-Coma-Score on admission.

Table 3. *Mechanisms of Cervical Spine Injuries with Head Trauma*

Mechanisms of injury	Head trauma			Total (dead)
	No	Yes		
		GCS* 3–8	GCS 9–15	
Motor-car	13	4 (1)	16	33 (1)
Motorcycle	1	3 (2)		4 (2)
Fall	6 (1)		5 (1)	11 (2)
Sport/leisure	6	3 (1)	2	11 (1)

* Glasgow-Coma-Score on admission.

Table 4. *Head Injuries in Cervical Spine Trauma Related to Neurological Findings (N = 59)*

Spinal neurological status	Head trauma			Total (dead)
	No	Yes		
		GCS* 3–8	GCS 9–15	
No deficit	9	3	9	21
Radicular deficit	5		7	12
Cord injury	10	3 (1)	4	17 (1)
incomplete	10	3 (1)	4	17 (1)
complete	2 (1)	4 (3)	3 (1)	9 (5)

* Glasgow-Coma-Score on admission.

Out of 4 motorcyclists, three had suffered a severe head trauma, and two died. Two swimming accidents caused by jumping head-first into unknown waters are worth mentioning: there was a fracture dislocation in one case and complete tetraplegia in the other case.

Morbidity and Mortality

The severity of head injury paralleled the extent of cord lesion (Table 4). The prognosis became more unfavorable with the severity of head injury. Out of 59 patients, six died, four with a severe head injury. A 76 year old female patient with a minor head trauma, but complete paraplegia, died on the 38th day of pneumonia. A 79 year old male, the only patient without head injury with tetraplegia, died on the third day, probably of embolism.

Time of Treatment

The spinal injury was diagnosed on the day of accident in 42 cases (within four hours in 18 cases) and reduction was carried out. The diagnosis was delayed in 17 cases: in minor head injuries in 10 cases and in severe head trauma only once. Additional reasons for a delayed diagnosis were an accident abroad in eight cases, diagnostic difficulties in demonstrating a cervical spine fracture in six cases and one case each of alcohol intoxication and amputation. In six of these injuries – of these, three with head trauma –, the spinal injury was only diagnosed following the development of neurological deficits.

Management

The 33 head injuries included only one open injury. Contusion foci were present three times, subdural haematomas and hygromas twice, a depressed skull fracture once, and scalp lacerations 12 times. The space occupying lesions were removed immediately. During operation, the head was immobilized and fixed in a Mayfield clamp. Afterwards, Crutchfield extension (Gardner-Wells tongs) was applied for reduction and realignment of dislocations and fractures. In two older fracture dislocations, the reduction was incomplete and surgical correction was required. An operative spinal stabilization procedure was regarded as necessary in 47 cases. As a rule, this was carried out in the second week after the accident, or even earlier in confused or comatose patients. Treatment was nonsurgical in 12 patients: fractures of only articular and spinous processes in five, early deaths in three, and in four cases halo devices or cervico-thoracic braces were used until stabilization was achieved.

The anterior approach was used with fusion according to Smith and Robinson 30 times, four times at two levels[61]. An anterior decompression with excision of the vertebral body and stabilization with an iliac crest graft to replace the resected vertebra were carried out in seven cases. A posterior approach with wiring and/or fusion was applied in 10 cases. This technique was performed in the sitting or prone position with the head fixed in the Mayfield clamp. A combined procedure was applied five times. After the stabilization operation, the Gardner-Wells tongs were removed.

Complications were a break of wire loops in one case and a wound infection in another. Twice, the horseshoe graft slid out, requiring a second operation. Since then, all patients received a light, individually adapted plastic head-cervico-thoracic brace which can be worn under their clothing (manufactured by Habermann, Orthopedic Workshop of the Frankfurt/Main University).

Outcome

Of 59 patients, six had died. All 21 patients who did not have any neurological deficits on admission also did not show any deficits in the follow-up examination (Table 4). Two patients with severe head injuries and fractures of articular and spinous processes survived with no spinal deficit. In 12 cases, there was a root lesion which recovered completely in five cases; in seven cases there was major improvement but without complete recovery. An incomplete lesion was present 17 times. There was an improvement in 11 patients, three were lost to follow-up; two showed no significant improvement. One patient died. There was complete cord lesion in nine cases. Five patients died, four of them with head injuries. The survivors showed no essential improvement.

Discussion

Pathological and biomechanical studies show that the head and neck are to be considered as an unit[10, 12, 57, 65, 74, 76]. It is hence comprehensible that cervical spine injuries are often combined with head injuries[1, 2, 10, 31, 47, 66] or facial injuries[40, 74]. In addition, cranial nerve lesions have been observed[12, 25, 55]. There is a close relationship with regard to the vascular supply due above all to the special course of the vertebral artery[11, 63, 64]. There are very divergent reports on the frequency of combination of head injuries and cervical spinal lesions. The percentage obviously depends on the selection of patients. In fatal spinal injuries, head injuries are especially frequent[2, 10]. In addition, the incidence evidently depends on the type, extent and localization of the injury. In spinal lesions with fractures and/or dislocations, 45% head injuries were found[1], and 56% in our patients. Only 6% head injuries were present among all whiplash traumas[66].

The mortality of spinal lesions generally ranges from 1 to 30%[1, 14, 22–24, 26, 41–43, 52, 74]. Besides the severity of the injury and the localization, the associated injuries appreciably influence the prognosis[74]. In our patients, four out of seven cases with severe head injuries died. The divergent mortality data in the literature evidently depends on the selection of injured persons and the proportion of associated injuries. Injury to the central nervous system essentially determines mortality[10].

The course of a cervical spinal injury is determined by early diagnosis and adequate therapy. Aebi *et al.* reported that reduction was carried out in the first hours in only 25%[1]. The reasons for a delayed treatment are seen to be head injuries with impaired consciousness, alcohol intoxication, other associated injuries and lacking or inadequate neuroradiological diagnostics. Bohlman specifies the frequency of cervical spine injuries with associated head trauma obscuring the diagnosis as being 19%[10]. We found that with one exception, correct and early diagnosis was made in comatose patients with head injuries, but was only made with a delay in every second patient with minor head injuries; there was a secondary deterioration in five cases. The standard plain radiodiagnostics were often not sufficient. Streitwieser *et al.* report the sensitivity as 93%[69]. The reasons for the problems in detection consist in the localization, in the absence of bony abnormalities, and in spotaneous reductions[27, 39, 40]. When spinal injury is suspected, plain films should be supplemented by CT and magnetic resonance imaging. Based on this experience, *it is strongly recommended that patients with head injuries should always be examined with regard to involvement of the cervical spine*[73].

The management of patients with cervical spine lesions in whom head injuries are also present depends on the type and extent of the injuries and is oriented in detail in accordance with familiar principles. First priorities in the acute phase are above all maintenance of optimal oxygenation, and restoration and support of circulation[31, 52, 66]. This is indispensable, since the injured nervous tissue is known to react especially sensitively to lack of oxygen and ischaemia[3, 30, 51, 52, 71]. In patients with potential cervical injury, the first management includes the adequate immobilization of the neck. The general goals of management of spinal and head injuries are above all precise identification of the injuries, if necessary a decompression as early as possible and additional stabilization in spinal injury with the lowest neurological trauma. It should be taken into account that the neurological findings and the vegetative parameters may change rapidly in the acute phase. Monitoring of neural functions and vegetative parameters in these patients is imperative[31, 52]. First surgical procedures include space occupying lesions, in particular extracerebral haematomas. In associated spinal fractures and dislocations, an attempt must be made to attain reduction. This reduction can be carried out by means of traction as described by Crutchfield. In accordance with our experience, it is effective to fix the head in a Mayfield clamp during craniotomy. This can be exchanged then for the Gardner-Wells tongs which we currently prefer.

Acute surgical decompression of the spinal canal is only indicated when spinal cord or nerve root are compressed by bone fragments, when there is a disc prolapse, or when an epidural haematoma or an open injury is present[3, 44, 52, 58, 59, 62]. The sequelae of injuries may first become manifest on the basis of secondary deterioration of the neurological status. However, neurological deterioration is not caused only by compression, but may also be due to ischaemia[22, 23]. In immobilization and/or traction, further diagnostic clarification may then follow[52].

The question of surgical or nonsurgical treatment of spinal injury continues to be controversial[4, 5, 18, 44, 45, 51, 58, 74]. In our opinion, the advantages of a surgical stabilization consist in the definitive restoration of spinal stability, reconstruction of anatomical structures, preservation of neurological functions and facilitation of mobilization[1, 5, 28, 39, 44, 51, 74, 78]. A large variety of surgical techniques of stabilization and fusion are recommended[6–8, 13, 15–21, 33–37, 46, 47, 50, 53, 54, 67, 68, 70, 75, 77]. Especially in patients with impaired consciousness, an early operative stabilization may be useful.

References

1. Aebi M, Mohler J, Zäch GA, Morscher E (1986) Indication, surgical technique and results of 100 surgically-treated fractures and fracture-dislocation of the cervical spine. Clin Orthop 203: 244–257
2. Alker GJ, Oh YS, Leslie EV, Lehotay J, Panaro VA, Eschner EG (1975) Postmortem radiology of head and neck injuries in fatal traffic accidents. Radiology 114: 611–617
3. Banik NL, Hogan EL, Powers JM *et al* (1982) Degradation of cytoskeletal proteins in experimental spinal cord injury. Neurochem Res 7: 1465–1475
4. Blencke BA (1986) Zur Operationsindikation bei traumatischen Querschnittslähmungen. Hessisches Ärzteblatt 47: 606 – 623
5. Böhler J (1971) Operative Behandlung von Halswirbelsäulenverletzungen. Hefte Unfallheilkunde 108: 132–137
6. Böhler J, Gaudernak T (1980) Anterior plate stabilisation for fracture-dislocation of the lower cervical spine. J Trauma 20: 203
7. Böhler J (1982) Anterior stabilisation for acute fractures and non-unions of the dens. J Bone Joint Surg 64 A: 18
8. Boestman O, Myllynen P, Risha EB (1984) Posterior spinal fusion using internal fixation with the Daab plate. Acta Orthop Scand 35: 310–314
9. Bohlman HH, Eismont FJ (1981) Surgical techniques of anterior decompression and fusion for spinal cord injuries. Clin Orthop 154: 57–67
10. Bohlman HH, Boada E (1983) Fractures and dislocations of the lower cervical spine. In: Bailey *et al* (eds) The cervical spine. Lippincott Comp, Philadelphia, pp 232–267
11. Bose B, Northrup BE, Osterholm JL (1985) Delayed vertebrobasilar insufficiency following cervical spine injury. Spine 10: 108–110
12. Breig A (1978) Adverse mechanical tension in the central nervous system. Almquist & Wiksell, Stockholm, pp 121–139
13. Bremer AM, Naguyen TQ (1983) Internal metal plate fixation

combined with anterior interbody fusion in cases of cervical spine injury: Neurosurgery 12: 649–653

14. Buchholz RW, Burkhead WZ (1979) The pathological anatomy of fatal atlanto-occipital dislocations. J Bone Joint Surg 61 A: 148–150
15. Bundens DA, Rechtine GR, Bohlman HH (1984) Upper cervical spine injuries. Orthop Rev 13: 556–565
16. Capen DA, Garland DE, Waters RL (1985) Surgical stabilization of the cervical spine. A cooperative analysis of anterior and posterior spine fusions. Clin Orthop 196: 229–237
17. Chan RL, Schweigel JF, Thompson GB (1983) Halo-thoracic brace immobilisation in 188 patients with acute cervical spine injuries. J Neurosurg 58: 508–515
18. Cahill DW, Bellegarrique R, Ducker TB (1983) Bilateral facet to spinous process fusion: a new technique for posterior spinal fusion after trauma. Neurosurgery 13: 1–4
19. Clark CR, White AA (1985) Fractures of the Dens. J Bone Joint Surg 67 A: 1340–1348
20. Cloward RB (1961) Treatment of acute fractures and fracture-dislocations of the cervical spine by vertebral body fusion. J Neurosurg 18: 201
21. Cooper PR, Maravilla KR, Sklar FH, Moody SF, Clark K (1979) Halo immobilization of cervical spine fractures. J Neurosurg 50: 603–610
22. Ducker TB, Russo L, Bellegarrique R, Lucas JT (1979) Complete sensori motor paralysis after cord injury: mortality, recovery and therapeutic implications. J Trauma 19: 837–840
23. Ducker TB, Bellegarrique R, Saleman W, Walleck C (1984) Timing of operative care in cervical spinal cord injury. Spine 9: 525–531
24. Frankel HL, Hancock DO, Hyslop G, Melsak J, Michaelis LS, Ungar GH, Vernon JDS, Walsh JJ (1969) The value of postural reduction in the initial management of closed injuries of the spine with paraplegia and tetraplegia. Paraplegia 7: 179–192
25. Grundy DJ, Weeney T, Jones HWF (1984) Cranial nerve palsies in cervical injuries. Spine 9: 339–343
26. Harris P, Karmi MZ, McClemont E, Matlhoko DA, Paul KS (1980) The prognosis of patients sustaining severe cervical spine injury (C 2–C 7 inclusive). Paraplegia 18: 324
27. Harris JH, Burke JT, Ray RD *et al* (1984) Low (Type III) odontoid fracture: A new radiographic sign. Radiology 153: 353–356
28. Harris JH, Edeiken-Monroe B, Kopaniky DR (1986) A practical classification of acute cervical spine injuries. Orthop Clin North Am 17: 15–30
29. Herkowitz HN, Rothman RH (1984) Subacute instability of the cervical spine. Spine 9: 348–357
30. Hossmann V (1985) Zerebrale Ischämie. In: Hossmann V *et al* (eds) Kalziumantagonisten und cerebrale Erkrankungen. Springer, Berlin Heidelberg New York, pp 63–88
31. Jennett B, Teasdale G (1981) Management of head injuries. Davis, Philadelphia, pp 77–90 and 211–252
32. Jofe MH, White AA, Panjabi MM (1983) Physiology and Biomechanics. In: Bailey RW *et al* (eds) The cervical spine. Lippincott, Philadelphia, pp 23–35
33. Johnson RM, Wolf JW (1983) Stability. In: Bailey RW *et al* (eds) The cervical spine. Lippincott, Philadelphia, pp 35–53
34. Kiwerski J (1982) Anterior interbody fusion in the treatment of cervical spine injury. Int Orthop 6: 255–258
35. Kleinfeld F (1981) Zur Behandlung von Frakturen der Halswirbelsäule mit dem Halo-Fixateur-externe. Unfallheilkunde 84: 161–167
36. Knöringer P (1985) Zur operativen Behandlung von Verletzungen der unteren Halswirbelsäule. Neurochirurgica 28: 213–220
37. Lesoin F, Cama A, Lores G *et al* (1984) The anterior approach and plates in lower cervical posttraumatic lesions. Surg Neurol 21: 581–587
38. Levine AM, Edwards CC (1986) Treatment of injuries in the C 1–C 2 complex. Orthop Clin North Am 17: 31–44
39. Levine AM, Edwards CC (1986) Complications in the treatment of acute spinal injury. Orthop Clin North Am 17: 183–203
40. Lewis VL, Manson PN, Morgan RF, Cerullo LJ, Meyer PR (1985) Facial injuries associated with cervical fractures: recognition, patterns and management. J Trauma 25: 90–93
41. Louis R (1979) Traumatismes du rachis cervical. II. Fractures et luxations. Nouv Presse Med 8: 1931
42. Lucas JT, Ducker TB (1979) Recovery in spinal cord injuries. Adv Neurosurg 7: 281–294
43. Lucas JT, Ducker TB (1977) Morbidity, mortality and recovery rates of patients with spinal cord injuries undergoing anterior decompressive procedures or fusion or both. Surg Forum 28: 451
44. Magerl F (1980) Operative Frühbehandlung bei traumatischer Querschnittslähmung. Orthopädie 9: 34–44
45. Meinecke FW (1986) Rückenmarksverletzungen – Geschichte, konservative Behandlung. In: Probst J (ed) Unfallheilkunde, pp 227–233
46. Murphy MJ, Daniaux H, Southwick WO (1986) Posterior cervical fusion with rigid internal fixation. Orthop Clin North Am 17: 55–65
47. Nachemson A (1960) Fracture of the odontoid process of the axis. Acta Orthop Scand 29: 185
48. O'Brien PJ, Schweigel JF, Thompson W (1982) Dislocations of the lower cervical spine. J Trauma 22: 710
49. Pech P, Kilgore DP, Pojunas KW, Haughton VM (1985) Cervical spinal fractures: CT detection. Radiology 157: 117–120
50. Ragozzino MW, Deluca SA (1985) Upper cervical spine trauma. Am Fam Physician 32: 113–119
51. Ransohoff J (1980) Surgical intervention after traumatic injury. In: Windle WF (ed) The spinal cord and its reaction of traumatic injury. M Dekker, New York, p 311
52. Ransohoff J (1983) Acute care: pathophysiology, current treatment and experimental approaches. In: Bailey RW *et al* (eds) The cervical spine. Lippincott, Philadelphia, pp 288–305
53. Rether JR, Otte D (1984) Verletzungen der Halswirbelsäule beim Verkehrsunfall. Unfallheilkunde 87: 524–530
54. Rogers VA (1957) Fractures and dislocations of the cervical spine. J Bone Joint Surg 39 B: 341
55. Rosa L, Carol M, Bellegarrique R, Ducker TB (1984) Multiple cranial nerve palsies due to hyperextension injury to the cervical spine. Case report. J Neurosurg 61: 172–173
56. Roy-Camille R, Saillant G, Judet T, Mammondy P (1982) Traumatismes récents des cinq dernières vertèbres cervicales chez l'adulte. Ann Chir 36: 735
57. Rüter A (1971) Zur operativen Behandlung von Verletzungen der Halswirbelsäule. Hefte Unfallheilkunde 108: 138–142
58. Ruge D (1977) Spinal cord injuries. In: Spinal disorders diagnosis and treatment. Lea & Febiger, Philadelphia
59. Ruidisch MH (1986) Rückenmarksverletzungen – operative Behandlung. In: Probst J (ed) Unfallheilkunde 1986. Demeter, Gräfeling, pp 235–239

60. Scher AT (1983) Hyperextension trauma in the elderly: An easily overlooked spinal injury. J Trauma 23: 1066–1068
61. Smith GW, Robinson RA (1958) The treatment of certain cervical spine disorders by anterior removal of the intervertebral disc and interbody fusion. J Bone Joint Surg 40 A: 607
62. Schmit-Neuerburg KP, Obertacke U (1986) Erstversorgung. In: Probst J (ed) Unfallheilkunde 1986, Demeter, Gräfeling, pp 83 – 88
63. Schneider RC, Schemm GW (1961) Vertebral artery insufficiency in acute and chronic spinal trauma. J Neurosurg 18: 348–360
64. Schneider RC, Livingston KE, Cave AJE, Hamilton G (1965) Hangman's fracture of the cervical spine. J Neurosurg 22: 141–154
65. Schneider RC, Crosby EC, Russo RH, Gosch HH (1973) Traumatic spinal cord syndromes and their management. Clin Neurosurg 20: 424–492
66. Soderstrom LA, Brumback RJ (1986) Early care of the patient with cervical spine injury. Orthop Clin North Am 17: 3–13
67. Southwick WO (1980) Management of fractures of the Dens (odontoid process). J Bone Joint Surg 62 A: 482–486
68. Stauffer ES (1986) Management of spine fractures C 3 to C 7. Orthop Clin North Am 17: 45–53
69. Streitwieser DR, Knopp R, Wales LR *et al* (1983) Accuracy of standard radiographic view in detecting cervical spine fractures. Am Emerg Med 12: 538–542
70. Svengaard NA, Gronquist S, Delgado T, Salford LG (1982) Treatment of severe cervical spine injuries by anterior interbody fusion with early mobilization. Acta Neurochir (Wien) 60: 91–103
71. Symon L (1982) Physiologie und Pathophysiologie der zerebralen Durchblutung. In: Dietz H, Umbach W, Wüllenweber R (eds) Klinische Neurochirurgie, Vol 1. Thieme, Stuttgart, pp 103 – 129
72. Tönnis W, Loew F (1953) Einteilung der gedeckten Hirnschädigungen. Ärztl Praxis 5: 13–14
73. Van der Walt PJ (1983) Pitfalls in the diagnosis of spinal injuries. SAFR Med J 64: 96–97
74. Vigouroux RP, Reynier Y, Baldini M (1983) Traumatic lesions of cranio-cervical junction. In: Samii M, Brihaye J (eds) Traumatology of the skull base. Springer, Berlin Heidelberg New York, pp 214–226
75. Walter J, Doris PE, Shaffer MA (1984) Clinical presentation of patients with acute cervical spine injury. Am Emerg Med 13: 512–515
76. White AA, Panjabi MM (1978) Clinical biomechanisms of the spine. Lippincott, Philadelphia, pp 115–190
77. Whitehill R, Schmidt R (1983) The posterior interspinous fusion in the treatment of quadriplegia. Spine 8: 733–740
78. Wolter D (1985) Vorschlag für eine Einteilung von Wirbelsäulenverletzungen. Unfallchirurg 88: 481 – 484

Address for correspondence: Dr. W. I. Steudel, Department of Neurosurgery, Johann-Wolfgang-Goethe-Universität, D-6000 Frankfurt/Main, Federal Republic of Germany.

Acta Neurochirurgica, Suppl. 43, 91–93 (1988)

New Methods of Treatment of Severely Injured Sciatic Nerve and Spinal Cord. An Experimental Study

S. Rochkind[1], L. Barr-Nea[2], A. Bartal[1], M. Nissan[3], R. Lubart[4], and N. Razon[1]

[1] Department of Neurosurgery, Tel-Aviv Medical Center, Ichilov Hospital, [2] Department of Cell Biology and Histology, Sackler School of Medicine, Tel-Aviv University, [3] Center for Technological Education, Holon, [4] Department of Physics, Bar-Ilan University, Ramat Gan, Israel

Summary

It has been demonstrated that low-energy laser irradiation (LELI) applied simultaneously to the injured sciatic nerve and the corresponding segment of the spinal cord, accelerates the process of regeneration of the injured peripheral nerve. A beneficial influence of LELI was also observed when it was applied to the spinal cord following transection and implantation of a segment of an autologous sciatic nerve, but further studies are necessary to evaluate if real regeneration or only earlier distal cord automatism occurred.

Both methods are proposed for treatment of peripheral nerve lesions (PNS) and spinal cord injuries.

Keywords: Laser irradiation; low-energy laser; LELI; peripheral nerve lesion; spinal cord lesion.

Introduction

Low-energy laser irradiation (LELI) is now used worldwide for research as well as for treatment[1, 2]. For the last eight years, our group has been conducting a study of the effects of the He-NE laser (632.8 nm) upon the nervous system, in an effort to increase the rate of recovery in injured peripheral nerve and spinal cord.

The results have been assessed both by electrophysiological and morphological means. Moreover, other wave lengths (465, 520 and 588 nm) have also been tested in order to define the best possible treatment.

Our previous results[3–6] have shown that:

1. LELI increases the action potential of the normal sciatic nerve of the rat, and mitigates degeneration of the muscle following the cutting of the nerve. This effect extends for at least one year.

2. LELI applied to crushed sciatic nerve, prevents the fall in action potential and scar formation, and accelerates the regeneration process; it also attenuates the degenerative changes in the corresponding segments of the spinal cord, which generally follow an injury to the sciatic nerve (submitted for publication).

3. Increased vascularization followed LELI application to nervous tissues.

4. A systemic effect has been observed in burns, wounds and injuries of peripheral and central nervous system (submitted for publication).

5. The beneficial effect on injured peripheral nerve and spinal cord diminishes with decreasing wavelength from 632 nm to 465 nm.

In this report we wish to present some of the results of LELI: (a) applied on the leg and the spinal cord, following a crush injury of the rat sciatic nerve (as an example of the effect of laser on PNS) and (b) applied to the spinal cord following transection and implantation of a segment of the sciatic nerve in dogs (as an example of laser effect on CNS).

Materials and Methods

1. Sciatic Nerve

Sprague-Dawley rats underwent a bilateral crush-injury of the sciatic nerve as previously described[4, 5]. The wound was closed and the test rats were treated with LELI, using a 16 mW He-Ne laser or a 40 mW C.W. Argon, or 20 mW pulsed dye-laser transcutaneously over 2 sites: the injured site of the right leg, and the corresponding segments of the spinal cord: L 4–S 2.

Starting immediately after the operation, LELI was applied daily for 20 consecutive days. The action potential of the crushed sciatic nerve of 7 laser-treated rats and 9 control non-irradiated rats, was recorded up to 180 days after operation. Thirty-five rats were sacrificed 90 days after operation for histological preparations. The animals were anesthetized with Nembutal, and perfused intracardially with 3% gluteraldehyde. The sciatic nerves were removed and divided into 3 segments: proximal, crushed (scar) region, and distal, and then put into fresh 3% glutaraldehyde. The spinal column was removed, the cord freed from the vertebrae and put into a fresh fixative. The nerve segments were routinely prepared for electron and light microscopy.

2. Spinal Cord

Seventeen dogs underwent laminectomy and transection of the cord at D 12–L 1. An autograft of the left sciatic nerve was then implanted into the injured area. Neurorrhaphy was performed on the right sciatic nerve. Seven dogs did not receive any additional treatment and served as control. The other 10 had the LELI (He-Ne) applied for 20 days transcutaneously over the operated area. Twenty-one days after operation, 3 animals were anesthetized with Nembutal and perfused intracardially with 10% formalin. The operated region of the spinal cord was removed and cut into three fragments: proximal, implant, and distal. The cord was then freed from the vertebrae and the 3 fragments were put into fresh fixative.

The tissues were prepared for histological examination: cut at 6 μ and stained with haematoxylin eosin, cresyl fast violet and luxol fast blue.

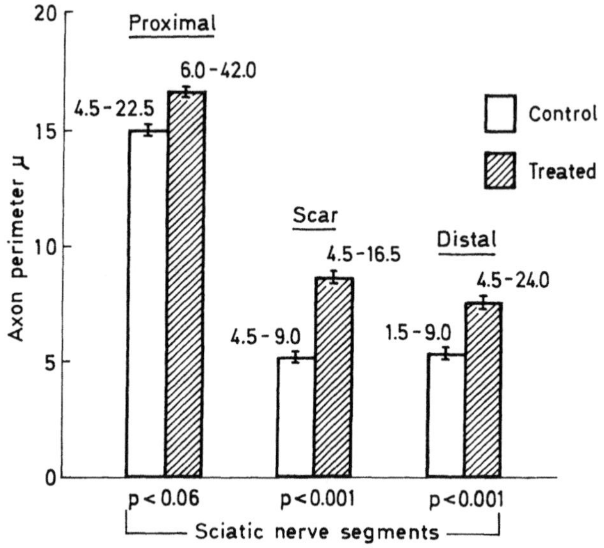

Fig. 2. Perimeters of the axons of the crushed sciatic nerve, mean and range values estimated in the three segments: proximal, crush region (scar) and distal

Results and Discussion

The effect of the combined laser irradiation (leg and spinal cord) on the action potential (AP) of the left (control) and right (He-Ne laser irradiated) sciatic nerve is presented in Fig. 1. The AP values recorded from the right leg are well above pre-crush (normal) values, while the AP of the left leg remained at approximately normal values. It is interesting to note that the AP of the left leg did *not* drop, as we had observed in rats which had received LELI on the right leg only and *not* on the spinal cord[5, 6]. Irradiation on the spinal cord maintains the action potential of the crushed non-irradiated sciatic nerve, and this explains the small difference between the values recorded in *both* legs.

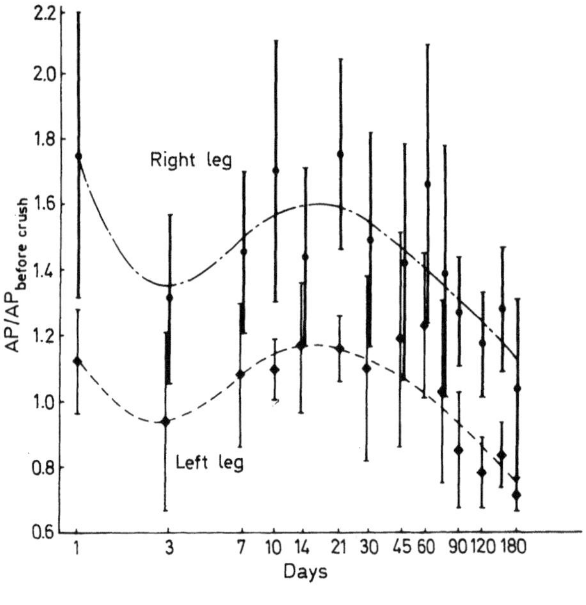

Fig. 1. Ratio of measured action potential of crushed nerve to the one observed before the crush injury – right leg: laser-irradiated; left leg: control, non-irradiated

Fig. 2 represents the mean perimeter of the three axonal segments: proximal, scar (crushed region), and distal in non-irradiated and irradiated sciatic nerves (He-Ne laser 16 mW) 90 days after crush. As could be expected, the difference between the two proximal groups is not significant (15/16.5 mμ), *i.e.* the retrograde degeneration was minimal in both cases. In the scar segment, the difference in size between the two groups of axons is the largest (5.2/8.5 μ), while the irradiated group contains much larger ones. This is primarily due to a thicker myelin envelope. In the distal segment, the situation is similar to that of the scar, with some very small axons in the control, representing degenerated ones. The beneficial effect diminishes with decreasing the wavelength from 632 nm to 465 nm.

The 7 dogs which underwent transection of the spinal cord, accompanied by implantation of a fragment of autologous sciatic nerve, became paralysed as was to be expected, and remained so for at least 6 weeks. The 10 dogs which underwent the same operation, but were irradiated immediately with LELI, were able to stand after 7–9 weeks and could walk a few steps between 9–12 weeks. An example of such a dog is presented in Fig. 3, but we are well aware that this as such does not constitute proof of a restored and functioning continuity of the spinal cord, but only of an earlier development of a distal cord automatism.

On the other hand, the histological picture obtained from the dogs which were sacrificed 21 days post-

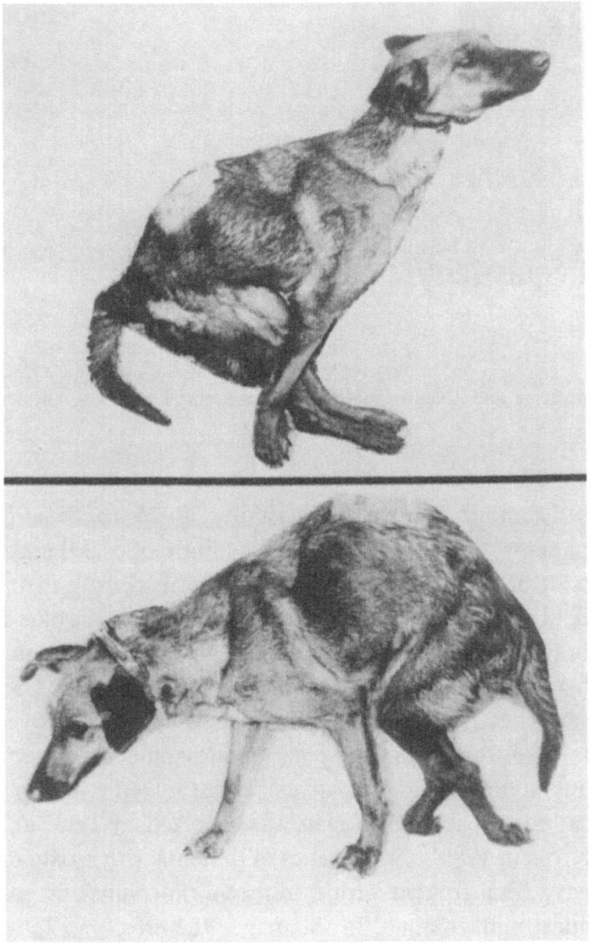

Fig. 3. Example of a dog which underwent spinal transection and implantation of a segment of the left sciatic nerve. Upper figure: two weeks postoperation, the dog is paralyzed. Lower figure: seven weeks postoperation, the dog is standing

operation, confirms and enlarges the clinical picture. No sign of rejection and no *prominent* scar tissue were observed at the site of contact between the spinal cord tissue and the transplanted nerve. Moreover, new axons and blood vessels originating from the spinal tissue, extended into the graft. These were seen only in the LELI dogs and not in the controls in which scar tissue had developed at the site of transection. Here again, the ability of the laser irradiation to promote and sustain regeneration in the nervous system is confirmed. But further experiments using histochemical markers for proof of possible functioning regeneration, are necessary.

In conclusion, *the data presented here indicates that* irradiation on both injured leg and spinal cord in rats induces better clinical results than the previously reported ones involving irradiation only to the injured leg. Comparing the results presented here with our previous ones[3–6], we would like to propose the following method of treatment in injuries of the PNS: (a) A single local injury should be treated by local LELI; (b) Multiple injuries: local LELI as much as possible, combined with central LELI.

We also wish to encourage further experimental studies on the possibility and effectiveness of treatment of spinal cord lesions by peripheral nerve transplantation in combination with laser irradiation.

Acknowledgement

We would like to thank Mr. N. Brusovanic for his invaluable technical assistance.

References

1. Letokhov VS (1985) Laser biology and medicine. Nature 316: 325–329
2. Mester E, Mester AF, Mester A (1985) The biomedical effects of laser application. Laser Surg Med 5: 31–39
3. Rochkind S (1978) Stimulating effect of laser energy on regeneration of injured peripheral nerves. Morphogenesis and Regeneration 73: 48–50 (in Russian)
4. Rochkind S, Nissan M, Razon N, Schwartz M, Bartal A (1986) Electrophysiological effect of He-Ne laser on normal and injured sciatic nerve in the rat. Acta Neurochir (Wien) 83: 125–130
5. Rochkind S, Nissan M, Barr-Nea L, Razon N, Schwartz M, Bartal A (1987) Response of peripheral nerve to He-Ne laser experimental studies. Laser Surg Med 7: 441–443
6. Rochkind S, Barr-Nea L, Razon N, Bartal A, Schwartz M (1987) Stimulatory effect of He-Ne laser on injured sciatic nerves of rats. Neurosurgery 20: 843–847

Address for correspondence: Dr. S. Rochkind, Department of Neurosurgery, Ichilov Hospital, Tel-Aviv Medical Center, 6 Weizman Street, Tel-Aviv 64239, Israel.

Acta Neurochirurgica, Suppl. 43, 94–99 (1988)

Implanted Pump Systems for Treatment of Spasticity

J. Zierski[1], **H. Müller**[2], **D. Dralle**[3], and **T. Wurdinger**[1]

Departments of [1] Neurosurgery, [2] Anesthesiology and Intensive Care Medicine and [3] Neuropediatrics, Justus-Liebig-University, Giessen, Federal Republic of Germany

Summary

In about one third of patients with violent spasticity due to spinal trauma, multiple sclerosis, and diffuse brain injury adequate control with oral antispastic medication cannot be achieved and successful rehabilitation is severely handicaped. In the past these patients were subjected to destructive chemical procedures or extensive surgery.

The authors present the results of management of uncontrollable spasticity by means of continuous intrathecal administration of baclofen with a totally implantable gas driven pump system (Infusaid). 30 patients were treated between June 1985 and January. 1987. The main indication was incapacitating spasticity resistant to oral treatment with baclofen and caused by spinal cord injury or lesion (11 patients), multiple sclerosis (11 patients), infantile cerebral palsy (3 patients) and cerebral injury, hypoxia or ischaemia (5 patients). Clinical assessment included spasticity scores, integrated electromyography (Iemg) and motography. Effective control for spasticity with mean reduction of Iemg by 55%, decrease of Ashworth's score from 3 to 0 and improvement of life quality was obtained in all patients with daily dose of 10–800 micrograms of Baclofen. Voluntary resting motoricity was not impaired and there were no untoward central side effects. The excellent effect of intrathecal baclofen in comparison with oral therapy is explained by local, spinal GABAergic inhibitory action of the drug which is delivered directly into spinal subarachnoid space. Dose finding and dose adjustment is performed prior to pump implantation by intermittend injections into a subcutaneous port. The complications of the procedure were minor (catheter displacement, disconnection) and easily correctable. The only limitation of this method is the inability to reduce spasticity in selective muscle groups. Intrathecal long term infusion of Baclofen is an alternative to surgery in treatment of spasticity.

Keywords: Spasticity; baclofen; pump system; intrathecal administration; complications; results.

Introduction

Successful rehabilitation and care of patients affected with handicapping spasticity due to spinal or cerebral lesion may be rendered difficult by lack of effective methods permitting permanent control of increased muscle tone and muscle spasms. Oral medication with antispastic agents and physiotherapy fail to provide satisfactory control of spasticity in 25 to 35% of patients[2, 8]. The main reason for failure of oral therapy with baclofen, the most often prescribed drug in the field of spasticity, is its limited passage across the blood-brain barrier[5, 6]. This disadvantage can be overcome by administering the drug directly into the spinal CSF spaces with the help of techniques which have proved to be effective during spinal opiate analgesia, *i.e.* via implantable delivery systems[18]. First intrathecal application of baclofen in man was reported by Penn and Kroin in 1985[12]. Small series of patients with spasticity have been treated with intrathecal morphine[4] or epidural midazolam[10]. In this paper the effects of long-term therapy with intrathecal baclofen infusion, using implantable pumps with continuous flow, will be described.

Patients and Methods

Forty-four patients (21 male and 23 female), aged 2–71 years, were treated with intrathecal baclofen in the period between May 1985 and August 1987. Selection criteria for this form of therapy included severe, handicapping spasticity and spasms interfering with daily care or physiotherapy and lack of effect of oral antispastic medication with baclofen, tizanidine, benzodiazepines and other antispastic agents. 28 patients were tetraplegic or tetraparetic, in 16 patients only the lower extremities were affected by spasticity. Five patients had preserved ability to stand or to walk with help. The categories of patients treated are shown in Table 1. The largest groups were patients in advanced stage of multiple clersosis and patients with sequelae of spinal trauma.

A trial of spinal baclofen was performed in all patients before implant of a pump for long-term infusion. For this purpose a totally implanted intrathecal silicone catheter was inserted percutaneously in the lumbar region and tunneled to a subcutaneous port in the upper abdomen. Bolus injections of baclofen in the range of 10–300 µg were used to assess the efficacy and duration of relief of spasticity. In a second operation the port was replaced by a constant flow implantable infusion pump (Infusaid, Model 400, Infusaid Corp., Norwood/MA, USA) delivering the drug in a daily volume of 1–2 ml. This pumping device is gas driven and requires no batteries. The rubber membrane of the drug chamber withstands app. 2,000 punctures. Depending on the fixed daily flow of a given pump the reservoir has to be refilled in 3–5 weeks intervals. This was done in our out-patient department or at the patient's home by a single percutaneous puncture of the pump.

The effect of intrathecal baclofen was evaluated by neurological examination, Ashworth score of spasticity, spasm score and clonus score, interrogation of the patient as well as the patient's family, assessment by physician and analysis of physiotherapy reports of individual patients. In 19 patients the effect of intrathecal baclofen upon spasticity was investigated by means of integrated electromyography using the method described by Schepelmann[14] and modified by the authors for study of spasms evoked by external and proprioceptive stimulation[11]. Videotapes of physiotherapy were taken before and after intrathecal application of baclofen in order to analyse the possible effect on the facilitation of passive joint movements. All patients were regularly seen in the out-patient department and neurological evaluation was performed on each occasion of pump refilling. Out of 44 patients receiving the trial bolus doses 37 cases were implanted with a continuous flow pump. The mean follow-up period for this group of patients is at present 10 months and ranges between 2 and 28 months. One third of the patients were followed for more than 12 months and over 500 ambulatory refills were performed.

Results

1. Dosage of Bolus Application

Application of baclofen intrathecally results in a dose-dependent reduction of spasticity, abolishment of spasms and loss of tendon reflexes. The effect is present within 1–2 hours after injection and may extend for a period of 4–48 hours. There was no relation between the dosage required to obtain a satisfactory effect and the degree of spasticity, the duration of spasticity or the period of oral antispastic treatment so far. There was, however, a relation between dosage and the local extent of central nervous lesion or spasticity. The average dosage in patients with tetraparesis or tetraplegia was higher than in patients with paraparesis or paraplegia. The doses necessary for a 24 hour suppression of spasticity varied between 25 and 425 µg (mean value: 166 µg).

2. Dosage During Continuous Infusion with Pump

The infusion dose necessary to obtain in a clinically satisfactory effect was higher than the bolus at a given duration of action and amounted to 236.5 µg ± 144 SD per 24 hours. During the follow-up period the dosage was reduced in a few patients, in the majority, however, an increase of dose was necessary suggesting possible tolerance. The range, mean, and median dosages required for achieving relief of spasticity via bolus injection, at the beginning of therapy and during follow-up with continuous infusion are shown in Table 1.

Table 1. *Aetiology of Spasticity and Dose Requirement of Intrathecal Baclofen per Day*

Aetiology of spasticity:		
1. Spinal spasticity		N = 32
Multiple sclerosis	17	
Trauma	11	
Vascular lesion	2	
Tumour	2	
2. Cerebral spasticity		N = 12
Infantile cerebral palsy	3	
Trauma	3	
Near drowning	3	
Encephalitis	2	
Basilaris artery thrombosis	1	
Dose requirement of intrathecal baclofen per day:		
1. Bolus injection		
Mean: 166 µg		
Median: 155 µg		
Range: 25–425 µg		
2. Start of infusion		
Mean: 236.5 µg		
Median: 240 µg		
Range: 15–600 µg		
3. During long-term infusion		
Mean: 350 µg		
Median: 333 µg		
Range: 10–900 µg		

3. Clinical Results

The results of neurological scoring (Ashworth score, spasm score, clonus score) before and after the start of intrathecal baclofen are shown in Table 2, which also include data of the opinion of patient and physician on the antispastic efficacy of this method. Complete abolishment of spasticity was achieved in almost 100% of the patients at the beginning of therapy. The average values of the scores during long-term therapy in the course of follow-up were slightly increased. This may be explained by the necessity to reduce the dosage immediately after the start of therapy and to accept some degree of spasticity and occasional spasm in order to preserve better posture or the ability to stand when supported.

Electromyographic results are shown in Fig. 1. Following intrathecal application of baclofen there was a significant reduction of electrical muscle activity at rest and during stimulation in the lower extremities and during repetitive external stimulation at the upper extremities.

Table 2. *Mean Values of Ashworth Score, Spasm Score, Clonus Score, Assessment by Patient or Physician before and after Start of Therapy as well as during Long-Term Therapy*

Ashworth score	
(1 (no increase of muscle tone) to 5 (extreme increase of muscle tone))	
Mean values	
before treatment (n = 44):	4.3 ± 0.80
after bolus injection (n = 44):	1.3 ± 0.75
during long-term infusion (n = 37):	1.5 ± 0.69
Spasm score	
(0 (no spasms) to 4 (spasms all the time))	
Mean values	
before treatment (n = 44):	2.8 ± 1.40
after bolus injection (n = 44):	0.06 ± 0.45
during long-term infusion (n = 37):	0.21 ± 0.41
Clonus score	
(0 (no foot clonus) to 2 (prolonged foot clonus))	
Mean values	
before treatment (n = 44):	1.6 ± 0.75
after bolus injection (n = 44):	0.09 ± 0.29
during long-term infusion (n = 37):	0.27 ± 0.50
Assessment by patient or family	
(1 (very good) to 4 (poor))	
Mean values	
after start of therapy (n = 44):	1.54 ± 0.73
during long-term infusion (n = 37):	1.37 ± 0.49
Assessment by physician	
(1 (very good) to 4 (poor))	
Mean values	
after start of therapy (n = 44):	1.40 ± 0.81
during long-term infusion (n = 37):	1.32 ± 0.58

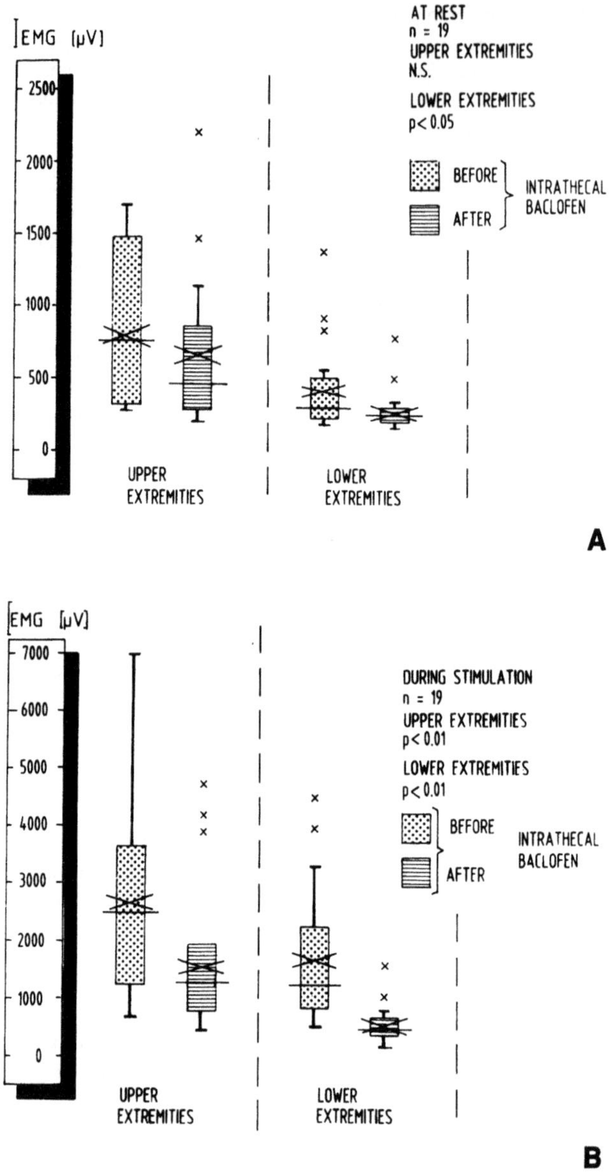

Fig. 1. Values of integrated electromyography recorded from six muscles of upper and lower extremities with epidermal electrodes. Recording at rest for 15 min and recording during stimulation for 30 min before and 2 h after an effective intrathecal baclofen bolus. All values presented as box-and-whisker-plot. Statistical analysis performed with Wilcoxon test

The changes in the motor situation could also be demonstrated by video or photography, as shown in Fig. 2. In patients with the ability to walk we performed motography, which also demonstrated functional improvement.

4. *Functional Results*

Out of 37 patients with implanted pumps 20 were permanently bedridden and unable to sit in a wheelchair because of severe spasms and contractures before start of therapy. In 14 of these patients transfer to wheelchair for several hours during daytime became possible. 6 patients in final stages of multiple sclerosis remained bedridden but nursing and general care became greatly facilitated due to the abolition of adductor spasticity. Out of 12 patients who had been confined to wheelchair 10 reported improved mobility. Facilitation of physiotherapy in terms of ease of passive movement of joints was achieved in all patients. The most striking effect was observed in patients with painful spasms; 13 out of 17 patients were completely relieved of pain. There was, however, no effect of intrathecal baclofen on pain not connected with spasms, and patients with painful arthropathy, causalgia-like pain or radicular pain reported no improvement of this symptom. In 3 patients respiratory function and feeding was impaired because of violent spasms of truncal and abdominal

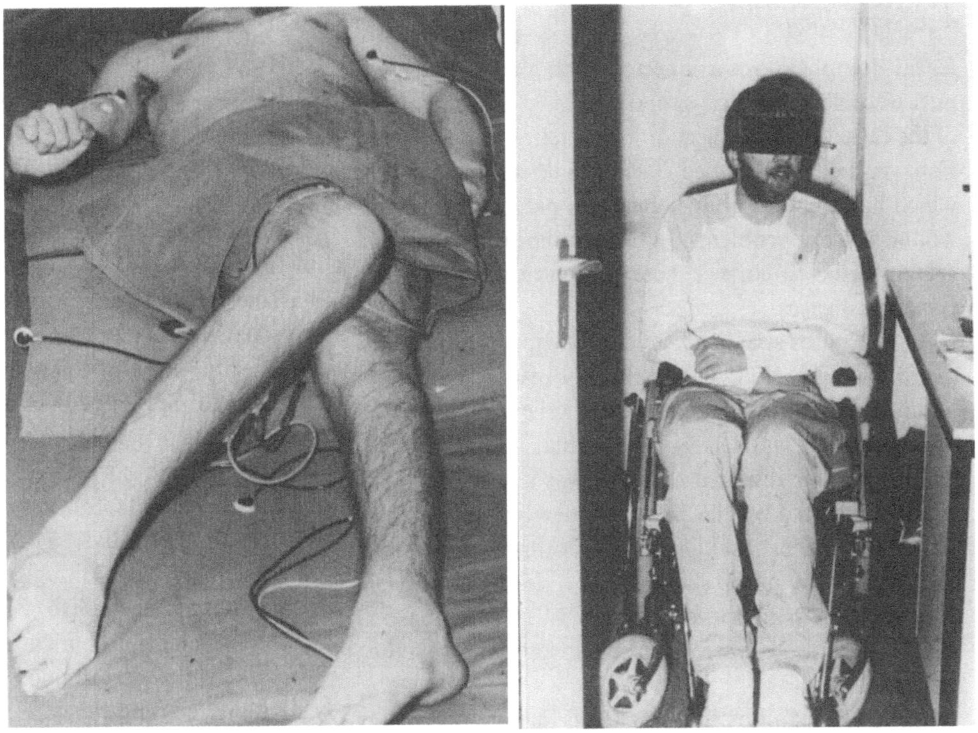

A B

Fig. 2. Photograph of a patient with spasticity due to spinal trauma before and after intrathecal baclofen therapy with an implanted pump

musculature, these functions were normalised with the abolition of spasms. 3 out of 5 patients with the ability to stand or walk reported improvement of their mobility and sometimes even learned to perform more complex movements, such as climbing stairs.

7 out of 44 patients were disqualified for permanent implantation after the initial trial of intrathecal baclofen via a port. In one patient with acute decerebration syndrome intrathecal baclofen was used to abolish decerebration episodes and was administered via external pump and implanted port for a restricted period of time of about 3 weeks. In 2 other patients with asymmetrical spasticity of the lower extremities the loss of muscle tone impaired walking and it was impossible to titrate the dosage in such a way as to abolish spasms on the one hand and to preserve the useful increased muscle tone on the other. In the remaining 4 patients other signs of their primary disease (pain, dystonia, ataxia) were more disturbing than the spasticity and the functional gain from long-term treatment seemed questionable.

5. *Side Effects*

The rationale behind the concept of intrathecal application of baclofen is to deliver the drug at its site of action, *i.e.* selectively to the spinal cord, thus reducing the dosage and possible cerebral side effects. As those side effects were usually only seen during bolus injection, this mode of application seems to be connected with a higher portion of drug reaching supraspinal centers than during continuous infusion.

Following implantation of the catheter – and therefore lumbar puncture – 22 patients complained of transient headache and 6 of nausea and vomiting. Increased somnolence after the first bolus injection was observed in 6 patients and was reversible after dose reduction or disappeared when continuous infusion was started. Other transient side effects of bolus application were bradycardia and moderate hypotension in 11 cases, or urinary retention lasting 24–48 hours in 6 cases. Severe sedation and bradypnoea occured in 3 patients and were clearly connected with bolus application of dosages exceeding 500 µg. Among 37 patients with long-term infusion increased somnolence was reported on 4 occasions only and occured during a daily infused dosage of more than 1,000 µg of baclofen. Sedation was reversed by dose reduction. One of these patients had additional nausea and vomiting. In one case heavy sedation and bradypnoea occured during continuous infusion, probably due to pump failure.

6. Complications

Local complications connected with the catheter or port were noted in 5 cases and consisted of dislocation of the catheter, disconnection, obstruction and leakage. Four patients developed haematoma of the pocket, which had to be aspirated and three patients had local wound healing problems. In total, minor repeated surgery in order to correct these complications was necessary in 8 cases.

In two cases spasticity suddenly reappeared during the follow-up period, 4 and 6 months respectively after pump implantation. Checking the patency and position of the catheter as well as the residual volume in the pump reservoir revealed that this was due to the lack of drug delivery by the pump. A new pump was inserted. In another case signs of overdosage occured and emptying of the pump showed that it delivered 9 ml of drug solution in 24 hours instead of 1.7 ml. The reason for this failure is presently being investigated by the manufacturer.

General complications included meningitis in 2 cases. Both patients had increased CSF count during the trials with bolus application and responded promptly to intrathecally administered gentamycin. Two patients developed urinary infection during the stay in hospital. In another two patients deep vein thrombosis occurred; we feel that the sudden loss of long-lasting spasticity may facilitate the development of this complication in a paraplegic patient. One of these patients died at home due to pulmonary embolism. There were two other deaths during the follow-up period, both unrelated to treatment modality (one case with pneumonia and the other with viral myocarditis). Two patients had transient psychotic symptoms after withdrawal or oral baclofen and start of intrathecal application. There were no infections connected with ambulatory filling procedures.

Discussion

Intrathecal administration of baclofen is a new method of treatment of severe spasticity. The results of our series of 44 patients, including 37 patients with implanted pumps for continuous long-term infusion of baclofen, confirm the preliminary results of earlier reports[3, 11, 12, 13]. In contrast to unconvincing results of spinal stimulative procedures for severe spasticity[7] the effect upon spasms and increased muscle tone is clinically spectacular, particularly for all forms of spinal spasticity. Good, but mostly less profound effects were observed in cases of cerebral spasticity. This may be explained by our selection of patients in this category of spasticity. In the patients with cerebral spasticity we have treated so far, arthropathy and deformities were far advanced and possibly the functional effects would have been better if this form of therapy had been started earlier.

Continuous application of baclofen with the help of fully implanted systems operating without batteries has the advantage of a theoretically life-long control of spasticity. Tolerance may lead to an increase of dosage, but possibly will also reduce the rate of cerebral side effects. Spinal baclofen as a functional and pharmacological procedure seems to be superior to extensive destructive surgery like longitudinal myelotomy and chemical rhizotomies[1, 15]. In our opinion it constitutes an alternative to percutaneous radiofrequency rhizotomy, although long-term results of this method still have to be established[9, 16]. A disadvantage of intrathecal baclofen application is a certain lack of selectivity, which is provided by the selective posterior rhizotomy of Sindou[17].

Our experience suggests that the main indication for intrathecal administration of baclofen are spasticity and painful spasms in patients with multiple sclerosis and spinal cord lesions. In these patients the complete abolition of handicapping spasticity can be achieved in all cases. As functional outcome depends on the existence of contractures and joint deformities it would be useful to include patients at an earlier stage of spastic syndromes, especially with spasticity due to diffuse cerebral damage, in our series.

In 18% of our patients we had minor mechanical complications connected with the implanted system. They usually occured at the beginning of our series and could be easily corrected. They seem to be avoidable by strict adherence to the surgical technique (correct positioning of the catheter, correct fixation etc.) and minor improvements in design of the catheter and connectors. In general, initial failure of this method is usually due to technical complications, which may be easily detected by injection of water-soluble contrast medium into the port or the side-port of the pump.

Trials of intrathecal baclofen by bolus application are part of the treatment regimen to demonstrate if the patient really profits from this therapy. They always should be done before the implant of the expensive pumping device. As the testing procedure may take some time, it is useful to implant a subcutaneous port first, which may help to reduce the hygienic risks. Bolus application should start with low dosage in the range of 50–100 µg of baclofen and at intervals of at least 24 hours. If necessary, the dosage may be gradually increased using the same intervals of injection. In some patients dose finding is only possible with the help of external pumps and continuous infusion, especially in patients with a small dose range. In those patients a definite dosage has to be found, which just abolishes spasms or reduces muscle tone without impairing residual motor functions, which very often depend on the preservation of a certain degree of increased muscular tone. In addition, bolus application may facilitate cerebral side effects. Our results suggest that for long-term therapy continuous application via implantable pump is preferable to repeated bolus application.

References

1. Bishof W (1951) Die longitudinale Myelotomie. Zbl Neurochir 2: 79 – 88
2. Corston RN, Johnson F, Godwin-Austen RB (1981) The assessment of drug treatment of spastic gait. J Neurol Neurosurg Psychiatry 44: 1035–1039
3. Dralle D, Müller H, Zierski J, Klug N (1985) Intrathecal baclofen for spasticity. Lancet II: 1003
4. Erickson DL, Blacklock JB, Michaelson MA, Sperling K, Lo JN (1985) Control of spasticity by implantable continuous flow morphine pump. Neurosurgery 16: 215–217
5. Faigle JW, Keberle H (1972) Metabolismus und Pharmakokinetik von Lioresal. In: Birkmayer W (ed) Aspekte der Muskelspastik. Huber, Bern Stuttgart Wien, pp 94–100
6. Faigle JW, Keberle H, Degen PH (1981) Pharmakokinetik von Baclofen (Lioresal). In: Bauer HJ, Koella WP, Struppler A (eds) Therapie der Spastik. Verlag für angewandte Wissenschaft, Munich, pp 137–147
7. Illis S, Read DJ, Sedgwick EM, Tallis RC (1983) Spinal cord stimulation in the United Kingdom. Review. J Neurol Neurosurg Psychiatry 46: 299–304
8. Jones RF, Burke D, Marosszeky JE, Gillies JD (1970) A new agent for the control of spasticity. J Neurol Neurosurg Psychiatry 33: 464–468
9. Kasdon DL, Lahti ES (1984) A prospective study of radiofrequency rhizotomy in the treatment of posttraumatic spasticity. Neurosurgery 15: 526–529
10. Müller H, Gerlach H, Boldt J, Börner U, Hild P, Oehler KU, Zierski J, Hempelmann G (1986) Spastik-Behandlung mit rückenmarksnaher Gabe von Morphin und Midazolam. In vitro-Experimente, Tierversuche und klinische Untersuchungen zur Verträglichkeit und Effektivität. Anaesthesist 35: 306–316
11. Müller H, Zierski J, Dralle D, Börner U, Hoffmann O (1987) The effect of intrathecal baclofen on electrical muscle activity in spasticity. J Neurol 234: 348–352
12. Penn RD, Kroin JS (1985) Continuous intrathecal baclofen for severe spasticity. Lancet II: 125–127
13. Penn RD, Kroin JS (1987) Long-term intrathecal baclofen infusion for treatment of spasticity. J Neurosurg 66: 181–185
14. Schepelmann F (1976) Human motor activity in decerebrate states and their sequelae. Acta Neurochir (Wien) 46: 185–217
15. Scott BA, Weinstein Z, Chiteman R, Puliam M (1985) Intrathecal phenol and glycerin in metrizamide for treatment of intractable spasms in paraplegia. J Neurosurg 63: 125–127
16. Müller H, Zierski J, Penn RD (eds) (1988) Local-spinal therapy of spasticity. Springer, Berlin Heidelberg New York Tokyo
17. Sindou M, Millet MF, Mortamis J, Eysette M (1982) Results of selective posterior rhizotomy in the treatment of painful and spastic paraplegia secondary to multiple sclerosis. Appl Neurophysiol 45: 335–340
18. Yaksh TL, Müller H (1982) Spinal opiate analgesia. Experimental and clinical studies. Springer, Berlin Heidelberg New York

Address for correspondence: Prof. Dr. J. Zierski, Neurochirurgische Abteilung des Krankenhauses Neukölln, Rudower Strasse 48, D-1000 Berlin.

VII. Basic Research in Neurosurgery

Acta Neurochirurgica, Suppl. 43, 103–106 (1988)

Ki-67 Immunoperoxidase Stain as Marker for the Histological Grading of Nervous System Tumours

T. Shibata[1], **P. C. Burger**[2], and **P. Kleihues**[1]

[1] Laboratory of Neuropathology, Institute of Pathology, University of Zürich, Zürich, Switzerland, [2] Department of Pathology (Neuropathology), Duke University Medical Center, Durham, North Carolina, U.S.A.

Summary

The monoclonal antibody Ki-67 recognizes a nuclear antigen which is expressed during the G_1, S, G_2 and M phases of the cell cycle. Immunostaining of frozen biopsy material thus presents a convenient and rapid method for the estimation of the growth fraction in human neoplasia. This report summarizes the results obtained in 178 neurosurgical biopsies. The highest incidence of Ki-67 positive nuclei was observed in 8 metastatic carcinomas (mean: 20%) and in 53 cases of glioblastoma multiforme (10%). Glioblastomas showed considerable variation, the fraction of stained nuclei ranging from 1 to 28%. For astrocytomas and oligodendrogliomas we found a close correlation between the Ki-67 index and the histological grade in agreement with known biological behaviour. In anaplastic gliomas (WHO grade III) the upper limit was 11%, in grade II gliomas, 6%. Juvenile pilocytic astrocytomas and pituitary adenomas showed mean staining indices of approximately 1%. In 13 meningiomas the mean Ki-67 index was 1% but rose to 5% in recurrent and anaplastic meningiomas. Assessment of the growth fraction in human brain tumours by immunostaining with Ki-67 could become an important tool in the prediction of the biological behaviour of nervous system neoplasms and the planning of adjuvant therapy.

Keywords: Brain tumours; monoclonal anitbody Ki-67; cell kinetics; growth fraction.

Introduction

In the histopathological examination of surgical biopsies, the main emphasis is placed on an exact disease classification. In neoplastic diseases, additional information is required on the prospective biological behaviour of a lesion since this is known to vary considerably, even among neoplasms belonging to the same histological type. The classical approach to this problem is the assessment of cellularity, nuclear pleomorphism, degree of differentiation, chromatin density and nucleolar size. In gliomas, the presence of vascular proliferation and necrosis closely correlated with progressive anaplasia and poor prognosis. Since the determination of the mitotic rate has proved to be of little use in the estimation of the proliferative potential, more precise methods have been proposed, including ^{3}H-thymidine autoradiography[7] and immunostaining for bromodeoxyuridine[6]. However, both methods require preoperative administration of the respective precursor. In 1983, Gerdes *et al.* reported the production of the mouse monoclonal antibody Ki-67 which recognizes a human nuclear antigen associated with cell proliferation[3]. Immunostaining with Ki-67 provides a convenient and rapid method for the estimation of the growth fraction in frozen tumour biopsies[2]. This technique has recently been applied to neoplasms of various tissues, including the nervous system[1]. It can be carried out intraoperatively and allows the investigation of a large number of biopsies without patient pretreatment or selection. We here report our experience with this technique and data obtained in 178 neurosurgical biopsies.

Materials and Methods

Investigations were carried out on a total of 178 neoplasms originating from the central or peripheral nervous system, the meninges and the pituitary gland (Table 1). Tissue preparation and immunohistochemistry were carried out as previously described[1]. Briefly, fresh biopsy tissue was frozen in isopentane suspended in liquid nitrogen. Cryostat sections (6–8 µm) or squash preparations were immersed in acetone for 10 minutes and subsequently dried for 20–30 minutes. Sections were immunostained with undiluted murine monoclonal antibody Ki-67 (Dianova GmbH, D-2000 Hamburg, F.R.G.) by the peroxidase-antiperoxydase method. Goat anti-mouse IgG and swine anti-goat IgG (Dako Corp., Santa Barbara, Ca., U.S.A.) were used at dilutions of 1 : 20. For quantitative evaluation, two large fields of each neoplasm were photographed and printed at a final magnification of × 390. Rather than determining the staining index in areas in which stained nuclei were especially prevalent,

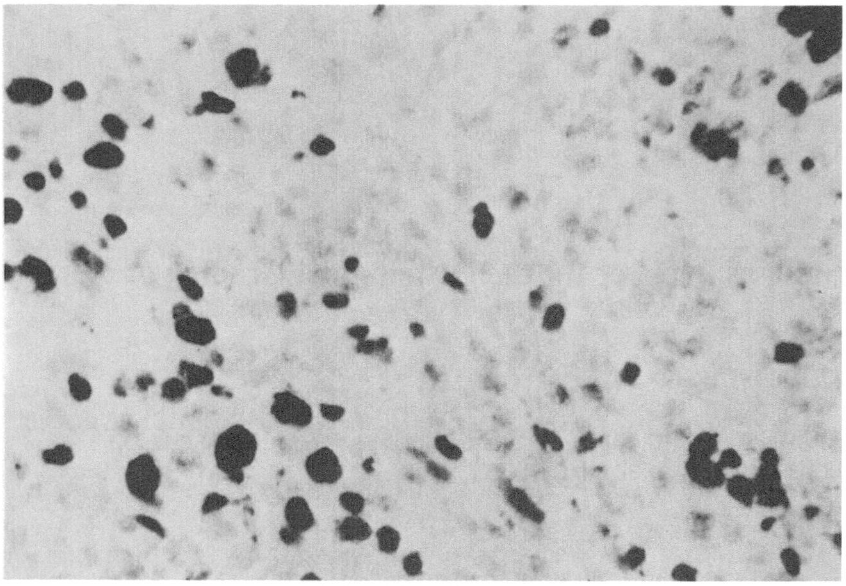

Fig. 1. Pleomorphic immunoreactive nuclei in a frozen section of a glioblastoma multiforme. Ki-67 staining index, 16% (PAP, × 275)

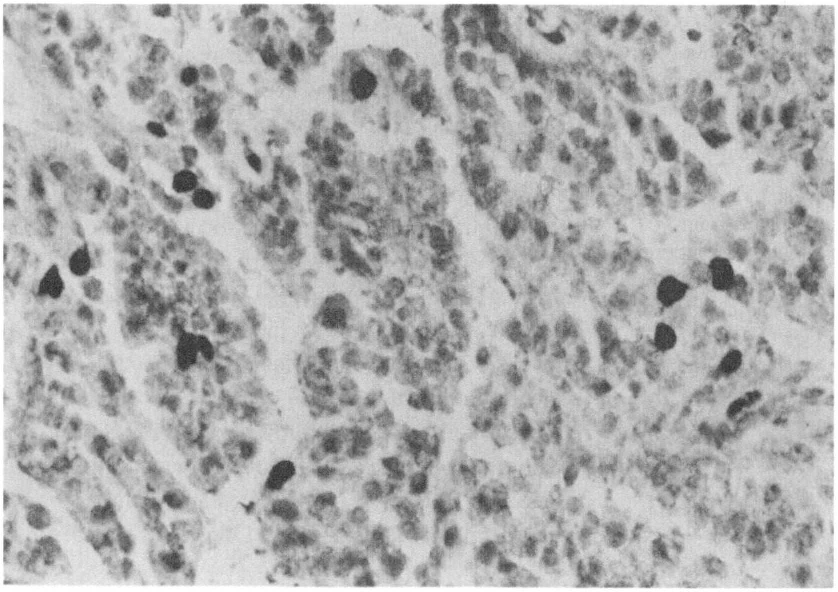

Fig. 2. Scattered Ki-67 positive nuclei in an anaplastic choroid plexus papilloma. Staining index, 2.9% (PAP, × 275)

an attempt was made to assess the mean Ki-67 index in representative tumour areas.

Results

The results obtained in 178 neurosurgical biopsies are shown in Table 1. For each type of neoplasm, the mean Ki-67 index and the range observed are reproduced. As expected, highest values were found in metastatic carcinomas although the Ki-67 index varied considerably in this group, ranging from 2.5 to 56%. This was similarly true for the most malignant type of human gliomas, the glioblastoma multiforme (Fig. 1). The mean staining index was below that of metastatic lesions but there was a broad variation ranging from 2–28%, with no evidence of bimodal distribution. A large growth fraction was also found in malignant pediatric brain tumours, *e.g.* medulloblastoma and malignant choroid plexus papilloma (Fig. 2). Among astrocytomas and oligodendrogliomas, the Ki-67 index was found to parallel histological grade as defined in the WHO classification of nervous system tumours[11]. The correlation between Ki-67 index and WHO grading was particularly evident in the case of astrocytomas, with mean staining indices of 4% for the anaplastic variant, 2% for differentiated astrocytomas (fibrillary, protoplasmic and gemistocytic) and 1% for the juvenile pil-

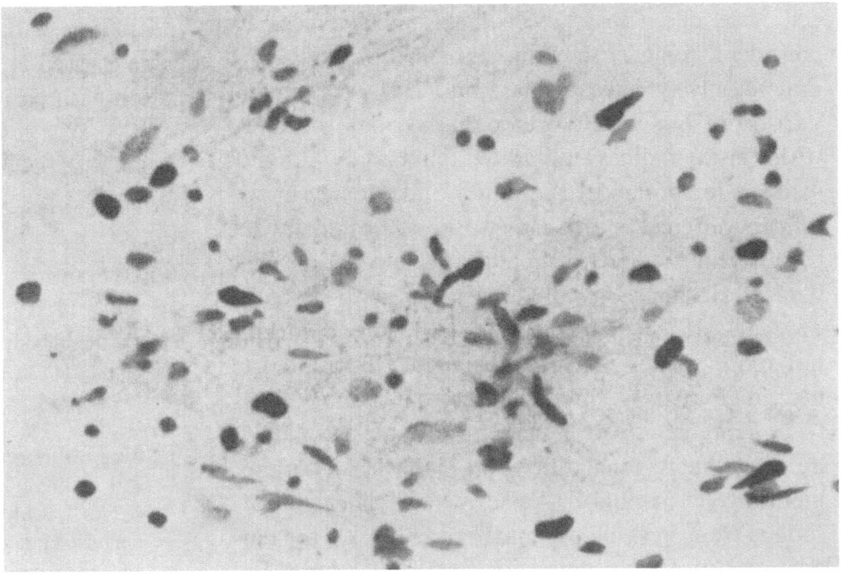

Fig. 3. Numerous immunoreactive nuclei in a squash preparation from a glioblastoma multiforme. Ki-67 staining index, 32% (PAP, × 275)

ocytic astrocytoma. Of the four ependymomas investigated, two were of the spinal myxopapillary type (WHO grade I) and had a very small growth fraction of only 0.2 and 1.1%. In meningiomas, highest values were found in the hemangiopericytic variant, in anaplastic meningiomas and in recurrent lesions. Among 34 pituiatry adenomas, an ACTH-producing adenoma associated with Cushing's disease showed the highest index (4%). In schwannomas, the staining index was generally low and labelled cells were usually found in tumour portions with Antoni A architecture rather than in the degenerative Antoni B areas.

In a total of 34 neoplasms, the Ki-67 index was determined both in frozen sections and squash preparations. Values obtained by either method varied little from each other, the correlation coefficient being 0.96%. The highest staining index of all squash preparations was 32% in a glioblastoma (Fig. 3) and this corresponded with a value of 26% in the frozen section.

Discussion

Since the monoclonal antibody Ki-67 recognizes a human nuclear antigen expressed in all continuously cycling cells, *i.e.* during G_1, S, G_2 and M phases of the

Table 1. *Ki-67 Immunoreactivity in 178 Neurosurgical Biopsies*

Tumour type	WHO grade	Number of cases	Ki-67 staining index	
			Mean	Range
Metastatic carcinoma		8	20.2%	(2.5–56.7)
Glioblastoma multiforme	IV	53	9.8%	(1.3–27.6)
Medulloblastoma	IV	1	10.2%	
Astrocytoma, anaplastic	III	6	4.3%	(1.6–9.0)
Oligodendroglioma, anaplastic	III	5	5.3%	(1.7–8.3)
Oligo-Astrocytoma, anaplastic	III	2	11.1%	(10.9–11.2)
Chor. Plexus papilloma, anapl.	III	2	10.1%	(2.9–17.4)
Astrocytoma	II	12	1.8%	(0.6–3.2)
Oligodendroglioma	II	12	2.2%	(0.5–6.0)
Oligo-Astrocytoma	II	4	1.9%	(1.3–2.9)
Ependymoma	I–II	4	0.9%	(0.3–1.6)
Astrocytoma, pilocytic	I	13	0.9%	(0.3–1.6)
Meningioma	I	13	1.4%	(0.2–3.2)
Meningioma	II–III	4	4.8%	(2.9–8.7)
Schwannoma	I	5	1.3%	(0.3–2.7)
Pituitary adenoma	I	34	1.1%	(0.2–3.7)

cell cycle, one would expect that staining indices are considerably higher than those obtained with methods that selectively detect S-phase nuclei. However, comparison of our results with published data obtained with ^{3}H-thymidine autoradiography[7] and immunostaining for bromodeoxyuridine[4-6] fails to reveal significant differences. In our view, this is most likely due to different methods of quantitative evaluation. The data of Hoshino and co-workers appear to be based on evaluation at high magnification, thus favouring hot spots whereas we have deliberately attempted to determine the Ki-67 index in large fields more likely to be representative for the entire neoplasm. We anticipated that regional differences in the density of proliferating cells would be less evident in smear preparations from fresh biopsy material. However, the correlation between both methods, *i.e.* frozen section and smear preparation is generally high. This makes the latter method particularly suitable for stereotactic surgery[9] and pituitary adenomas removed by transsphenoidal surgery[8]. It requires very small amounts of tissue and allows assessment of the growth fraction in several portions of the neoplasm without undue increase in workload.

Although immunostaining for Ki-67 represents a convenient and rapid method for the estimation of the growth fraction in human central nervous system tumours, its future routine application would only be meaningful if the results obtained are of prognostic value and helpful in the planning of postoperative chemo- and radiotherapy. In meningiomas, a correlation between clinical behaviour and Ki-67 index is emerging both from our data and those of Roggendorf *et al.*[10]. In both studies, recurrent and anaplastic variants showed a considerably higher staining index. A similar correlation seems to exist in pituitary adenomas[8]. These benign neoplasms had Ki-67 values ranging from 0.1 to 3.7%. Invasive adenomas demonstrated significantly higher values (mean, 1.2%) than noninvasive adenomas (0.6%). However, gliomas represent a greater challenge for the prediction of biological behaviour. Although data on a great number of astrocytic neoplasms and glioblastomas have accumulated, the time elapsed is still too short for a meaningful correlation with patient survival.

Acknowledgement

We thank Mrs. B. Walker for excellent technical assistance.

References

1. Burger PC, Shibata T, Kleihues P (1986) The use of the monoclonal antibody Ki-67 in the identification of proliferating cells: application to surgical neuropathology. Am J Surg Pathol 10: 611–617
2. Gerdes J (1985) An immunohistological method for estimating cell growth fractions in rapid histopathological diagnosis during surgery. Int J Cancer 35: 169–171
3. Gerdes J, Schwab U, Lemke H, Stein H (1983) Production of a mouse monoclonal antibody reactive with a human nuclear antigen associated with cell proliferation. Int J Cancer 31: 13–20
4. Hoshino T, Nagashima T, Cho KG, Murovic JA, Hodes JE, Wilson CB, Edwards MSB, Pitts LH (1986) S-Phase fraction of human brain tumours in situ measured by uptake of bromodeoxyuridine. Int J Cancer 38: 369–374
5. Hoshino T, Nagashima T, Murovic JA, Wilson CB, Davis RL (1986) Proliferative potential of human meningiomas of the brain. A cell kinetics study with bromodeoxyuridine. Cancer 58: 1466–1472
6. Hoshino T, Nagashima T, Murovic JA, Wilson CB, Edwards MSB, Gutin PH, Davis RL, DeArmond SJ (1986) In situ cell kinetics studies on human neuroectodermal tumours with bromodeoxyuridine labeling. J Neurosurg 64: 453–459
7. Hoshino T, Wilson CB (1979) Cell kinetic analyses of human malignant brain tumours (gliomas). Cancer 44: 956–962
8. Landolt AM, Shibata T, Kleihues P (1987) Growth rate of human pituitary adenomas. J Neurosurg 67: 803–806
9. Ostertag C, Volk B, Shibata T, Burger PC, Kleihues P (1987) The monoclonal antibody Ki-67 as a marker for proliferating cells in stereotactic biopsies of brain tumours. Acta Neurochir (Wien) 89: 117–121
10. Roggendorf W, Schuster T, Peiffer J (1987) Proliferative potential of meningiomas determined with the monoclonal antibody Ki-67. Acta Neuropathol (Berl) 73: 361–364
11. Zülch KJ (1979) Histologic classification of tumours of the central nervous system. International Histological Classification of Tumours, No 21, World Health Organization, Geneva (1979)

Address for correspondence: Dr. P. Kleihues, Abteilung Neuropathologie, Universitätsspital, CH-8091 Zürich, Switzerland.

Acta Neurochirurgica, Suppl. 43, 107–113 (1988)

Glioma Biology in vitro: Goals and Concepts*

M. Westphal, M. Hänsel, H. Nausch, E. Rohde, J. Köppen, M. Fiola, F. Hölzel, and **H.-D. Herrmann**

Laboratory for Brain Tumour Biology, Department of Neurosurgery, University Hospital Eppendorf, Hamburg, Federal Republic of Germany

Summary

Gliomas are heterogeneous in their cellular composition, affecting therapeutic efforts such as surgical removal, radiotherapy, chemotherapy and immunotherapy. 106 gliomas were taken into culture in our laboratory and 12 cell lines could be established there from. Experiments were carried out in as many early cultures as possible and with the constant experimental system of the cell lines.

To subdivide and possibly classify the heterogeneous group of gliomas the following approaches emerged:

– Immunostaining of cells for glial markers such as GFAP, A4, A2B5, Leu 7 as well as fibronectin will allow one to distinguish different groups of glial cultures.

– Performance of growth factor sensitivity tests allows the assessment of major aspects of growth control in cultured gliomas.

– Cytogenetic evaluation in early cultures and correlation with the expression of oncogenes may yield useful information on mechanisms of escape from normal growth control.

– One of our cell lines (NCE-G 28) in which cells switch from GFAP to fibronectin expression and transiently express the x-hapten may serve as a model to study crucial aspects of cellular differentiation.

– Using different extracellular matrices for the initiation of cultures even from very benign lesions with low proliferative potential it is possible to include these into comparative studies with glioblastomas.

Keywords: Glioma; tissue culture; glial markers; growth factors; oncogenes; cellular differentiation.

Introduction

Cell culture is a valuable tool in the exploration of the biology of human malignancies. The in vitro study of human neoplasms is often hampered by their heterogeneous nature, difficulties to establish cells in culture and the adaptation of cultured cells to in vitro conditions which make it difficult to extrapolate to the in vivo situation. In this brief conceptual overview we intend to show which approaches may lead to results which are useful in the understanding of the biology of gliomas despite the above mentioned setbacks of in vitro studies.

In contrast to meningeomas which usually present few problems in cell culture, glial neoplasms are known to be often of a complex cellular composition of biologically different cell populations. Growth failure in vitro is therefore a frequently encountered phenomenon especially with more malignant lesions. Initiation of cell cultures derived from gliomas has been facilitated by the routine use of an extracellular matrix (ECM) which in our laboratory is derived from bovine corneal endothelium[1] but other matrices[2] might be as suitable. Using this ECM it was possible to cultivate more than 90% of the biopsy specimens. It was obvious that the initial cultures showed more attached cells and greater degree of differentiation as well as heterogeneity on ECM when compared to plastic alone (Fig. 1). The usefulness of ECM after adaptation of the cells to in vitro conditions is limited to those cases where cells maintain a tendency to attach only loosely to plastic (less than 1% in our series) and to the reduction of serum requirements for experiments which require long-term culture in the absence of serum. However it has to be conceded that the employment of ECM introduces an additional variable to in vitro studies and in this context it could recently be shown that an ECM derived from human arachnoid cells might induce cells to differentiate and slow their growth rate[3].

After cultures are established, they can be divided into two groups – those cultures which still show senescence and have only a limited in vitro life-span and those cultures which go on to become established cell lines with infinite growth. In those cultures which cannot be maintained for more than 15 to 20 in vitro passages, the possibility of terminal differentiation has to be investigated. In 20 cases of glioma-derived cultures, we immunostained cultures at the end of their

* Supported by the Deutsche Forschungsgemeinschaft.

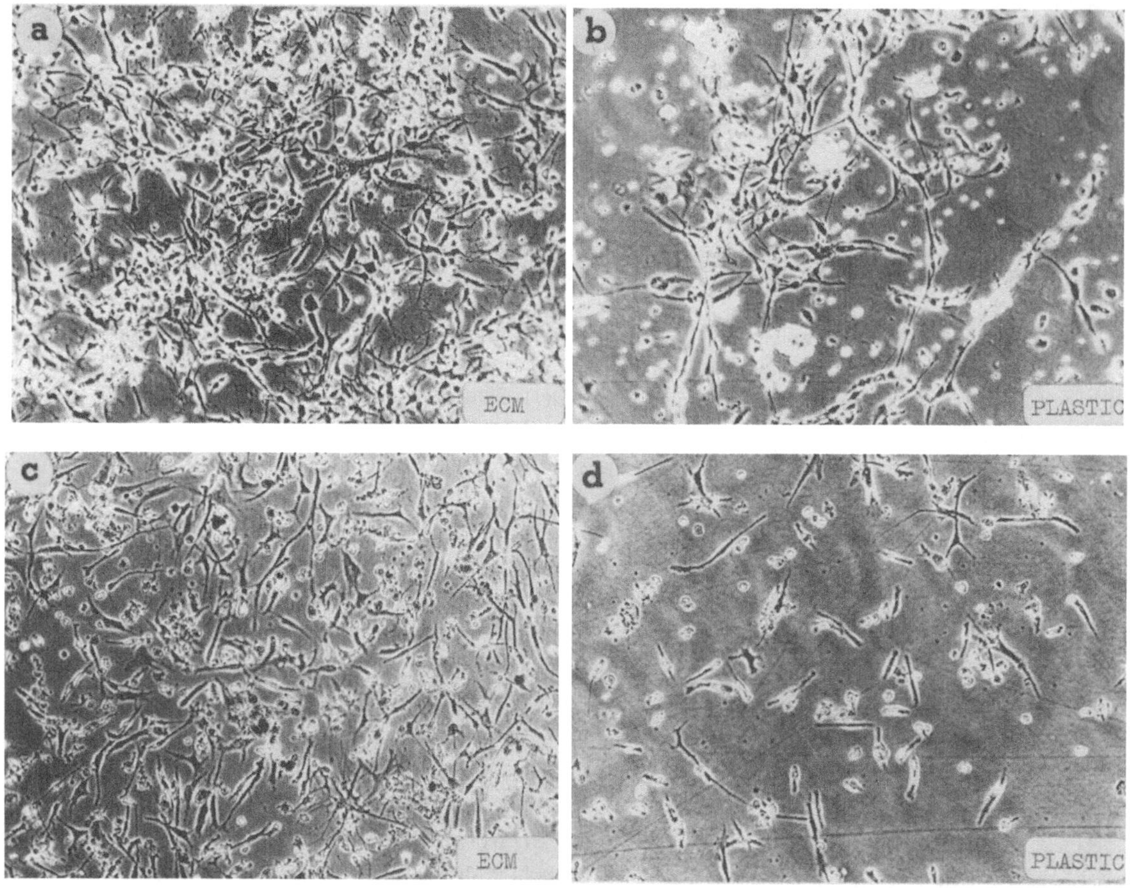

Fig. 1. Cells from two different gliomas which were initiated on extracellular matrix (a, NCE-G 98, recurrent glioblastoma, c, NCE-G 102, oligodendroglioma D) as well as on uncoated dishes (b, d)

life span (cessation of mitotic activity) with antisera detecting glial fibrillary acidic protein (GFAP), fibronectin (FN), galactocerebroside (galc), Leu 7 or A2B5. Despite the presence of highly differentiated looking cells in the cultures, we found no case in which astrocytoma cells would eventually turn positive for GFAP if they were negative before. Neither were we able to detect cells which were positive for galc in oligodendroglioma derived cells. The most prevalent marker in all cases was FN. As for those cultures which went on to become cell lines (12 cases from 106 biopsies), markers were checked as well. Only one of the cell lines was positive for GFAP; this cell line was derived from a gliosarcoma and produces tumours in nude mice upon injection[4]. The other cells lines are positive for FN but had GFAP expression in early culture.

Cell lines are to be distinguished from senescent cells by their autocrine potential by which we understand mechanisms of secretion of growth factors by the cultured cells which enable them to grow independent from outside stimuli. These mechanisms can be further subdivided into those which are activated during in vitro culture and have to be regarded as reactive or those which are already detectable in early passages and can be constitutional and thereby of actual relevance to the in vivo biology of the tumour. Autocrine mechanisms can be investigated indirectly by several approaches including radioreceptorassays for growth factors, proliferation assays with growth factors, isolation of growth factor activities from conditioned media and analysis of oncogene expression of which some are related to growth factors or their receptors[5–7].

Epidermal growth factor receptors (EGF-R) in gliomas are of interest because they have been reported in these tumours[8] as well as the overexpression of the erb B protooncogene[9] encoding an EGF-R related protein[5] and the concomitant overrepresentation of chromosome 7[7, 9] which carries the gene for the EGF-R[10]. Furthermore TGFα, a growth factor secreted by many malignant solid tumours[11] acts via the EGF-R[12]. Using 125-I labeled EGF, we investigated 39 cultured gliomas for the presence of EGF-R. The 39 gliomas consisted

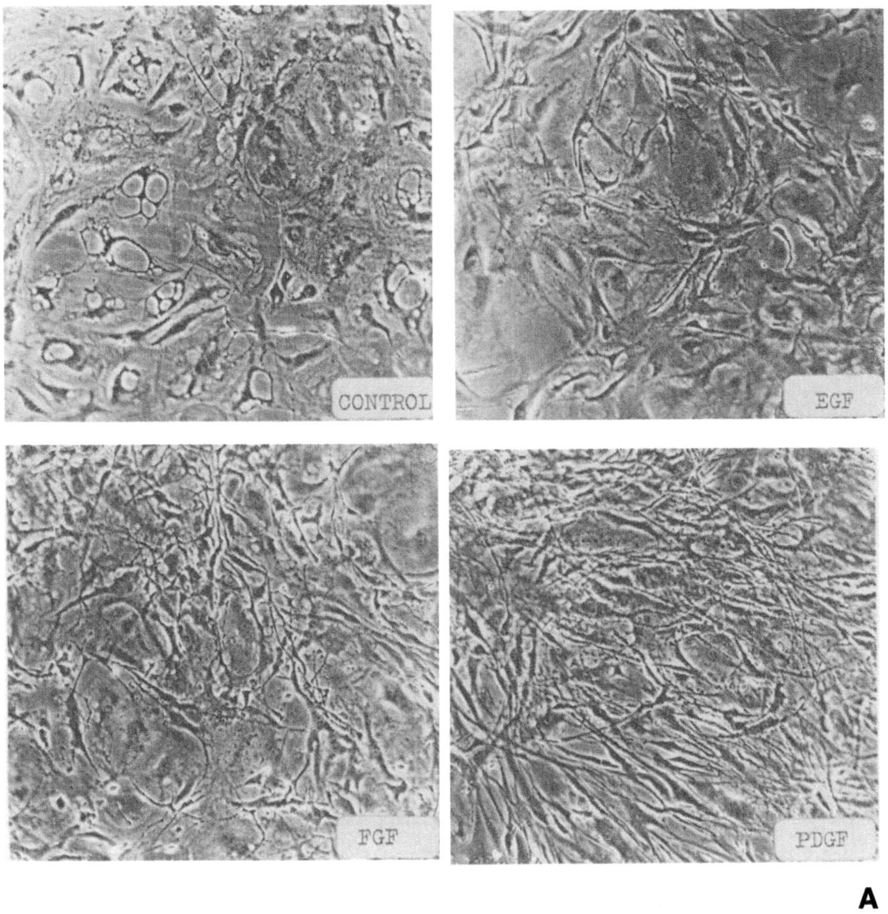

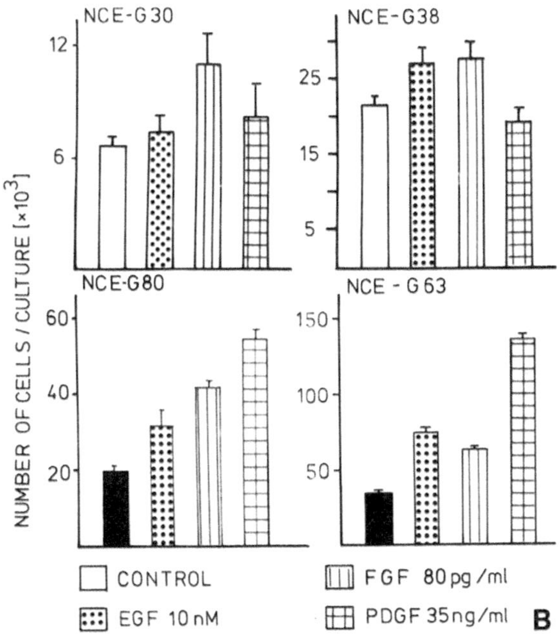

Fig. 2. A) Phase contrast photomicrographs from one growth experiment in which serum-free cell cultures from a glioma, NCE-G 80 (glioblastoma multiforme) were incubated with EGF (10 nM), FGF (80 pg/ml) and PDGF (35 ng/ml) as indicated. B) Corresponding cell numbers plus three more examples from identical growth experiments using three further cases of cultured glioma cells

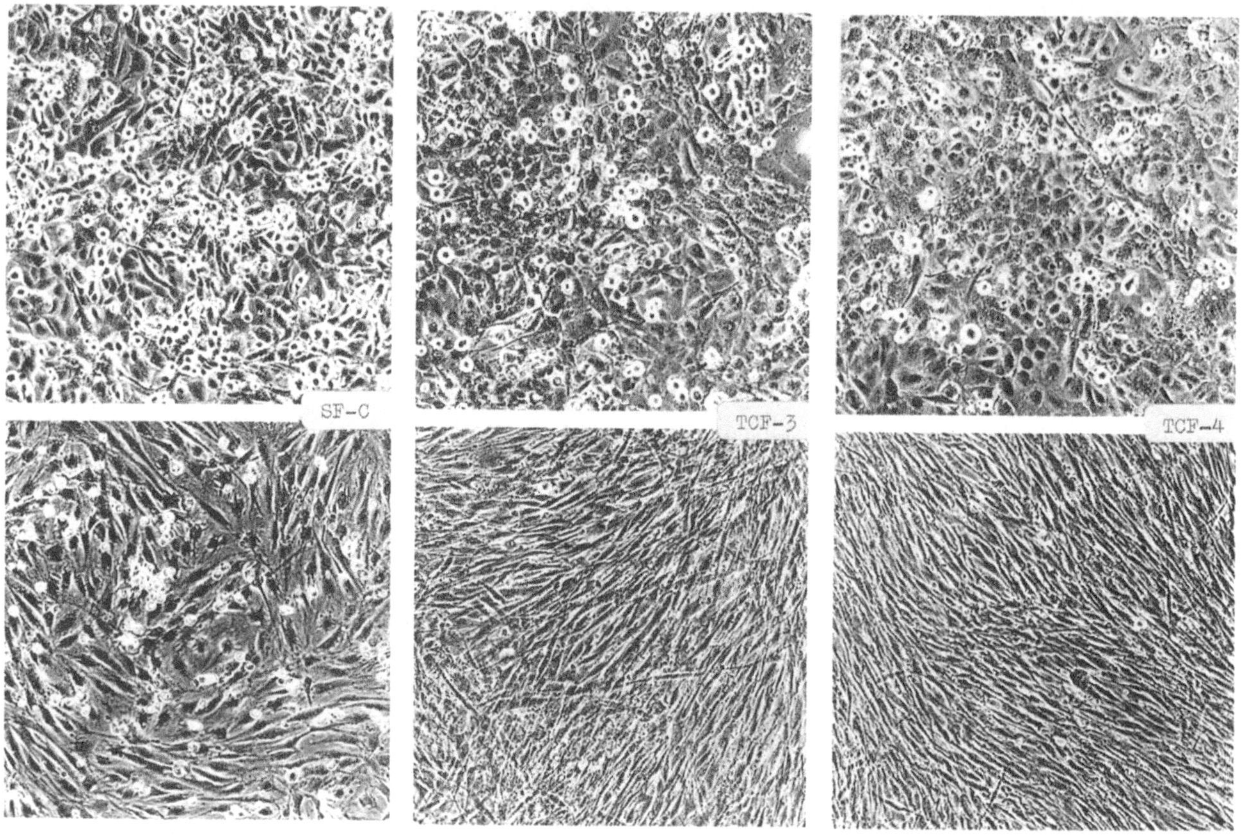

A

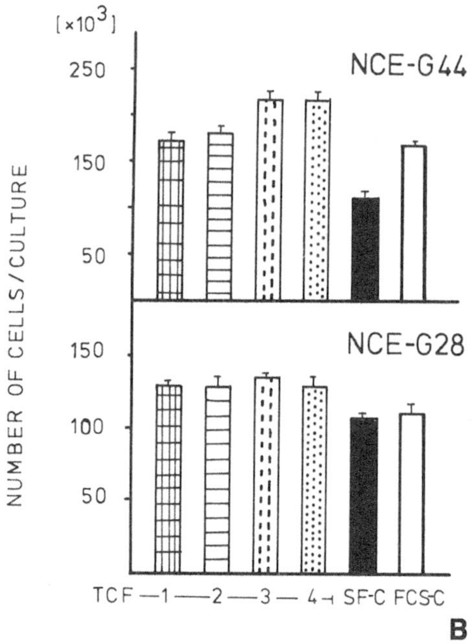

B

Fig. 3. A) Photomicrographs from an experiment in which tumour cyst fluids were added to serum-free cultures of two different gliomas, NCE-G 28 (gliosarcoma), upper row and NCE-G 44 (glioblastoma) lower row. 10 µl of cyst fluids were added every second day, after six days of incubation the cells were enzymatically detached and counted in a coulter counter. The respective cell numbers are contained in panel B. B) Cell counts from growth experiments with cyst fluids added to serumfree cultures. FCS-C: control with 10 µl fetal calf serum (FCS) added every second day. SF-C serum-free control. TCF-1 was from a mammary carcinoma metastasis, TCF-2 oligodendroglioma D, TCF-3 glioblastoma and TCF-4 from oligodendroglioma C of 17 malignant astrocytomas/glioblastomas, 11 benign astrocytomas and 11 oligodendrogliomas of which only one was malignant. In those cultures which eventually did not establish permanently, EGF-R was demonstrated in amounts of up to 50,000 receptors per cell. EGF-R binding on cultured cells decreased with senescence. We never found a cell turning EGF-R positive during culture periods, indicating that the expression of EGF-R is not a cell culture artifact. Ten cell lines which were analyzed expressed EGF-R to various degrees. Nine cell lines express up to 300,000 R per cell but one cell line shows only barely detectable levels of binding despite vigorous uninterrupted in vitro growth for two years, indicating that high levels of EGF-R are not essential to becoming an established cell line. In general we observed the highest receptor levels in the more malignant lesions, the cell lines with the highest levels being derived from a gliosarcoma, two glioblastomas and one malignant astrocytoma. Extracts from ten cell lines were analyzed for TGFα immunoreactive material but no immunoreactivity corresponding to the adequate molecular weight (approx. 6,000 D) was detected by Western blot analysis but only material with much higher mass (30,000 D) of which the nature is still to be defined (Antiserum obtained from Biotope, Bellevue, Wash., U.S.A.).

The elucidation of gliopoesis including the immunocytological characteristics of the cellular lineages and the growth factors involved in their development has been started with the rat optic nerve[13]. In these studies it was demonstrated that different kinds of glia develop from common precursor cells and then can differentiate definitively depending upon a variety of external stimuli and thereby are also accessible to manipulation in vitro[13, 14]. In a series of experiments we examined whether the gliopoetic principles referred to may be applicable to gliomas by assessing the response pattern of a series of cultured human gliomas to EGF, basic fibroblast growth factor (FGF) and platelet derived growth factor (PDGF). Cell proliferation, cellular morphology and staining pattern with four different antibodies against the afore mentioned markers were used as endpoints. In these experiments we found that in early passages the response patterns to the growth factors were different among the gliomas (Fig. 2). In general, EGF had potent proliferation inducing activity only in one case. PDGF was the broadest acting factor. FGF seemed to be of lesser relevance in the cases which were of more malignant histological appearance. In no case we were able under serumfree conditions with the addition of the factors to induce the expression of GFAP. Mostly the cells which reacted to the growth factors stained positive for fibronectin. The ratio between GFAP and FN cells did not change indicating that all cells took part in the proliferative response.

A more specific analysis of the results obtained with PDGF and FGF allows some speculations on the relevance of autocrine mechanisms in gliomas. PDGF is present in serum and therefore cells responding to PDGF may have been selected during the culture procedure. On the other hand PDGF, the c-sis oncogene product[15] is produced by many gliomas in a autocrine fashion[15] and we have shown that c-sis related products are indeed present in NCE-G 28 a cell line of only weak response to growth factors[4]. It may therefore be that there are three groups of gliomas in relation to the response to PDGF. One group does not respond because of maximal autocrine self-stimulation. Another group is weakly responsive because PDGF is of lesser importance than FGF or EGF-(TGFα). A third group responds well because PDGF is of relevance to growth control but is not produced sufficiently by the cells themselves. FGF has recently been shown to be secreted by capillary endothelial cells[16]. It is assumed that it may serve as a self-stimulating factor for these cells but also may act as a tumour maintenance factor by first causing capillary invasion and then stimulation of the surrounding target cells which include glia[17]. In this respect it may be possible to distinguish gliomas. We found that in the more malignant gliomas, there seems to be less dependence on FGF. Clinically this receives an interesting aspect in the observation that these tumours very often have a central necrosis by outgrowing their blood-supply. This leads us to speculate that in these tumours FGF may not be the major driving force of tumour cell division otherwise capillary growth and tumour expansion would have to be expected to be more coordinated. If however the central necrosis is due to some other energy problem, our findings might indicate that these tumours make some FGF like substance themselves rendering them less dependent in culture. Production of such factors by brain tumours has been demonstrated and named tumour angiogenesis factor[18].

Another way to detect possible autocrine mechanisms which may be active within a glioma is the analysis of products secreted in cell culture as well as the analysis of tumour cyst fluids (TCFs). Production of a mitogenic activity was demonstrated in a gliosarcoma derived cell lines, NCE-G 28. This activity is mitogenic in a broad spectrum of cells[4]. Cyst fluids which were obtained from gliomas during surgery were analyzed in cell culture experiments for their mitogenic potential. The TCFs clearly contain factors which will stimulate glioma cell proliferation in vitro (Fig. 3). These factors are barely active on cells with substantial autocrine potential of their own (NCE-G 28). Other cells which despite all characteristics of an established cell line require serum in their medium for optimal growth (NCE-G 44) will respond well to the addition of TCFs. Some response to the same TCFs is obtained in those cell cultures which heavily depend on serum and fail to grow well in vitro. These different responses indicate marked differences between tumours in the degree of activation of autocrine mechanisms and the subsequent ability to be responsive.

In the chromosomal analysis of gliomas only very general indications have been obtained so far[7, 19]. From our studies we can confirm, that the chromosome 7 bearing the gene for the EGF-receptor is overrepresented in all malignant gliomas investigated. This polysomy 7 in gliomas has also been found by others[7, 10, 21] and may be related to the EGF-R/erb B expression mentioned above. The most consistent chromosomal aberrations found in gliomas are loss of one gonosome and losses of chromosomes 10 and 22. An important cytogenetic finding was provided by the analysis of NCE-G 28 and three clones derived thereof. The parental cell line is GFAP positive (50% of the cells), two clones are negative but stain 100% with FN. A cell line obtained from a nude mouse tumour explant is positive for GFAP only for 5% of its cells and stains with FN otherwise. Between these different cell lines are four shared marker chromosomes, clearly indicating that within one glioma there may be cells with different staining patterns but close chromosomal relatedness.

In vitro studies of gliomas can help us to understand mechanisms which sustain tumour growth in vivo. The interaction of glioma cells with the extracellular matrix is part of the control mechanisms for cell proliferation and possibly differentiation. Analysis of growth factors and growth factor receptors provides insights into cellular growth control and autocrine mechanisms. The actual role of these aspects for tumourigenesis remains to be answered but their elucidation allows one to investigate new therapeutic approaches as well as to introduce biological parameters to glioma classification.

Acknowledgements

This study was supported by the Heinrich Bauer Stiftung, Hamburg and the Deutsche Forschungsgemeinschaft, Bonn – Bad Godesberg.

References

1. Westphal M, Hänsel M, Brunken M, König A, Köppen JA, Herrman HD (1987) Initiation of primary cell cultures from human intracranial tumours on extracellular matrix from bovine corneal endothelial cells. Exp Cell Biol 55: 152–163
2. Gospodarowicz D, Cohen D, Fujii DK (1982) Regulation of cell growth by the basal lamina and plasma factors: relevance to embryonic control of cell proliferation and differentiation. In: Cold Spring Harbor conference on cell proliferation, vol 9. Cold Spring Harbor Laboratories, Cold Spring Harbor, pp 95–124
3. Rutka JT (1986) Effects of extracellular matrix proteins on the growth and differentiation of an anaplastic glioma cell line. Can J Neurol Sci 13: 301–306
4. Westphal M, Hänsel M, Müller D, Laas R, Kunzmann R, Rohde E, König A, Hölzel F, Herrmann H-D (1988) Biological and karyotypic characterisation of a new cell line derived from human gliosarcoma. Cancer Res 48: 731–740
5. Downward J, Yarden Y, Mayes E (1983) Close similarity of epidermal growth factor receptor an v-erb-B oncogene protein sequences. Nature 307: 605–608
6. Doolittle RF, Hunkapiller MW, Hood LE (1983) Simian sarcoma virus onc gene, v-sis, is dervied from the gene (or genes) encoding a platelet-derived growth factor. Science 221: 275–277
7. Shapiro JR (1986) Biology of gliomas: Chromosomes, growth factors and oncogenes. Seminars in Oncology 13: 4–15
8. Libermann TA, Nusbaum HR, Razon N, Kris R, Lax I, Soreq H, Whittle N, Waterfield MD, Ullrich A, Schlessinger J (1985) Amplification, enhanced expression and possible rearrangement of EGF receptor gene in primary human brain tumours of glial origin. Nature 313: 144–147
9. Henn W, Blin N, Zang KD (1986) Polysomy of chromosome 7 is correlated with overexpression of the erb B oncogene in human glioblastoma cell lines. Human Genet 74: 104–106
10. Robson EB, Meera Khan P (1983) Report of the committee on the genetic constitution of chromosomes 7, 8 and 9. In: Human gene mapping 6. Cytogenet Cell Genet 32: 144–152
11. Nickell KA, Halper J, Moses HL (1983) Transforming growth factors in solid human malignant neoplasms. Cancer Res 43: 1966–1971
12. Massague J (1983) Epidermal growth factor-like transforming growth factor. II. Interaction with epidermal growth factor receptors in human placenta membranes and A 431 cells. J Biol Chem 258: 13614–13620
13. Raff MC, Miller RH (1984) Glial cell development in the rat optic nerve. Trends in Neurosci 7: 469–472
14. Raff MC, Miller RH, Noble M (1983) A glial progenitor cell that develops in vitro into an astrocyte or an oligodendrocyte depending on culture medium. Nature 303: 390–396
15. Betsholtz C, Heldin C-H, Nister M, Ek B, Wasteson A, Westermark B (1983) Synthesis of a PDGF-like growth factor in human glioma and sarcoma cells suggests the expression of the cellular homologue to the transforming protein of simian sarcoma virus. Biochem Biophys Res Commun 117: 176–182
16. Schweigerer L, Neufeld G, Friedman J, Abraham JA, Fiddes JC, Gospodarowicz D (1987) Capilary endothelial cells express basic fibroblast growth factor, a mitogen that promotes their own growth. Nature 325: 257–259

17. Gospodarowicz D, Neufeld G, Schweigerer L (1986) Molecular and biological characterization of fibroblast growth factor, an angiogenic factor which also controls the proliferation and differentiation of mesoderm and neuroectoderm derived cells. Cell Diff 19: 1–17
18. Klagsbrun M, Knighton D, Folkman J (1976) Tumour angiogenesis activity in cells grown in tissue culture. Cancer Res 36: 110–114
19. Bigner SH, Mark J, Bullard DE, Mahaley MS, Bigner DD (1986) Chromosomal evolution in malignant human gliomas starts with specific and usually numerical deviations. Cancer Genet Cytogenet 22: 121–135

Address for correspondence: Dr. M. Westphal, Universitätskrankenhaus Eppendorf, Abteilung für Neurochirurgie, D-2000 Hamburg 20, Federal Republic of Germany.

Acta Neurochirurgica, Suppl. 43, 114–117 (1988)

Gene Dosage Effect in Cells with Monosomy of Chromosome 22 Derived from Human Meningiomas

R. Casalone[1], **A. Minelli**[1], **G. Butti**[2], **P. Gaetani**[2], **V. Silvani**[2], **C. L. Solero**[3], and **C. Danesino**[1]

[1] Biologia generale e Genetica Medica, Università di Pavia, [2] Clinica Neurochirurgica, Università di Pavia, [3] Instituto Neurologico C. Besta, Milano, Italy

Summary

Enzymatic determinations of four enzymes, Arylsulphatase A (ARSA), α-iduronidase (IDUA), β-galactosidase (GLB) and hexosaminidase were performed in 19 cases of meningioma. Gene dosage effect was demonstrated for ARSA and IDUA; preliminary evidence for mapping a "protective protein" of GLB in the region 22pter → q11 was also obtained.

Keywords: Meningioma; chromosome; gene dosage effect.

Introduction

Gene dosage effect (GDE) has been extensively studied in cells from patients with constitutional chromosomal imbalance and has contributed information on gene mapping and on the pathogenesis of the phenotypic consequences of chromosomal abnormalities[7].

Only a few reports have been published concerning the study of GDE in neoplastic cells with acquired chromosomal imbalance[4, 11]. We report the results of the study of four enzymes, Arylsulphatase A (ARSA) and α-iduronidase (IDUA) whose genes are mapped on chromosome 22, and β-galactosidase (GLB) and hexosaminidase (HEX) whose structural genes are on other chromosomes in cultured cells from human meningiomas showing monosomy for chromosome 22.

Material and Methods

Detailed cytogenetic findings of 10 of the cases used in this study have been published by Casalone *et al.*[2]. These are cases 5, 21, 22 (monosomy of chromosome 22), case 16 (inverted duplication of 22 (pter → q11)), cases 23, 24, 25, 27, 28, 29 (normal karyotype or anomalies not involving chromosome 22). In addition 8 new cases with normal karyotype and 1 case with monosomy 22 were used for enzymatic assays. Essential information on new cases are reported in Table 1.

The cases studied have been divided as follows: in group A the cases with monosomy 22, group B includes only case 16, while in group C are listed control cases (normal chromosomes or anomalies not involving chromosome 22).

Table 1. *Summary of Histopathological and Chromosomal Findings in Cases Not Included in Casalone et al. 1987*

Case no.	Age/Sex	Histological diagnosis	Site of tumor origin[a]	No. of days in vitro	Karyotype and/or clonal rearrangements (No. of cells)
32	35/m	Endotheliomatous type	B	19	46, XY (20)
33	45/m	Endotheliomatous type	B	24	46, XY (9)
34	42/m	Endotheliomatous type	B	14	46, XY (32)
35	38/f	Endotheliomatous type	B	13	45, XX,-22 (10) 44, XX,-22, + random losses (8)
36	43/f	Meningotheliomatous type	B	18	46, XX (22)
37	52/f	Meningotheliomatous type	C	9	46, XX (10)
38	56/f	Transitional type	B	10	46, XX (20)
39	50/m	Meningotheliomatous type	C	16	46, XY (15)
40	48/m	Fibroblastic type	SC	35	46, XY (11)

[a] C (convexity); B (base); SC (spinal cord)

Table 2. *Levels of Activity of the Enzymes Indicated in 4 Cases with Monosomy 22 (A), 1 Case with inv dup(22)(pter → q11) (B) and 14 Cases with Normal Karyotype or Anomalies Not Involving Chromosome 22 (C)*. All activities are expressed as nM/mg/hr

Case no.	ARSA	IDUA	GLB	HEX
A 5	98.18	110.01	212.88	2010.22
21	33.51	13.5	95.08	1969.20
22	59.93	13.1	87.64	3409.63
35	49.95	51.35	471.51	3346.65
(Mean)	(60.39)	(46.96)	(216.77)	(2683.92)
(SD)	(27.44)	(45.70)	(179.24)	(802.19)
B 16	68.52	93.67	575.13	3550.46
(Mean A + B)	(62.01)			(2857.23)
(SD A + B)	(24.04)			(795.49)
C (n)	(14)	(14)	(14)	(14)
(Mean)	(172.46)	(64.44)	(330.35)	(4135.32)
(SD)	(73.80)	(32.83)	(145.60)	(2238.43)

Table 3. *Results of t test and Wilcoxon Rank Sum Test Obtained Comparing Cases With and Without Monosomy 22 for Each Enzyme Assayed* A, B, C as in Table 2

t test—t values	df	ARSA	IDUA	GLB	HEX
A/C	16	2.92^{b}	0.86^{ns}	1.30^{ns}	1.25^{ns}
A + B/C	17	3.23^{c}	—	—	1.22^{ns}
Wilcoxon rank sum test—p values					
A/C		0.017	0.500	0.157	0.157
A + B/C		0.010	—	—	0.328

a = $p < 0.05$, b = $p < 0.01$, c = $p < 0.005$, ns = no significant, df = degrees of freedom

The methods for cell culture and chromosome analysis are as in Casalone *et al.*[2]. The methods for cell preparation, protein determination and enzymatic assays are as in Galjaard[8], with minor modifications.

The enzymes assayed were: Arylsulphatase A, whose gene is mapped on chromosome 22 in the region q13.31 → qter; α-iduronidase (22 pter → q11), β-galactosidase, whose structural gene is on chromosome 3, but requires for its full expression a "protective protein" whose locus is mapped on chromosome 22, and β-hexosaminidase with loci for α and β subunits of hexosaminidase A and B on chromosome 15 and 5[10].

Results and Discussion

In Table 2 A and 2 B respectively are reported the results of the assay of the four enzymes in the 4 cases with monosomy 22, with their means and standard deviation and the values obtained in case n. 16; Table 2 C includes the means and standard deviations of the 14 cases with normal karyotype or anomalies not involving chromosome 22.

In Table 3 are entered the results of the statistical analysis of the data; cases with monosomy of chromosome 22 were compared with other cases both with the t test and the wilcoxon rank sum test using the EPISTAT program by T. L. Gustafson.

Arylsulphatase A

Each of the four cases with monosomy 22 had an activity lower than the mean of the cases of group C; case 16 also has a low level of ARSA; due to the presence of the inverted duplication of the region pter → q11, it is in fact monosomic for the distal region of chromosome 22 where the locus for ARSA is mapped. It can thus be added to the cases with full monosomy when considering this enzyme. The ratio between the means of group A + B and C is 0.36 (expected in case of GDE 0.50); the difference between the two groups is significant with the two statistical tests (see Table 3). Our results demonstrate clearly the presence of an uniplex GDE for ARSA in cultured cells from meningiomas with monosomy for chromosome

22. GDE for ARSA was already reported in single cases for leukocytes[1] and chorionic villi[5].

The low levels of ARSA are unlikely to cause relevant metabolic changes in meningiomas as the enzyme is mainly relevant in myelin metabolism. In particular no storage of sulfatides, the natural substrates of ARSA, is expected in cultured meningioma cells as a consequence of its reduced activity as it is not observed in fibroblasts of heterozygotes for metachromatic leukodistrophy who have similarly reduced activities. They show however minor but statistically significant changes in I.Q. and reaction time[3].

α-iduronidase

The results obtained for this enzyme are not homogeneous three out of the four cases with monosomy 22 have, as expected, activities lower than the mean of cases with normal karyotype, while the fourth case (case n. 5 in Table 2) has the 2nd highest value of all cases including group C.

Case n. 16, is trisomic for the region including the gene for IDUA and the ratio to group C cases is 1.45, nearly equal to the expected 1.5 caused by a triplex GDE. The demonstration of a GDE for IDUA in meningiomas needs further investigation: in fact case 16 behaves as expected and trisomy for the region including the gene is followed by an increased activity with the expected 1.5 ratio to controls. In the case with monosomy however, the presence of the high activity of case 5 makes the difference with controls not significant even if the other cases show reduced activities. Meningiomas were reported to be more frequent and to develop in younger age in cats with iduronidase deficiency[9]. We tested IDUA in peripheral leukocytes of three patients with meningiomas finding activities within the range of controls (data not shown). Conversely, in three patients with IDUA deficiency diagnosed in our laboratory (2 cases with Scheie syndrome and one with Hunter syndrome respectively of age 10 yr, 14 yr and 18 months) no meningioma was observed. Thus the presence of a relationship, if any, between IDUA deficiency and meningiomas is still to be demonstrated.

β-galactosidase

GLB, whose structural gene is on chromosome 3, was chosen to be tested for the presence on chromosome 22 of a "protective protein"[12] necessary for its normal activity. Thus variations in GLB activity may represent a direct consequence of GDE effect for this protein.

In case 16 the trisomy for the region 22pter → q11 is followed by an increased activity of GLB with a ratio to the mean of group C of 1.74 (expected for triplex GDE 1.5) (see Table 2). This result, if interpreted as demonstration of GDE, could narrow the mapping of the protein to 22pter → q11 but obviously more cases are needed for confirmation. In the group of cases with monosomy, three out of four cases show low activities, as expected in case of GDE, while the fourth (case 35 in Table 2) is comparable to the highest values of group C. This makes the difference between cases with and without monosomy 22 not significant (see Table 3).

β-hexosaminidase

Total activity of HEX was used as control enzyme: as expected, all cases with monosomy 22 as well as case 16 fell within the range of cases of group C and the difference are not significant (see Table 3).

Dutrillaux and Muleris[6] proposed a model for carcinogenesis in colorectal carcinomas in which the various chromosomal anomalies found, induced by GDE a modification of the balance between de novo and salvage pathway of nucleotide synthesis in favour of the latter. The same model can not be applied to meningiomas, where the analysis of cultured cells shows anomalies in 40–70% of cases only, and monosomy of chromosome 22 is often the only anomaly present. The enzymes we tested have known catabolic activity on lipids, glicolipids and glicosaminoglicans, and their complete loss of activity causes specific storage disease.

Thus no storage of these substances is expected to follow the low activities found (36 to 60% of controls) in our cases.

Acknowledgements

This work was partially supported by Associazione per lo Studio delle Malformazioni ASM-Milano and by grant Regione Lombardia 850 to C.D.

We thank Prof. M. Fraccaro for his advice in the preparation of the manuscript.

References

1. Cantu JM, Hernandez A, Vaca G, Plascencia L, Martinez-Basalo C, Ibarra B, Rivera H (1981) Trisomy 22q12 → qter: "aneusomie de recombinaison" of a pericentric inversion. Ann Genet 24: 37–40
2. Casalone R, Granata P, Simi P, Tarantino E, Butti G, Buonaguidi R, Faggionato F, Knerich R, Solero L (1987) Recessive cancer genes in meningiomas? An analysis of 31 cases. Cancer Genet Cytogenet 27: 145–159
3. Christomanou H, Martinius J, Jaffè S, Betke K, Förster C (1980) Biochemical, psychometric and neurophysiological studies in heterozygotes for various lipidoses. Hum Genet 55: 103–110

4. Danesino C, Pasquali F (1983) Gene dosage effect for β-glucuronidase (GUSB) in monosomy 7 cells of patients with myeloproliferative disorders. Acta Anthropogenetica 7: 133–142
5. Danesino C, Minelli A, Rossella F (1985) Gene dosage effect in chorionic villi. First trimester fetal diagnosis. In: Fraccaro M *et al* (eds). Springer, Berlin Heidelberg New York, pp 246–250
6. Dutrillaux B, Muleris M (1986) Induction of increased salvage pathways of nucleotide synthesis by dosage effect due to chromosome imbalances may be fundamental in carcinogenesis: the example of colorectal carcinoma. Ann Genet 29: 11–15
7. Epstein CJ (1986) The consequences of chromosome imbalance. Cambridge University Press
8. Galjaard H (1980) Genetic metabolic disease. Elsevier/North Holland Biomedical Press, Amsterdam New York Oxford
9. Haskins ME, McGrath J (1983) Meningiomas in young cats with mucopolysaccharidosis I. J Neuropathol Exp Neurol 42: 664–670
10. Human Gene Mapping 8 (1985) Eight International Workshop on Human Gene Mapping. Cytogenet Cell Genet 40: Nos 1–4, 1985
11. Sadamori N, Takei H, Yao E, Nagamine M, Tomonaga Y, Tagawa M, Kusano M, Ichimaru M (1984) Erythrocyte glutathione reductase activity and acquired trisomy 8 in various haematologic disorders. Cancer Genetics and Cytogenetics 12: 255–259
12. Sips HJ, de Wit-Verbeek HA, de Wit J, Westerveld A, Galjaard H (1985) The chromosomal localization of human β-galactosidase revisited: a locus for β-galactosidase on human chromosome 3 and for its protective protein on human chromosome 22. Hum Genet 69: 340–344

Address for correspondence: C. Danesino, C.P. 217, I-27100 Pavia, Italy.

Acta Neurochirurgica, Suppl. 43, 118–120 (1988)

The Effect of Type Beta Transforming Growth Factor on Proliferation of Clonogenic Cells from Human Gliomas

E. Helseth, G. Unsgaard[1], **A. Dalen**[2], and **R. Vik**

Brain Tumour Research Laboratory, The Institute of Cancer Research, [1] Department of Neurosurgery, [2] Department of Virology, University Hospital of Trondheim, Trondheim, Norway

Summary

Type beta transforming growth factor (B-TGF) is a potent regulator of cell growth and differentiation. Growth of many human tumour cell lines are inhibited by B-TGF.

The effect of B-TGF on proliferation of clonogenic cells from 12 human glioma biopsies was registrated in a thymidine incorporation assay. B-TGF appeared to be a potent growth inhibitor for some gliomas, while it had no effect on others. Maximum inhibition was about 60%. Though not significant, glioblastomas appeared to be less sensitive to inhibition by B-TGF than astrocytomas and oligodendrogliomas.

Very little is known about the growth inhibitory action of B-TGF. The negative autocrine growth theory for cancer cells postulates that a reduced production, or production of a defective growth inhibitor normally found in the cell, may account for the autonomous nature of some cancer cells. In view of this theory, we searched for B-TGF in protein extracts from a glioblastoma cell line, T-MGl, whose growth was inhibited by B-TGF.

Protein extract from T-MGl cells was analysed for B-TGF activity using a soft agar colony formation assay with normal rat kidney fibroblasts. B-TGF was not found in the extract. Since, according to the literature B-TGF has been found in all other cell lines examined, we believe that there may be a lack of B-TGF or an altered B-TGF in the gliomas whose growth are inhibited by B-TGF. This problem will be studied further.

Keywords: Growth factor; Beta transforming growth factor; glioma; glioblastoma; astrocytoma; oligodendroglioma.

Introduction

Sporn and Todaro proposed in 1980 the concept of autocrine secretion to account for the relatively autonomous nature of the cancer cell[3]. The original autocrine hypothesis suggested that one way a cell could become independent of external control was by synthesis of its own growth factors, for which it also possessed functional receptors. It is now clear that the original concept of autocrine secretion is too restrictive. An extended autocrine hypothesis should also account for the situations where the autonomous growth is caused by enhanced expression of growth factor receptors or increased activity in the postreceptor signalling pathways. However, normal cellular proliferation is a delicate balance between growth stimulating and growth inhibiting factors. A deleted or a defective growth inhibitor normally produced by the cell, can therefore also account for the autonomous nature of cancer cells. This is the negative hypothesis for autonomous growth of cancer cells.

Type beta transforming growth factor (B-TGF), a hormonally active polypeptide found in normal and transformed tissues, is a potent regulator of cell growth and differentiation[2]. The name "transforming growth factor" was given to B-TGF because of its ability to transform normal rat fibroblasts (NRK-49 F) to anchorage independent growth. Anchorage independent growth in semisolid medium is considered the best in vitro correlate to tumourigenicity. With regard to growth the effects of B-TGF are bifunctional, and whether its action is growth stimulating or growth inhibiting depends on the cell type, growth conditions and the simultaneous action of other growth factors. Interestingly, B-TGF fits into both the positive and negative autocrine hypothesis for autonomous growth of cancer cells.

Since B-TGF is a potent inhibitor of many human tumour cell lines, we found it relevant to register the effect of B-TGF on proliferation of clonogenic cells from freshly resected human gliomas. We have further studied if there is a deleted or defective production of B-TGF in gliomas sensitive to B-TGF, which could explain the autonomous nature of these tumours. For this purpose we have used a human glioblastoma cell line which is strongly inhibited by B-TGF. Primary cultures are not suitable for mechanistic studies.

Table 1. *Effects of B-TGF on (^3H)-thymidine Incorporation in Clonogenic Cells from Human Brain Tumours (% of Control)*

	Sex	Age	Hist. diagnosis	B-TGF[1]
A	f[2]	34	glioblastoma	72%
B	m[3]	36	oligodendroglioma	42%
C	m	36	oligodendroglioma	52%
D	m	63	oligodendroglioma	49%
E	f	67	glioblastoma	116%
F	m	71	glioblastoma	101%
G	f	51	astrocytoma	76%
H	m	41	astrocytoma	71%
I	f	50	mixed astrocytoma—oligodendroglioma	67%
J	m	67	glioblastoma	180%
K	m	44	astrocytoma	72%
L	m	61	oligodendroglioma	41%

Mean of one experiment done in triplicate
[1] B-TGF 10 units/ml.
[2] f—female.
[3] m—male.

Materials and Methods

Cells

The glioblastoma cell line, T-MGl, was established in our laboratory. The glial origin of the cell line was confirmed by demonstration of the glial fibrillary acidic protein. T-MGl cells were maintained in medium containing 10% human heat inactivated AB Rh+ serum. The rat fibroblast cell line NRK-49 F (ATCC CRL 1570) was maintained under the same conditions as T-MGl, except that 5% fetal calf serum was used instead of 10% human serum.

Biopsies

Freshly resected tumour pieces were disaggregated mechanically to single cell suspensions in medium containing 20% human inactivated AB Rh+ serum (G. Unsgaard, in press).

Type beta transforming growth factor (B-TGF)

Human B-TGF (purity = 95%) isolated from blood platelets was purchased form R and D systems (Minneapolis, U.S.A.). In addition we used B-TGF produced in our laboratory as described earlier (E. Helseth, in press). So far we have not detected any difference in biological effects between the purchased B-TGF and the B-TGF produced in our laboratory. One unit of B-TGF activity was defined as the concentration of B-TGF which gave half maximal (^{3}H)-thymidine incorporation (ED_{50}) in a NRK-49 F soft agar assay in the presence of 2 ng/ml epidermal growth factor.

Agarose (^3H)-thymidine incorporation assay

Agarose permits the proliferation of malignant cells, but inhibits the growth of normal cells. The agarose isotope assay with T-MGl cells and single cell suspensions from freshly resected biopsies was performed as described earlier (G. Unsgaard, in press). Cells cultured in agarose were pulsed with (^{3}H)-thymidine 20 h before harvesting and registration of (^{3}H)-thymidine incorporation on the seventh day of culture. The cells were continuously exposed to B-TGF.

Protein extraction from T-MGl cells and detection of B-TGF in T-MGl protein extracts

The protein extraction procedure and the assay for detection of B-TGF activity have been described earlier (E. Helseth, in press). Briefly, proteins were extracted from T-MGl cells with 1 M acetic acid. The extracted material was dialyzed against 1 M acetic acid in a membrane tube with molecular cut off 3,500 daltons. The dialyzed material was lyophilized, resuspended in a small volume of 1 M acetic acid and run on a Bio-Gel P-30 column. The fractions from the P-30 column were lyophilized and utilized for detection of B-TGF

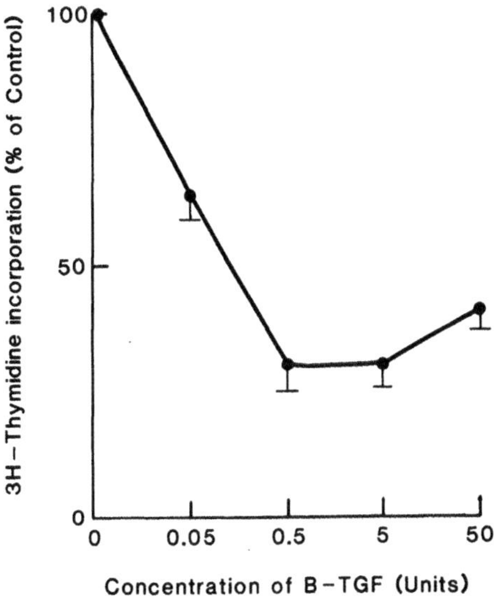

Fig. 1. Inhibition of soft agar growth of T-MGl cells by continuous exposure to different concentrations of B-TGF for 7 days as per cent of (^{3}H)-thymidine incorporation in control cultures. Mean ± SD of three experiments done in triplicates with the same batch of B-TGF

activity in the NRK-49 F agarose assay and in the T-MGl agarose assay. For NRK-49 F agarose assay 2 ng/ml EGF was added.

Results

Effects of B-TGF on growth of clonogenic cells from human brain tumours in agarose

B-TGF inhibited most of the patient biopsies. There were, however, vast individual variations. Glioblastomas seemed to be less sensitive to inhibition by B-TGF than astrocytomas and oligodendrogliomas (Table 1).

Effect of B-TGF on clonogenic growth of T-MGl cells in agarose

B-TGF had a biphasic effect on the soft agar growth of T-MGl cells. Small doses inhibited the growth in a dose-dependent manner, while inhibition was reduced at very large doses (Fig. 1). Maximum inhibition was about 70%.

Search for B-TGF activity in T-MGl protein extracts

We did not detect any B-TGF activity in protein extracts from T-MGl cells, neither measured by colony formation of NRK-49 F cells in agarose in the presence of 2 ng/ml EGF nor by inhibition of colony formation of T-MGl cells in agarose. We have used the same procedures for extraction and detection of B-TGF as we used when we produced B-TGF from human platelets. With reference to these data we have estimated that the amount of B-TGF activity in 1 gm wet weight T-MGl cells, if present at all, can be at most 1 : 1,000 of the amount found in 1 gm wet weight platelets.

Discussion

B-TGF is a potent inhibitor of many human tumour cell lines. In our study, we found that clonogenic growth of single cell suspensions from human gliomas showed great individual variation in sensitivity to B-TGF. Thus, growth inhibition by B-TGF was not a common feature for gliomas. This reflects that gliomas are a very heterogeneous group of tumours[1]. Based on these results, together with the fact that B-TGF is a potent suppressor of the immune system, we find it unlikely that B-TGF will be an effective agent in the treatment of gliomas. However, further studies on gliomas sensitive to B-TGF may reveal the mechanism for growth inhibition by B-TGF and this may give a lead as to why the cells have become cancer cells.

Reduced production or production of a defective growth inhibitor normally found in the cell may contribute to the autonomous growth of cancer cells. B-TGF acted as a growth inhibitor to T-MGl cells and we searched for B-TGF in protein extracts from T-MGl cells. B-TGF could not be detected in these extracts. The interpretation of these data was somewhat difficult, mainly due to the sensitivity of the assay. According to our estimates the level of B-TGF activity in T-MGl cells, if present at all, must have been less than 1/1,000 of that found in human platelets (activity/gm wet weight). This estimation was based on the B-TGF activity found in protein extracts from human platelets using the same extraction procedure as for T-MGl cells. B-TGF has so far been found in every transformed and nontransformed cell line examined[2]. The amount of B-TGF found in these cell lines is more than 1/1,000 of that found in human platelets. With the reservation of the sensitivity of our B-TGF assay, we are at the time being tempted to suggest that a reduced/absent production or production of a defective B-TGF may contribute to the autonomous growth of T-MGl cells and of gliomas sensitive to B-TGF. This problem will be studied further when more sensitive assays for B-TGF is available.

References

1. Bigner DD, Schold C, Bigner SH, Bullard DE, Wickstrand C (1981) How heterogeneous are gliomas? Cancer Treatment Reports [Suppl 2] 65: 45–49
2. Roberts AB, Sporn MB (1985) Transforming growth factors. Cancer Surveys 4: 683–707
3. Sporn MB, Todaro GJ (1980) Autocrine secretion and malignant transformation of cells. New Engl J Med 303: 878–880

Address for correspondence: E. Helseth, M.D., The Institute of Cancer Research, University of Trondheim, Regionsykehuset, N-7000 Trondheim, Norway.

Acta Neurochirurgica, Suppl. 43, 121–125 (1988)

Human Glioma Cell Lines: Tumour Associated Antigens Distribution and Sensitivity to Antibody-Toxin or Ligand-Toxin Conjugates. A Preliminary Report

M. Colombatti, M. Bisconti, P. Lorenzi, G. Stevanoni, B. Dipasquale, M. Gerosa[1], and **G. Tridente**

Istituto di Scienze Immunologiche and [1] Dipartimento di Neurochirurgia, University of Verona, Verona, Italy

Summary

We have investigated the phenotype of seven human glioma cell lines established in vitro from primary tumour explants. Indirect immunofluorescence and flow cytofluorimetry revealed a heterogeneous distribution of surface GE 2 and CG 12 Tumour Associated Antigens (TAA). In one group of cell lines TAA were detected both at the cell surface and in the cytosol, whereas in a second group of glioma cell lines TAA were found only in the cytosol. We have also investigated the sensitivity of glioma-derived cell lines to antibody-toxin and ligand-toxin conjugates (Immunotoxins). Monoclonal antibodies anti GE 2 antigen linked to ricin toxin A subunit (RTA) showed poor cytotoxicity, which increased about 50 fold when the whole toxin was linked to anti GE 2 monoclonals. Treatment with human recombinant interferon γ (IFN-γ) greatly augmented the percentage of HLA-DR$^+$ cells and the amount of HLA-DR antigens per cell. IFN-γ treatment resulted in a net increase of sensitivity to anti HLA-DR Immunotoxins (IT). Human diferric transferrin linked to RTA exhibited a potent cytotoxic effect against human glioma-derived cells when used in the presence of the lysosomotropic carboxylic ionophore monensin.

Keywords: Gliomas; tumour antigens; immunotoxins.

Abbreviations: FCS fetal calf serum, mAb monoclonal antibody, RTA ricin A chain, Ig immunoglobulins, FITC fluorescein isothiocyanate, BSA bovine serum albumin, IT immunotoxins, MBS m-maleimidobenzoyl-N-hydroxisuccinimide ester, SPDP N-succinimidyl-3-(2-pyridyldithio)-proprionate, TR transferrin, HPLC high performance liquid chromatography, IFN-γ interferon γ, NP-40 Nonidet P 40, IC 50 dose inhibiting 50% protein synthesis.

Introduction

Human gliomas are composed of heterogeneous cell populations. Heterogeneity has been suggested by their pathological appearance, karyotypic variability and sensitivity to chemotherapeutic agents[16, 17]. The monoclonal antibody technology has greatly improved the identification of tumour cell populations based on antigen expression[9, 14]. Monoclonal antibodies of Tumour Associated Antigens expressed by human gliomas have been developed by several laboratories[1, 8, 15, 18], to help recognize tumour cells in vitro in tissue sections and for in vivo immunolocalization and immunotherapy. Experimental immunotherapy approaches make use of hybrid molecules obtained by chemically linking monoclonal antibodies directed against surface antigens or ligands recognizing specific surface receptors to potent cytotoxins[20]. One of the most widely used toxins is ricin an its A chain. Ricin is constituted of two functional subunits: the A chain, an enzyme able to inhibit the protein synthesis of the target cell, and the B chain, which binds to ubiquitous structures of the cell surface and helps the A chain to reach the cytosol[11]. Great differences in the cytotoxic potency of various immunotoxins have been described[20]. This variable cytotoxicity has been related to the presence or absence of the B chain in the Immunotoxin, to the properties of the toxin-linked monoclonal or ligand and to the intracellular distribution of the cell surface-bound Immunotoxin.

In our experiments we have observed heterogeneity of Tumour Associated Antigens expression by human glioma-derived cell lines and identified two predominant phenotypes. In addition, human glioma cell lines were highly sensitive to Immunotoxins constituted of human transferrin linked to ricin A chain when used in the presence of the lyosomotropic agent monensin.

Material and Methods

Reagents

Chemicals, analytical grade, were from Pierce (Holland). Human recombinant interferon γ was a gift from Cytotex S.A. (Martigny, Switzerland). Human transferrin was from Sigma (St. Louis, MO) and the goat anti-mouse Ig conjugated to FITC was from Becton and Dickinson (Mountain View, CA). Anti CG 12 and GE 2 monoclonals were kindly supplied by Dr. S. Carrel, Ludwig Institute for Cancer Research (Lausanne, Switzerland). Monoclonal D 1/12 anti

Table 1. *TAA Expression by Human Glioma Cell Lines*

cell line	passage	GE 2+ cells (percentage)		CG 12+ cells (percentage)	
		Fresh cells	Fixed cells	Fresh cells	Fixed cells
Hu 126[a]	9	48	n.d.[b]	21	n.d.
	33	91	99	76	95
Hu 195	11	36	80	33	n.d.
	30	76	94	74	85
Hu 112	8	51	n.d.	73	n.d.
	41	95	81	96	66
Hu 104	13	4	7	4	24
	31	8	64	4	20
Hu 197	10	13	70	15	70
	33	3	6	9	35
Hu 70	9	40	n.d.	6	n.d.
	34	5	41	5	90
Hu 228	3	49	n.d.	9	n.d.
	28	12	55	27	54
	31	2	2	9	4

[a] Cell lines with "positive" phenotype are boxed.
[b] n.d., not determined.

HLA-DR was a gift of Dr. R. Accolla, Ludwig Inst. (Lausanne). RTA was supplied by Dr. P. Casellas, Clin Midi (Montpellier, France).

Cell lines

Human glioma cell lines were obtained as previously described[13]. Briefly, tumour biopsies were minced and then disaggregated with exposure to an enzyme cocktail. Cells were then filtered through a wire mesh and the single cell suspension cultured at 37 °C in a humidified atmosphere with 5% CO_2, 95% air in RPMI medium additioned with 15% FCS.

Cytofluorimetry

For cell surface analysis of TAA, cells in logarithmic growth phase were harvested and stained using indirect immunofluorescence techniques with saturating amounts of protein A purified[5] anti GE 2 or anti CG 12 monoclonal antibodies (mAb) and FITC conjugated goat anti-mouse Ig antibody. After staining the cells were run through an Epics V cytofluorimeter with an exciting wavelength of 488 nm at 200 mW power. Photomultiplier gains remained constant during analysis of similarly treated samples. Intracellular antigens were detected by staining cells fixed with 4% paraformaldehyde (10 min, r.t.) and permeabilyzed with 70% methanol and 0.1% Tween 20 in PBS-BSA 1%.

Conjugate synthesis

Ricin was purified as described[10]. Synthesis of toxic conjugates (Immunotoxins) was performed as described previously[3]. Whole ricin toxin was linked to anti Ge 2 mAb with the cross-linker m-maleimidobenzoyl-N-hydroxisuccinimide ester (MBS). Ricin A chain was linked to human transferrin (TR) or to mAb D 1/12 with N-succinimidyl-3-(2-pyridyldithio)-propionate (SPDP). Hybrid molecules were separated from unreacted reagents by gel filtration using a TSK 3,000 SW HPLC column. Conjugates with a stoichiometry of 1:2–3 (ligand/mAb : toxin) were pooled, concentrated by vacuum dialysis and filtred through 0.22 µm Millipore filters.

Protein synthesis assay

Protein synthesis was assayed essentially as described[3] by evaluating incorporation of ^{14}C-leucine. Samples containing ricin or ricin Immunotoxins (IT) were assayed in the presence of 0.1 M lactose to prevent nontarget cell binding of ricin. To assess the sensitivity of IFN-γ treated cells to anti HLA-DR-RTA IT, cells were treated with 500 IU/ml IFN-γ for 48 hr before adding the IT to the cell cultures. The specificity of the IT was determined by comparing the IT effect to the effect of RTA alone (anti GE 2, HLA-DR, TR) or to ricin + 0.1 M lactose (anti GE 2-ricin). The concentration of RTA required to inhibit the protein synthesis of treated cells to 50% of control (IC 50) was taken as 1.0.

Results

TAA distribution in human glioma cell lines

The expression of GE 2 and CG 12 TAA was studied in human glioma cell lines at various times in culture. Fresh cells or fixed cells were treated with mAb to the different TAA, stained with a second step fluorescent antibody and the cells analyzed with a flow cytofluorimeter to determine the percentage of positive cells. Cell surface expression of TAA was investigated in fresh cells, whereas the presence of intracellular antigens could be estimated in fixed and permeabilyzed cells. As shown in Table 1, 3 cell lines (Hu 70, Hu 104 and Hu 197) showed very low percentage of TAA^+ cells (between 2% and 9% at the latest culture passages). These cell lines were therefore considered "negative" as regard surface TAA expression. One cell line (Hu 228) showed initially 49% of $GE\,2^+$ cells and 9% of $CG\,12^+$ cells, but the percentage of TAA^+ cells was reduced to less than 10% at the latest culture passages (Table 1). Three other human glioma cell lines (Hu 126, Hu 195 and Hu 126) were instead composed of a high percentage (76% to 96%) of TAA^+ cells and conserved this phenotype throughout a long culture period. When we examined the expression of TAA in fixed cells, a very large proportion (66% to 99%) of fixed cells of the "positive" cell lines showed intracellular GE 2 and CG 12 antigens (Table 1) and, unexpectedly, also cells with a surface "negative" phenotype showed intracellular TAA (6% to 90%). In two "negative" cell lines, Hu 197 and Hu 228, the percentage of intracellularly TAA^+ cells decreased with time in culture (Table 1). It must be mentioned, however that the TAA positivity in fixed cells with a "negative" cell surface can only reflect the presence of truly intracellular antigens, whereas the TAA detected in fixed cells with a "positive" phenotype could represent the sum of surface and intracellular antigens. The limit of detection of the cytofluorimetric analysis is about $3–5 \times 10^3$ molecules per cell, therefore cells with a surface "negative" phenotype expressed less than $3–5 \times 10^3$ TAA molecules per cell.

Sensitivity of human glioma cells to Immunotoxins

To investigate the sensitivity of human glioma cell lines to IT, cells were treated for 24–48 hr with toxins (RTA, ricin) or with IT (anti GE 2-RTA, anti GE 2-ricin, D 1/12-RTA and TR-RTA). The cytotoxic effect of the various IT was compared to the effect of RTA. RTA is toxic only by nonspecific mechanisms at high concentrations, since it does not bind the cell surface. After incubation the protein synthesis was estimated by ^{14}C-leucine incorporation. Ricin B chain binds terminal galactose residues present on all cells[11]. IT containing whole ricin are made cell-specific by the addition of 0.1 M lactose to the culture medium to prevent binding of ricin B chain to nontarget cells[21]. As shown in Table 2, anti GE 2-RTA was only about 13 times more cytotoxic for glioma cells than RTA. Inhibition of 50% protein synthesis (IC 50) was attained by RTA at 2×10^{-7} M in the absence and at 10^{-8} M in the presence of 80 nM monensin. These were taken as 1.0 for comparison with the IT. The carboxylic ionophore monensin has been shown to potentiate the cytotoxic effect of IT, probably by interfering with the intracellular degradation of internalized IT[2, 12]. Treatment with 80 nM monensin, however, did not increase significantly the cytotoxic effect of anti GE 2-RTA IT (Table 2). Unlike what described for lymphocyte antigens (*i.e.* T 3, CALLA) treatment with anti GE 2 mAb did not result in modulation and subsequent internalization of GE 2 antigens (M. Colombatti *et al.*, in preparation).

Table 2. *Effect of Immunotoxin Treatment on Human Glioma Cells*

Toxin or Immunotoxin	Cytotoxic index[a]
RTA	1
RTA + 80 nM monensin	1
Ricin	4000
Ricin + 0.1 M lactose	67
anti GE 2-RTA	13
anti GE 2-RTA + 80 nM monensin	2
anti GE 2-ricin	670
D 1/12-RTA	20
D 1/12-RTA + IFN-γ	130
TR-RTA	10
TR-RTA + 80 nM monensin	4100

[a] The cytotoxic index was calculated based on the concentration of RTA required to inhibit 50% protein synthesis (IC 50). This concentration was arbitrarily given the value of 1. This is taken as a reference value to which the cytotoxic effect of the various IT is compared.

Whole ricin IT are often more potent than RTA IT[20]. Linkage of anti GE 2 mAb to ricin resulted in a 50 to 300 fold increase in cytotoxicity as compared to anti GE 2-RTA IT (Table 2). Ricin was about 4,000 fold more toxic than RTA. Ricin blocked with 0.1 M lactose was 60 fold more toxic than RTA and about 10 fold less toxic than anti GE 2-ricin IT in the presence of 0.1 M lactose (Table 2). The gain in cytotoxicity by anti GE 2-ricin IT is due to ricin linkage to the specific mAb.

In the attempt to find a reagent capable of killing human gliomas, we treated glioma cells in culture with 500 IU/ml IFN-γ. This resulted in a considerable increase of $HLA\text{-}DR^+$ cells and of the number of surface HLA-DR antigens per cell (Table 3). The increase of surface HLA-DR antigens rendered the target cells about 6 fold more sensitive to the anti HLA-DR IT D 1/12-RTA as compared to IFN-γ untreated cells.

Human transferrin linked to RTA (TR-RTA) was about 10 fold more toxic than RTA alone. However, TR-RTA was over 4,000 fold more potent than RTA alone in killing human glioma cells when 80 nM monensin was added to the medium (Table 2).

Table 3. *Effect of IFN-γ on HLA-DR Antigen Expression by Human Glioma Cells*

Time (hr)[b]	— IFN-γ HLA-DR⁺ cells (percentage)	+ IFN-γ[a] HLA-DR⁺ cells (percentage)	MFI ($\times 10^2$)[c]
24	11	42	2.5
48	9	89	4.3
72	8	91	4.9
96	3	97	6.3

[a] Cells were treated with 500 IU/ml IFN-γ.

[b] Cells were cultured in the presence or in the absence of IFN-γ for the times indicated and then analyzed for HLA-DR antigen expression by indirect immunofluorescence and flow cytofluorimetry.

[c] MFI = mean fluorescence intensity, calculated by computer assisted statistics on flow cytofluorimeter analyzed cells.

Discussion

We have investigated the expression of TAA by human glioma-derived cells and their sensitivity to several IT.

Two different phenotypes were observed as regard TAA expression. One group of cell lines showed surface and intracellular GE 2 and CG 12 TAA, whereas a second group of human glioma cell lines expressed TAA only in the cytosol. These two phenotypes did not change significantly during culture and they may represent different cell subpopulations existing in vivo. This may be important for tumour cell localization in vivo and for immunotherapy. The existence of surface TAA^- cell populations may limit the use of mAb in vivo. Immunodiagnosis is usually performed on fixed tissue sections using immunoperoxidase techniques. Positivity for TAA in fixed cells can be mistakenly interpreted as cell surface positivity. Only techniques that allow a distinction between surface and intracellular antigens may help in determining the feasibility of the immunolocalization and immunotherapy approach in gliomas.

The results of experiments carried out with IT indicated that GE 2 antigen is not a good IT target, probably due to its intrinsic properties. Treatment with monensin failed to enhance the cytotoxicity of anti GE 2-RTA IT. This is consistent with lack of modulation and internalization of surface-bound anti GE 2 IT. Antigens or receptors known to be modulated and internalized upon binding the specific antibody or ligand effectively mediate IT cytotoxicity[2, 12].

Increase of antigen positive cells and of antigen density on the cell surface resulted in a higher sensitivity to IT, as demonstrated by experiments with anti HLA-DR IT. Use of anti HLA-DR IT cannot be envisaged in brain tumours due to HLA-DR antigens presence on normal cells in the brain[6]. Anti HLA-DR IT used against IFN-γ stimulated cells might represent, however, a good working model for IFN-γ inducible tumours arising in tissues where expression of HLA-DR antigens is weak or not inducible by IFN-γ.

Transferrin enters the cell via receptor mediated endocytosis through acidified organelles[4]. The potent cytotoxic effect of TR IT used in the presence of lysosomotropic substances has been described in human leukemic T cells[12]. Receptors for transferrin are expressed in high numbers by normal proliferating cells and by growing tumour cells[19]. The cytotoxic effect of RTA-IT is weak as compared to the effect of RTA alone when assayed in the absence of monensin, but increases over 4,000 fold when cells are treated in the presence of monensin. The in vivo use of monensin chemically linked to carrier proteins to potentiate IT has been described in mice[7]. The potent antiglioma cell activity demonstrated by TR IT in our experiments is promising and certainly deserves further investigation to assess their possible use as anti-tumour reagents.

Strategies for IT use in vivo will probably require the combined use of different IT; clinical application of IT, however, cannot be approached without taking into consideration several aspects related to in vivo pharmacokinetics, toxicity, specificity, etc. that still need to be fully evaluated.

Acknowledgements

We thank Drs. S. Carrel and R. S. Accolla for the gift of monoclonal antibodies and Dr. P. Casellas for generously supplying RTA. This work was supported in part by the Italian National Research Council (C.N.R.), Oncology Research Project.

References

1. Cairncross JG, Mattes MJ, Beresford HR, Albino AP, Houghton AN, Lloyd KO, Old LJ (1982) Cell surface antigens of human astrocytoma defined by mouse monoclonal antibodies: Identifications of astrocytoma subsets. Proc Natl Acad Sci USA 70: 5641–5645

2. Casellas P, Bourrie PJ, Gros P, Jansen FK (1984) Kinetics of cytotoxicity induced by immunotoxins. Enhancement by lysosomotropic amines, and carboxylic ionophores. J Biol Chem 259: 9359–9363
3. Colombatti M, Greenfield L, Youle RJ (1986) Cloned fragment of diphteria toxin linked to T-cell-specific antibody identifies regions of B chain active in cell entry. J Biol Chem 261: 3030–3035
4. Dautry-Varsat A, Ciechnover A, Lodish HF (1983) PH and the recycling of transferrin during receptor-mediated endocytosis. Proc Natl Acad Sci USA 80: 2258–2262
5. Ey PL, Prouse SJ, Jenkins CR (1978) Isolation of pure IgG_1, IgG_{2A} and IgG_{2B} immunoglobulines from mouse serum using protein A-sepharose. Immunochemistry 15: 429–435
6. Fontana A, Mc Adam KPWJ, Kristensen F, Weber F (1983) Biological and biochemical characterization of interleukin l-like factor from rat C_6 glioma cells. Eur J Immunol 13: 685–689
7. Jansen FK, Blythman HE, Casellas P, Derocq JM, Dussussoy D (1987) Enhancement of ITs: In vitro and in vivo. In: Bonavida B, Collier RJ (eds) Membrane-mediated cytotoxicity, vol 45. Liss AR, Inc, New York, pp 167–173
8. Kemshead JT, Fritschy J, Garson JA, Allan PM, Coakham HB, Brown S, Asser U (1983) Monoclonal antibody 127.11 detects A 2,020–240,000 mol. WT. Glycoprotein present on a subset of neuroectodermally derived cells. Int J Cancer 31: 187–195
9. Koprowski H, Steplewski Z, Herlyn D, Herlyn M (1978) Study of the antibodies against human melanoma produced by somatic cells hybrids. Proc Natl Acad Sci USA 75: 3405–3409
10. Mise T, Funatsu G, Ishiguro M, Funatsu M (1977) Isolation and characterization of ricin E from castor beans. Agric Biol Chem 41: 2041–2048
11. Olsnes S, Pihl A (1982) Toxic lectins and related proteins. In: Choen P, Van Heyningen S (eds) Molecular action of toxins and viruses. Elsevier, North Holland, Amsterdam, pp 51–105
12. Raso V, Lowrence J (1984) Carboxylic ionophores enhance the cytotoxic potency of ligand – and antibody – delivered ricin A chain. J Exp Med 160: 1234–1240
13. Rosemblum ML, Kuebel KT, Vasquez DA, Wilson CB (1976) Clonogenic tumour cell kinetics following 1,3 bis (2-chloroethyl)-1-nitrosourea brain tumour therapy. Cancer Res 36: 3718–3725
14. Schlom J (1986) Basic principles and applications of monoclonal antibodies in the management of carcinomas: The Richard and Hinda Rosenthal foundation award lecture. Cancer Res 46: 3225–3238
15. Schnegg JF, Diserens AC, Carrel S, Accolla RS, de Tribolet N (1981) Human glioma-associated antigens detected by monoclonal antibodies. Cancer Res 41: 1209–1213
16. Shapiro JR, Yung W-KA, Shapiro WL (1981) Isolation karyotype and clonal growth of heterogeneous subpopulations of human malignant gliomas. Cancer Res 41: 2349–2359
17. Shapiro WR, Boisler GA, Chernik MN, Postner JB (1979) Human brain tumour transplantation into nude mice. J Natl Cancer Inst 62: 447–453
18. Sikora K, Alderson T, Phillips J, Watson JV (1982) Human hybridomas from malignant gliomas. Lancet 1: 11–17
19. Sutherland R, Delia D, Schneider C, Newman R, Kemshead J, Greaves M (1981) Ubiquitous cell – surface glycoprotein on tumour cells is proliferation – associated receptor for transferrin. Proc Natl Acad Sci USA 78: 4515–4519
20. Youle RJ, Colombatti M (1986) Immunotoxins: Monoclonal antibodies linked to toxic proteins for bone marrow transplantation and cancer therapy. In: Roth J (ed) Monoclonal antibodies for the diagnosis and therapy of cancer. Futura Publishing Co, Kisco MT, New York, pp 173–213
21. Youle RJ, Neville DM Jr (1980) Anti-thy 1.2 monoclonal antibody linked to ricin is a potent cell-type-specific toxin. Proc Natl Acad Sci USA 77: 5483–5486

Address for correspondence: Dr. M. Colombatti, Istituto di Scienze Immunologiche, University of Verona, Policlinico B. go Roma, Verona, Italy.

Acta Neurochirurgica, Suppl. 43, 126–129 (1988)

Ganglioside Content and Composition in Human Gliomas

S. M. Gaini[1], **L. Riboni**[2], **C. Cerri**[3], **N. Grimoldi**[1], **E. P. Sganzerla**[1], and **B. Berra**[4]

[1] Institute of Neurosurgery, [2] Department of Medical Chemistry and Biochemistry, [3] Institute of Internal Medicine and Physiopathology, School of Medicine, [4] Institute of General Physiology and Biochemistry, School of Pharmacy, University of Milano, Italy

Summary

Many alterations of ganglioside content and distribution have been described in human and experimental tumours. Our previous data showed the presence of lipid alterations in meningiomas, in particular an increased monosialilganglioside content. Therefore we analyzed the distribution and content of gangliosides in various gliomas. The data show that ganglioside content is inversely porportional to tissue malignancy and that the ganglioside pattern can be described as lacking of polisialilgangliosides with increased GD_3 content. The amount of GD_3 (as percent of total gangliosides sialic acid) increases from 15% in astrocytomas grade I to 60% in grade IV. The GD_3 increase seems to be almost specific of glioma. Because anti-GD_3 antibodies could be used to localize immunohistochemically the ganglioside and to help the tumour grading, we used a purified preparation of GD_3 to produce monoclonal antibodies in balb/c mice. But because some clones did produce anti-GD_3 antibodies the low yield requires further experiments to obtain an antibody useful for this purpose.

Keywords: Ganglioside; cerebral gliomas; monoclonal antibodies.

Introduction

In recent years several studies on tumour-associated gangliosides on both tumour tissues and tumour cell lines have been reported[1–2]. Specific monoclonal antibodies directed against these gangliosides in human tumours have been produced[3–5], and current studies are investigating their possible role as effective tumour targets as well as useful tools in the diagnosis and/or therapy of lesions such as melanomas, breast cancer, adenocarcinomas and small cell lung cancer.

Little work has been undertaken on the gangliosides of intracranial human malignancies.

In previous studies we observed that dramatic ganglioside changes, possibly consequent upon incomplete synthesis, occur in different histological types of human intracranial tumours such as meningiomas, pituitary adenomas, neurinomas and gliomas; these variations result in the accumulation of a selective ganglioside species (different in different neoplasias) which could be a well defined tumour target.

On these bases we have focused our interest on human gliomas, aiming to confirm and extend previous studies on a larger casuistry, to isolate the main glioma associated ganglioside (GD_3) and to evaluate its potential role as glioma specific antigen.

Materials and Methods

Tumours – Tumour specimens were obtained from 43 patients aged from 24 to 73 years. Small pieces of the tumours were examined for their histological features: all the tumours were astrocytomas graded according to the WHO[6] as: grade I (3 cases), grade II (6 cases), grade III (20 cases) and grade IV (14 cases).

Immediately after operation, all macroscopically visible necrotic parts were removed and samples were stored at −20 °C until processed.

DNA extraction and determination – DNA was extracted from the fresh tissue with perchloric acid[7] and it was quantified by the method of Burton[8] as modified by Karlsson and Svennerholm[9].

Ganglioside extraction and purification – Extraction and purification of the crude ganglioside mixture from each sample of tumour was attained as described by Tettamanti *et al.*[10]. Briefly, tissues were homogenized in 10 mM phosphate buffer, pH 6.8/tetrahydrofuran (1 : 4, by vol) and submitted to centrifugation. The extraction was repeated in the same conditions three more times. The total lipid extracts so obtained were then submitted to a two-stages partitioning with diethylether and water respectively. The final extracts, after dialysis and liophylization were resuspended in chloroform/methanol (2 : 1, by vol) and centrifuged at 10,000 × g for 5 min to remove unsoluble material.

Total and individual ganglioside quantification – Total gangliosides were quantified as lipid-bound sialic acid by the resorcinol-HCl method[11] as modified by Miettinen and Takki-Lukkainen[12].

The ganglioside distribution was assessed in each individual sample by monodimensional thin-layer chromatography (TLC) with different solvent system: chloroform/methanol/0.2% $CaCL_2$ = 55 : 45 : 10, by vol; propanol/32% NH_4OH/water = 6 : 2 : 1 by vol; chloroform/methanol/0.2% KCl = 60 : 35 : 8 by vol. After development, gangliosides were detected on the plate by spraying it with a p-dimethylaminobenzaldheide reagent[13] followed by heating at 120 °C for 10 min and quanitified by densitometry.

Isolation and characterization of GD_3 ganglioside from astrocytomas – In order to isolate and purify GD_3 ganglioside, the ganglioside mixtures extracted from the tumours were pooled, submitted to column chromatography on a Silica Gel 100 column (40 × 0.7 cm) equilibrated and eluted at a flow rate of 35 ml/h using chloroform/methanol/water (60 : 35 : 1.5 by vol) as solvent at 18–20 °C. The elution pattern was monitored by TLC and the fractions containing GD_3 were collected, dried and rechromatographed under the above conditions. The homogenous material thus obtained was stored in chloroform/methanol, 2 : 1 by vol at −20 °C. The purity of the isolated ganglioside, assessed by TLC was higher than 98% in different solvent systems. The characterization of GD_3 ganglioside was attained by determination of its carbohydrate composition, and by partial acid hydrolysis and sialidase treatment, both followed by TLC recognition of the formed neutral glycosphingolipids[10].

Immunological procedures – Six Balb/c mice were given 100 µg of purified GD_3 ganglioside suspended in 50% Freund adjuvant intraperitoneally four times at 30 days interval. The mice were boosted e.v. with GD_3 in saline 3 days before sacrifice and the spleen cells were removed and used for fusion with P 3 murine myeloma according to Schnegg *et al.*[14]. After fusion, hybrid lines producing antibodies (detected with rabbit anti mouse $F(ab)_2$ antibodies) were cloned in a 96 well plate. Culture fluid from the hybrids were tested with the agar immunodiffusion method of Outcherlony against the gangliosides.

Results and Discussion

Table 1 shows the content of DNA and of total gangliosides in the different histological classes of human astrocytomas. The increase of malignancy, accompanied by an elevation of DNA content, is characterized by the drop of total lipid-bound sialic acid content. On comparing the relative mean values, a significant exponential relationship ($p < 0.01$) between DNA and total gangliosides was found (Fig. 1).

As for the ganglioside pattern, in human astrocytomas and, above all in the malignant ones (grade III and IV), a peculiar distribution, different from all the other intracranial tumours investigated, i.e. meningiomas, neurinomas, pituitary adenomas (15–16 and unpublished results) was present.

A representative TLC pattern of the four histological classes is shown in Fig. 2. The most significant evidence appears to be the selective predominance of one ganglioside species, migrating on TLC as a doublet; this ganglioside was isolated from the tumours and characterized by enzymatic and chemical analyses as ganglioside GD_3. Its content increased from nearly

Table 1. *Content of DNA (mg/g fresh tissue) and of Total Gangliosides (µ Moles Liquid-Bound Neu/AC/mg DNA) in the 4 Histological Types of Human Astrocytomas*

	Grade I (3)	II (5)	III (17)	IV (13)
DNA	1.215 ± 0.120	1.734 ± 0.206	2.312 ± 0.202	3.463 ± 0.197
Total gangliosides	1.304 ± 0.161	0.828 ± 0.106	0.395 ± 0.058	0.163 ± 0.022

Data are expressed as mean ± SD. The number of cases is given in parentheses.

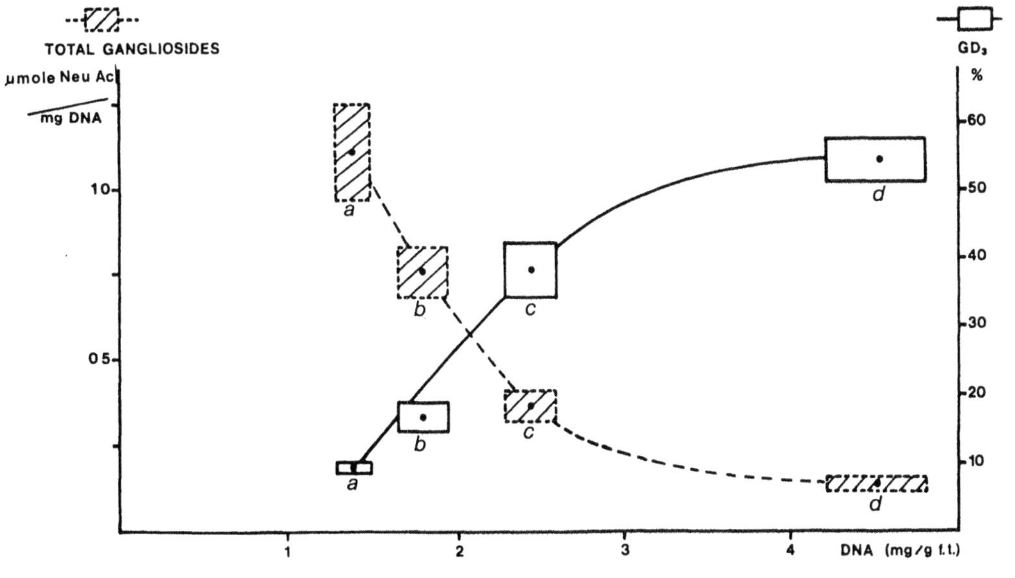

Fig. 1. Plots of DNA content versus total gangliosides and GD_3 (expressed as % of lipid-bound Neu/Ac). *a, b, c, d*: mean content ± SD of the two parameters in I, II, III, IV grade astrocytomas respectively. *f.t.* = fresh tissue

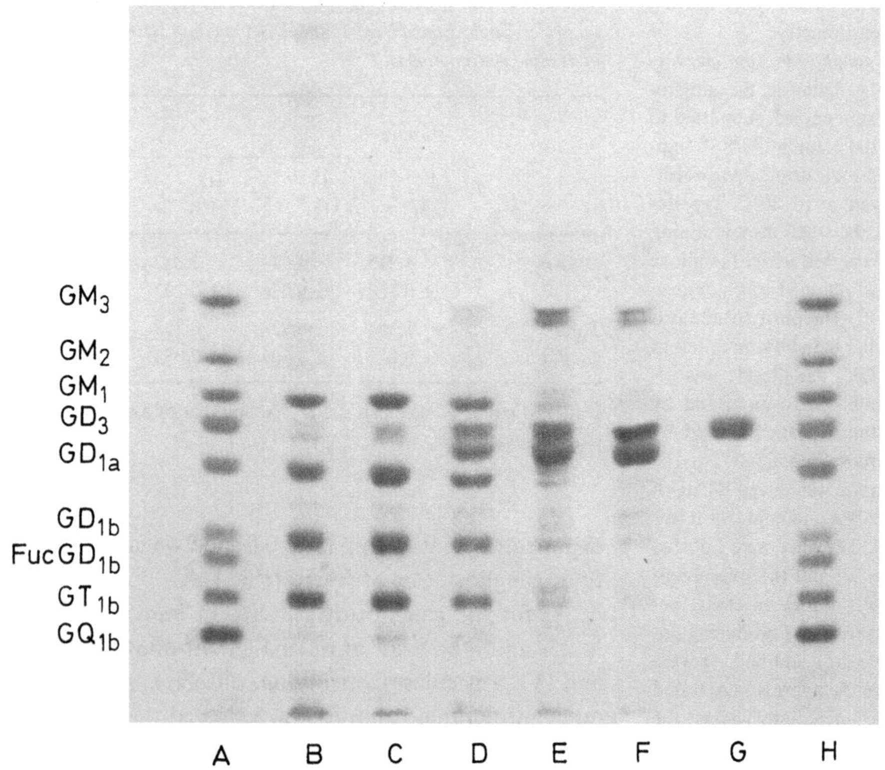

Fig. 2. Thin layer chromatography pattern of total gangliosides in human astrocytomas. A standard ganglioside mixture; B normal human brain; C through F astrocytomas grade I to IV; G GD_3 standard

10% of total ganglioside-bound sialic acid in grade I tumours to more than 60% in grade IV lesions. The bivariate curve fitting analysis demonstrated a significant exponential relationship ($p < 0.01$) between DNA and GD_3 percentage (Fig. 1). This correlation suggests the possibility of classifying astrocytomas also on the basis of their ganglioside features and suggests GD_3 as a potentially effective astrocytoma target.

Another evident feature of the ganglioside composition of human astrocytomas is the progressive disappearance with malignancy of the more complex species (polysialilgangliosides) (Fig. 2). In particular the GD1a, GD1b, GT1b and GQ1b contents progressively decrease in higher grade tumours, being almost absent in the grade IV ones. In normal human CNS GD_3 represents a minor ganglioside[17] in astrocytes or glial cells. However, studies on developmental changes of brain gangliosides have demonstrated that GD_3 is the predominant ganglioside in embryonic cerebrum[18] and its content decreases with a concomitant increase of polysialilgangliosides. The drop of GD_3 and the appearance of polysialilgangliosides are synchronized with the differentiation of proliferating cells into postmitotic neurones. The changes we observed in human astrocytomas, in agreement with previous studies on other tumours, suggested that changes in the absolute amounts and the relative proportions of various carbohydrate moieties (in this case of gangliosides) are precisely programmed and genetically predictable[19].

Although the splenic cells from the immunized mice fuse with the expected yield, very few cell lines produced antibodies which reacted with GD_3 ganglioside. Furthermore in order to give a visible precipitation band all reacting lines needed very high ratio of antibodies to antigen.

These results suggest that antibodies produced by hybridomas obtained from mice immunized with purified GD_3 human astrocytomas are directed against a scarcely immunoreactive component of the ganglioside.

The low activity of our monoclonal antibodies make it difficult to use these antibodies for immunochemical detection of GD_3 ganglioside in histological tumour sections. On the other hand our data showed the possibility of producing monoclonal antibodies against specific components of a brain tumour, and we hope to be able to produce a more specific and immunoreactive antibody. Work is in progress in our laboratory with the aim of obtaining an astrocytoma-directed antibody whose role in the diagnosis and immunotherapy of these tumours in the future could be of considerable clinical value.

Acknowledgement

The authors express their thanks to Miss Rita Paglino for invaluable technical assistance in the preparation of all microscopical slides.

References

1. Hakomori SI (1981) Glicosphingolipids in cellular interaction, differentiation and oncogenesis. Ann Rev Biochem 50: 733–764
2. Hakomori SI, Kannagi R (1983) Glycosphyngolipids as tumour-associated and differentiation markers. J Natl Cancer Inst 71: 231–251
3. Tai T, Paulson JC, Cahan LD, Irie RF (1983) Ganglioside GM_2 as a human tumour antigen (OFA-I-1). Proc Natl Acad Sci (USA) 80: 5392–5396
4. Hellström I, Brankovan V, Hellström KE (1985) Strong anti-tumour activities of IgG_3 antibodies to the human melanoma-associated ganglioside. Proc Natl Acad Sci (USA) 82: 1499–1502
5. Mansson JE, Fredman P, Nilsson O, Lindholm L, Holmgren J, Svennerholm L (1985) Chemical structure of carcinoma ganglioside antigens defined by monoclonal antibody C 50 and some alied gangliosides of human pancreatic adenocarcinoma. Biochim Biophys Acta 834: 110–117
6. Zülch KJ (1980) Principles of the New World Health Organization (WHO) classification of brain tumours. Neuroradiology 19: 59–66
7. Zamenhof S, Bursztyn H, Rich K, Zamenhof PJ (1964) The determination of deoxyribonucleic acid and of cell number in (calf) brain. J Neurochem 11: 505–509
8. Burton K (1956) A study of the conditions and mechanisms of the diphenylamine reaction for the colorimetric estimation of deoxyribonucleic acid. Biochem J 62: 315–323
9. Karlsson I, Svennerholm L (1978) Biochemical development of rat forebrains in severe protein and essential fatty acid deficiencies. J Neurochem 31: 657–662
10. Tettamenti G, Bonali F, Marchesini S, Zambotti V (1973) A new procedure for the extraction, purification and fractionation of brain gangliosides. Biochem Biophys Acta 296: 160–170
11. Svennerholm L (1957) Quantitative estimation of sialic acid II. A colorimetric resorcinol-hydrochloric acid method. Biochem Biophys Acta 24: 604–611
12. Miettinen T, Takki Lukkainen IT (1959) Use of butyl acetate in determination of sialic acid. Acta Chem Scand 13: 856–858
13. Partridge SM (1948) Filtre-paper filtration chromatography of sugars. General description and application of qualitative analysis of sugars in apple juice, eggwhite and foetal blood of sheep. Biochem J 42: 238–248
14. Schnegg JF, Diserens AC, Carrel S, Accolla RS, De Tribolet N (1981) Human glioma-associated antigens detected by monoclonal antibodies. Cancer Res 41: 1209–1213
15. Traylor TD, Hogan EL (1975) Gangliosides of human cerebral neoplasia. Trans Amer Soc Neurochem 6: 147
16. Berra B, Riboni L, De Gasperi R, Gaini SM, Ragnotti G (1983) Modifications of ganglioside pattern in human meningiomas. J Neurochem 40: 777–782
17. Abe T, Norton WT (1974) The characterization of sphyngolipids from neurons and astroglia of immature rat brain. J Neurochem 23: 1025–1036
18. Rösner H, Al-Aqtum M, Henke-Pahle S (1985) Developmental expression of GD_3 and polysialogangliosides in embrionic chicken nervous tissue reacting with monoclonal antigangliosides antibodies. Developmental Brain Res 8: 85–95
19. Feizi T (1984) Monoclonal antibodies reveal saccharide structures of glycoproteins and glycholipids as differentiation and tumour associated antigens. Biochem Soc Trans 12: 545–550

Address for correspondence: S. M. Gaini, Institute of Neurosurgery, University of Milano, Via F. Sforza, 35, I-20122 Milano, Italy.

Acta Neurochirurgica, Suppl. 43, 130–134 (1988)

The Response of Peripheral Blood Mononuclear Cells of Glioma-Bearing Patients to Stimulation with Microbial Antigen and IL-2: Proliferation and IFN-Gamma Production

A. Maleci, C. Ausiello[1], F. M. Gagliardi, B. Guidetti, and **A. Cassone[2]**

Dipartimento di Scienze Neurologiche, Universitá di Roma "La Sapienza", [1] Istituto CNR di Tipizzazione Tissutale e Problemi della Dialisi, L'Aquila, [2] Istituto Superiore di Sanitá, Laboratorio di Batteriologia e Micologia Medica, Roma, Italy

Summary

Peripheral blood mononuclear cells (PBMC) of malignant glioma-bearing patients were stimulated in vitro with a glucomannan-protein antigen of Candida albicans (GMP) or Interleukin-2 (IL-2), then assayed for proliferation and production of Interferon gamma. PBMC of healthy, age and sex matched subjects were the controls. PBMC from glioma-bearing patients did not differ, as a whole, from PBMC of healthy donors in IL-2 or GMP-induced proliferation. However, they showed a definitely lesser ability to produce IFN. The results are discussed in the framework of the impairment of immune responses known to affect glioma-bearing patients.

Introduction

Malignant glioma is one of the most frequent brain tumours, the prognosis of which has only slightly improved by combined surgery and radio- or chemotherapy approaches. Immunomanipulation has also been attempted in glioma-bearing patients and in other tumour patients, as well. The results, however, have not been very encouraging[1, 2]. Obviously the efficacy of immunotherapy is severely affected by the host's immune system and, in this connection, several immune responses have been reported to be depressed in glioma-bearing patients, particularly those concerning T-cell function[3, 4] also in relation to the site of primary tumour[5] and to the immunosuppressive effects of glioma cells, serum factors[6, 7] and supportive therapy[8]. In particular, the ability of circulating lymphocytes to respond to common microbial antigens by proliferating in vitro, or as skin testing in vitro, has been reported by several authors to be greatly diminished in glioma-bearing patients. This view, however, was not confirmed in a recent report by Blomgren *et al.*[5], who showed that the response to PPD was affected only in those patients who had brain stem gliomas. In general, the researches quoted above did not concern other parameters connected with antigen stimulation of the immune system, among which production of lymphocytokines such as Interleukin-2 (IL-2) or Interferon (IFN) gamma are essential for a correct activation of those immunoeffectors which are potentially important in controlling tumour growth (Natural killer cells, Lymphokine-activated killers or other cells). Therefore, in this study we re-examined the in vitro proliferative response of PBMC of glioma-bearing patients following stimulation with a powerful, common microbial antigen such as the glucomannan-protein complex of Candida albicans. We also explored the GMP ability of inducing the production of IL-2 or IFN gamma in PBMC cultures.

Materials and Methods

1. Patients

All patients who entered this study had a primary malignant glioma of the cerebral hemisphere and had not previously received radio- or antiblastic therapy. Peripheral blood was taken before surgery and, at that time, all patients were under treatment with corticosteroids (Betametasone, 8 mg/day) and anticonvulsants (Phenytoin, Phenobarbital and Mephobarbital 100 mg each/day). Haematological routine examination showed that all patients had no significant alteration in the haematocrit or in the leukocyte formula when they entered this study. PBMC of age, sex-matched healthy donors were used as controls.

2. Candida Preparation (GMP) and Interleukin-2 (IL-2)

The glucomannan protein component of the cell wall of Candida albicans (GMP), constituting the major antigenic determinant of this microorganism, was obtained from washed yeast-form of Candida albicans, strain BP, serotype A, as described elsewhere[9]. As already shown, this preparation contained phosphorylated cell wall components of different size and charge, consisting predominantly of mannan (about 80% of dry weight) and protein (about 8%) and residual amount of glucan and ribonucleic acid[9]. The preparation was used at a final concentration of 50 µg/ml. IL-2 was a commercial purified preparation (Cellular Products, Buffalo, NY, USA) obtained from PBMC stimulated with Phytohaemoagglutinin (PHA). The purification scheme removes the inducing agent as well as B-cell growth factor and immune interferon. It was used at 5% v/v final concentration corresponding to 20 IU as standardized against a recombinant IL-2 preparation (Hoffman La Roche, Basel, Switzerland).

3. PBMC Preparation and Proliferation Assay

PBMC collection from peripheral blood, culture and proliferation assay were as reported elsewhere[9]. GMP and IL-2 were added at the beginning of PBMC cultures at the final concentration of 50 µg/ml and 5% v/v respectively. All tests were performed in triplicate. Trays were incubated in 5% CO_2 at 37 °C and harvested on 7th day of culture. Incorporation of radiolabelled Thymidine added eighteen hours before harvesting, at a concentration of 0.5 µCi/well was measured as reported previously[9]. The background (unstimulated) ^{3}H-Thymidine incorporation was subtracted to the test (GMP- or IL-2 stimulated) incorporation and the proliferative response was taken as positive only when it exceeded 1,500 cpm (mean value).

4. Assay of Interleukin-2

Murine (C 57 BL/6) tumour specific cytotoxic T cell line (CTLL 2) was used for IL-2 assay. The assay was carried out as reported by Gillis *et al.*[10].

5. IFN Assay and Characterization

IFN titre was determined by a previously described assay for the inhibition of Sindbis virus haemoagglutinin yield after a single growth cycle[11]. IFN activity was expressed as international units (IU). The IFN that was produced in these experiments was characterized as IFN gamma. Specifically, this IFN was destroyed by acid treatment, neutralized by antibody to IFN gamma and was not affected by antibodies to IFN alfa or beta.

6. Statistical Analysis

The significance of the differences in the mean values of proliferation induced by GMP or IL-2 in normal controls versus glioma-bearing patients were analyzed by the pooled or paired Student's "t" test. Differences in the ratio responders/not responders to any treatment involving PBMC of healthy donors and glioma patients were analyzed by the Fisher's exact test.

Results

1. Lymphoproliferation in Response to GMP or IL-2

PBMC from 19 glioma-bearing patients or 20 control subjects were stimulated in vitro with GMP, at the previously established optimal dose (50 µg/ml) to induce lymphoproliferation, then lymphoproliferation assayed by ^{3}H-Thymidine incorporation. Other PBMC cultures were treated with IL-2 (20 IU/ml). In both cases, ^{3}H-Thymidine incorporation was assayed on day seven of culture. The results are reported in Table 1. PBMC of all healthy donors extensively proliferated in response to either GMP or IL-2, although the individual response varied with the single donor tested and was, as average, higher with IL-2 than GMP ($p < 0.05$, paired "t" test). Lymphocytes of glioma-bearing patients also proliferated on addition of GMP or IL-2, the difference in the mean proliferation induced by each stimulant being, in this case, not statistically significant ($p > 0.05$ paired "t" test). On a yes/no basis, the response was seen to be positive in 12 out of 13 PBMC cultures exposed to IL-2 and in 16 out of 19 cultures exposed to GMP ($p > 0.35$, Fisher's exact test). Comparing the intensity of proliferative response between patients and controls, we observed that both antigen- and IL-2-induced proliferations were higher in healthy subjects; however the difference was never such as to reject the null hypothesis at a probability level of 0.05 (not significant). Nor was any statistically significant difference observed between healthy donors and patients (whatever the stimulant) assuming a yes or no proliferative response and employing the Fisher's exact test.

2. Lymphokine Production in Response to GMP

In most cases, microbial antigen induced proliferation is accompanied by production and release of soluble mediators of immunity, such as IL-2 or IFN gamma, which may play a fundamental role in the activation of cytotoxic antitumour immunoeffectors. It was of interest, therefore, to examine whether, first, GMP-stimulated cultures were able to produce the afore mentioned lymphokines and, second, whether there was any difference between tested patients and controls in this aspect. Table 2 shows the data obtained with our groups of subjects. Significant IL-2 production with respect to the unstimulated counterpart was detected in 5 out of 14 PBMC of controls and only in 2 cultures out of 12 tested in the glioma-bearing patient population ($p > 0.20$, Fisher's exact test). On the other hand, PBMC of all normal subjects (twelve) tested, produced elevated IFN gamma levels (ranging from 30 to 1,000 IU), following GMP stimulation whereas only 6 PBMC cultures of glioma-bearing patients produced IFN gamma in a sample of 12 tested patients ($p = 0.0068$, Fisher's exact test). It is also relevant that some of the highest IFN gamma producers following GMP stimulation in vitro (pts 8, 10, 13) were not among those who most extensively proliferated in response to antigen stimulation.

Table 1. *PBMC Proliferation Induced by GMP or IL-2 in Glioma-Bearing Patients and in Age and Sex Matched Normal Subjects*[a]

Patients	PBMC proliferation after stimulation with[b]:		Controls	PBMC Proliferation after stimulation with[b]:	
	GMP	IL-2		GMP	IL-2
1.	40.4 ± 0.7	23.9 ± 0.8	1.	14.1 ± 2.9	27.1 ± 0.7
2.	92.0 ± 11.9	66.2 ± 0.9	2.	7.8 ± 0.1	78.6 ± 1.6
3.	5.2 ± 2.2	4.4 ± 1.5	3.	14.2 ± 1.1	11.0 ± 0.6
4.	9.4 ± 1.7	21.4 ± 5.0	4.	28.9 ± 1.5	27.1 ± 0.2
5.	31.2 ± 1.5	17.1 ± 4.6	5.	2.4 ± 0.8	10.7 ± 0.4
6.	1.4 ± 0.1	N.D.[c]	6.	56.5 ± 1.9	11.6 ± 0.2
7.	18.4 ± 4.8	10.5 ± 3.4	7.	12.8 ± 0.8	17.3 ± 0.3
8.	6.7 ± 0.8	0.5 ± 0.7	8.	17.5 ± 1.9	17.0 ± 0.3
9.	0.7 ± 0.2	23.9 ± 9.0	9.	8.1 ± 2.9	61.3 ± 1.4
10.	0.5 ± 0.1	10.2 ± 0.3	10.	5.4 ± 0.1	63.1 ± 5.2
11.	1.6 ± 0.3	115.7 ± 1.8	11.	28.2 ± 0.9	N.D.
12.	1.6 ± 0.1	7.6 ± 0.3	12.	25.4 ± 4.1	29.1 ± 5.2
13.	2.3 ± 0.1	14.8 ± 0.9	13.	14.2 ± 1.4	57.9 ± 2.4
14.	18.0 ± 0.9	16.3 ± 0.6	14.	35.2 ± 0.8	36.9 ± 5.3
15.	5.5 ± 0.6	N.D.	15.	48.6 ± 6.4	29.8 ± 7.7
16.	15.9 ± 3.3	N.D.	16.	32.2 ± 1.2	69.5 ± 11.8
17.	2.3 ± 0.4	N.D.	17.	44.2 ± 5.9	12.0 ± 3.1
18.	36.4 ± 9.3	N.D.	18.	43.8 ± 10.0	58.1 ± 4.5
19.	29.4 ± 1.2	N.D.	19.	26.6 ± 2.4	43.5 ± 7.0
			20.	35.1 ± 0.8	36.9 ± 5.3
Mean ± SD	16.5 ± 22.1	25.6 ± 31.5	Mean ± SD	25.1 ± 15.5	36.7 ± 22.0

[a] PBMC (2×10^5/well) were cultured for seven days with GMP (50 μg/ml) or IL-2 (20 IU).
[b] Measured by ^{3}H-Thymidine incorporation, expressed as cpm $\times 10^{-3}$ ± SE.
[c] Not done.

Table 2. *GMP-Induced Interferon Gamma and IL-2 Production in PBMC of Glioma-Bearing Patients and in Normal Subjects*[a]

Patients	Unstimulated		GMP-stimulated		Controls	Unstimulated		GMP-stimulated	
	IFN	IL-2	IFN	IL-2		IFN	IL-2	IFN	IL-2
4.	< 3	20	3	17	1.	< 3	5	30	8
6.	< 3	5	30	5	2.	< 3	1	100	4
7.	3	3	30	4	3.	< 3	4	30	5
8.	< 3	12	10	11	4.	10	5	100	6
9.	< 3	10	< 3	7	5.	3	11	30	24
10.	3	8	100	10	6.	< 3	20	100	15
12.	< 3	36	< 3	42	7.	< 3	12	30	14
13.	10	27	300	26	8.	10	8.	300	8
14.	3	6	300	61	9.	< 3	10	30	18
16.	< 3	7	< 3	5	10.	< 3	12	1000	14
18.	< 3	2	< 3	16	11.	< 3	9	1000	26
19.	< 3	3	< 3	3	12.	< 3	3	300	11
					13.	N.D.	6	N.D.	16
					14.	N.D.	2	N.D.	24

[a] PBMC (2×10^5/well) were cultured for seven days with GMP (50 μg/ml) and supernatants were assayed for presence and type of IFN and IL-2 produced, as described in Materials and Methods.
N.D. = Not done.

Discussion

Cell-mediated immune responses are generally believed to be impaired in glioma-bearing patients, although results which challenge such an assumption do indeed exist[12, 5]. In this study, we have examined the proliferative response of PBMC of glioma-bearing patients to stimulation with the dominant antigen of C. albicans, a human indigenous microorganism which is capable of inducing high level proliferative responses in all normal, healthy subjects[9]. This response has also been shown to be paralleled by production of IFN gamma and generation of cytotoxic cells, probably in the NK-LAK lineage, which prove highly effective against tumour targets in vitro[13]. A comparison between GMP and IL-2 induced proliferation has also been attempted here, in view of the well known importance of IL-2 in the control of proliferation and cell-mediated cytotoxicity generation. The results we obtained do not support the idea that glioma-bearing patients have a low lymphocyte proliferative potential as compared to healthy subjects. While it is true that the average lymphoproliferation values, whatever the inducer, was generally higher in PBMC of healthy subjects as compared to PBMC of glioma patients, the difference never reached a statistically significant level. It should be noted here that the extent of ^{3}H-Thymidine incorporation is affected by a number of variables and the overall quantitative measurements show a high standard deviation. In view of this, the simple assessment of positive versus negative responses may be more reliable given that a sufficiently high threshold of positivity has been assumed. By this approach, and assuming a not parametric test for assessment of statistical significance, it was seen that patients did not differ significantly from healthy subjects in the occurrence of positive proliferative responses to both GMP or IL-2. On the contrary, what appears to be significantly depressed in glioma patients is their capacity to produce IFN gamma and, perhaps, IL-2 in response to the antigen stimulation. This partial inability may well play a role in insufficient defense against cancer, considering the importance of both IFN gamma and IL-2 in the generation/recruitment of antitumour effectors[14]. In this context, it may be of interest to note that no or low level cytotoxicity is generated in PBMC cultures of glioma patients in response to the antigenic stimulation (Ausiello *et al.*, in preparation). Our patients were being treated with corticosteroids and anticonvulsants at the time of PBMC activation in vitro, therefore the possibility that reduced IFN-gamma production is due to underlying treatment must be considered. Nonetheless, several facts argue against medication – and in particular that with corticosteroids – explaining our results; i) lymphoproliferation was normal in our patients; ii) PBMC were washed free of patient serum before incubation, therefore the drugs would not be expected to be present to any significant concentration in our cultures; iii) the six patients whose PBMC culture produced IFN-gamma after GMP stimulation were receiving comparable (dose and duration) drug treatment at the time of testing, with respect to the unresponsive patients; iv) finally, a few patients who underwent anticonvulsants and corticosteroid treatments for CNS disorders other than glioma, did not show impaired IFN-gamma production when their PBMC were stimulated by GMP in vitro (data not shown).

Overall, these results encourage further studies aimed at dissecting the immune responses triggered by the microbial antigen in PBMC of glioma patients. This might help in outlining the immune profile of these patients and devising new immunotherapeutic approaches.

Acknowledgement

This work was supported by CNR (Italy) under special project ONCOLOGIA (grants n. 86.02680.44 and Assegnazione n. 44.85068.88) to C. Ausiello. The authors wish to thank the Hoffmann-La Roche Company for kindly providing rIL-2.

References

1. Ommaya AK (1986) Immunotherapy of gliomas: a review. Adv Neurol 15: 337–354
2. Terry WD, Rosemberg SA (eds) (1982) Immunotherapy of human cancer. Elsevier, North-Holland, New York
3. Brooks WH, Caldwell HD, Mortara RH (1974) Immune responses in patients with glioma. Surg Neurol 2: 419–423
4. Mahaley MS, Brooks WH, Roszman TL, Bigner DD, Dudka L, Richardson S (1977) Immunobiology of primary intracranial tumours. Part I: Studies of cellular and humoral general immune competence of brain-tumour patients. J Neurosurg 46: 467–479
5. Blomgren H, Blom U, Ullen H (1986) Relation between the site of primary intracranial tumours and mitogenic responses of blood lymphocytes. Cancer Immunol Immunother 21: 31–38
6. Roszman TL, Brooks WH, Elliot LH (1982) Immunobiology of primary intracranial tumours. VI Suppressor cell function and lectin-binding lymphocyte subpopulations in patients with cerebral tumours. Cancer 50: 1273–1279
7. Brooks WH, Netsky MG, Normansell DE, Horwitz DA (1972) Depressed cell-mediated immunity in patients with primary intracranial tumours. J Exp Med 136: 1631–1647
8. Newelt EA, Kikuchi K, Hill S, Lipsky P, Frenkel EP (1985) Immune responses in patients with brain tumours: factors such as anticonvulsants that may contribute to impaired cell-mediated immunity. Cancer 51: 248–255

9. Ausiello C, Spagnoli GC, Boccanera M, Casalinuovo I, Malavasi F, Casciani CU, Cassone A (1986) Proliferation of human peripheral blood mononuclear cells induced by Candida albicans and its cell wall fractions. J Med Microbiol 22: 195–202
10. Gillis S, Ferm MM, Ou W, Smith KA (1978) T cell growth factor: parameters of production and a quantitative microassay for activity. J Immunol 120: 2077–2085
11. Stanton GJ, Langford MJ, Dianzani F (1981) Virus yield reduction assay for IFN by titration of Sindbis virus haemagglutinin. Methods Enzymol 78: 351–357
12. Ullen H, Blom U, Blomgren H (1986) Natural killer activity of blood lymphocytes in patients with primary intracranial tumours. Correlation of histological tumour type and anatomical site. Eur J Cancer Clin Oncol 10: 1239–1245
13. Ausiello C, Spagnoli GC, Antonelli G, Malavasi F, Dianzani F, Casciani CU, Cassone A (1987) Generation and characterization of cytotoxic activity against tumour cell lines in human peripheral blood mononuclear cells stimulated in vitro by a glucomannan-protein preparation of Candida albicans. J Biol Reg Homeost Agents 1: 1
14. Rosemberg S (1985) Lymphokine-activated killer cells: a new approach to immunotherapy of cancer. JNCI 75: 595–603

Address for correspondence: A. Maleci, M.D., Dipartimento di Scienze Neurologiche, Sezione di Neurochirurgia, Universitá di Roma "La Sapienza", Roma, Italy.

Acta Neurochirurgica, Suppl. 43, 135–139 (1988)

Immunological and Immunohistochemical Studies in Brain Tumours

B. L. Imieliński[1], **J. Borowska-Lehman**[2], and **A. Żółtowska**[3]

[1] Department of Neurosurgery, [2] Department of Pathomorphology, [3] Immunology Laboratory, Medical Academy Gdańsk – School of Medicine, Gdańsk, Poland

Summary

The authors have studied different brain tumours with immunohistochemical technique using fluorescence and peroxidase methods. On frozen sections reactions for fibronectin (FN), prostaglandin E 2 (PGE 2), carcinoembryonic antigen (CEA) with polyclonal antibodies and for lymphocyte subpopulations B, NK and Pan-T with monoclonal antibodies have been investigated.

Both primary and metastatic tumours contained fibronectin in various intensity and showed increased reactions with PGE 2 antibodies comparing with surrounding tissues. A well developed FN positive fibrillary network in tumour tissue confirmed the presence of angiogenesis and visualised abnormalities of the vessel walls with focal disruptions.

The intracytoplasmatic location of PGE 2, more intense than in surrounding tissues revealed in this study, may indicate that tumour cells are capable of its production. Since prostaglandin inhibits lymphocyte cytotoxicity against tumour cells in vitro, this may be suggestive for its activity in tumour defence reaction.

CEA was not found in primary brain tumours by immunohistochemical procedure, either fluorescence or peroxidase method. It has been detected only in metastatic carcinomatous tumours in moderate intensity.

Immunostaining for subpopulations of lymphocytes in tumour slices with monoclonal antibodies showed only small number of cells in some specimens. The reasons for this failure are discussed.

The meaning of all these findings mentioned above for clinical purposes is considered. The application of immunohistochemical staining may be useful in diagnosing tumour of vague or doubtful microscopic appearances.

Keywords: Brain tumour; immunohistochemical methods; fibronectin; postaglandin E 2; carcinoembryonic antigen; lymphocyte subpopulations.

Introduction

The immunological phenomena are uniformly accepted in tumour pathology. In many experiments it has been established that tumour-bearing individuals initiate immune response to tumour associated neoantigens (Brooks *et al.* 1975, Wilson *et al.* 1974). These responses are demonstrable by in vitro analysis of cell mediated and humoral immune reactions. However, the majority of monoclonal antibodies which are believed to be specific for human neoplasms has reacted positively with normal cells and tumour cells of other neoplasms (Schnegg *et al.* 1985).

The experience of the present authors over the past 15 years consisted of assessment of round-cell infiltrations within the tumour as a sign of cellular immune response, as well as skin reactions on challenging with dinitrochlorobenzene (DNCB). The humoural response has been evaluated by the use of direct and indirect immunofluorescence technique (contacting the tumour with autologous and control sera), detection of immunoglobulin activity in the serum as well as autoantibodies and C 3 component of the complement (Imieliński 1980).

Recently, great attention has been paid to the possibilities of detection of various antigens in tissue slices. The techniques which have been employed until quite lately were based on immunofluorescent method, but they are being outsted by immunoperoxidase technique. Using it, one is able to localise the antigen not only in the light microscope, but also in the electron microscope (EM).

Material and Methods

The subject of investigation were specimens of tumour from 16 patients, freshly obtained from the neurosurgical operation theatre. As a control, the normal human brain tissue was used, taken from temporal lobectomies of 2 non-neoplasmatic patients.

For immunohistochemical procedure blocks of tissue were frozen in liquid nitrogen and then kept at −25 centrigrades until used.

The remaining tissue was processed in the usual way for routine microscopic studies in the light microscope.

The immunohistochemical techniques comprised fluorescence procedure and peroxidase conjugate (PC), or a peroxidase-antiperoxidase technique (PAP), as applied by Taylor and Burnes (1974). In all cases sections have been stained for fibronectin (FN), pros-

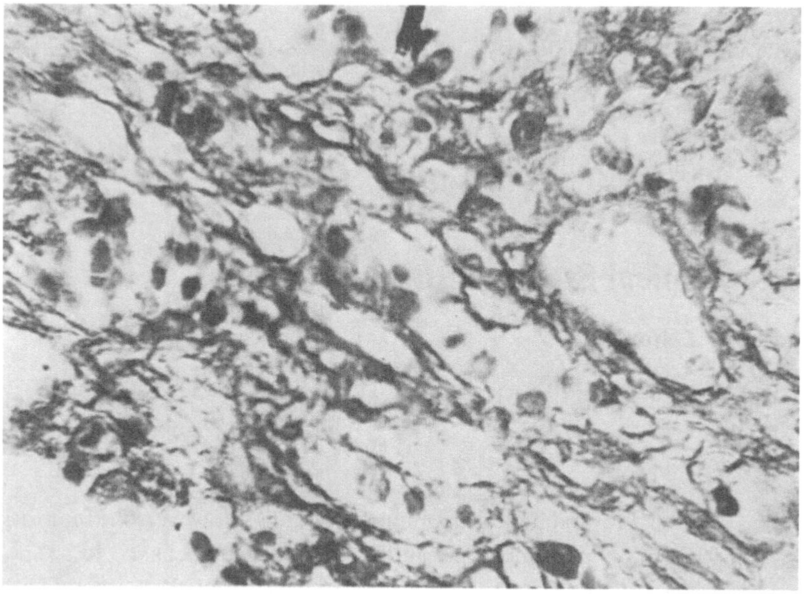

Fig. 1. Medulloblastoma, case 7. Fibronectin positive network around neoplastic foci and along the basement membrane of the capillary vessels. Frozen section. Immunoperoxidase × 320

taglandin E 2 (PGE 2), carcinoembryonic antigen (CEA) and lymphocytes B, NK and Pan-T.

Concerning PGE 2 the reaction was assessed as positive when the staining of neoplastic cells was more intense than that of the stroma. The immunohistochemical reagents which have been in use, were as follows: Behring OSIE 04/05, BMA-020, BMA 030, BMA-070, Miles 61-355-1, Dakopatts P-167, P-260, Z-113, Z-196 and Heintel A-1080.

Results

Fibronectin was detected in all the tissue specimens examined. Usually it was visible in markedly thickened vessel walls and in the basement membrane of the capillaries. Very often this basement membrane showed focal disruption by infiltrating neoplastic cells. The cell bodies of the astrocytes and their fibres were also positively stained for FN. Some of the astrocyte processes were shortened and thickened. In the papilloma of the chorioid plexus a positive FN immunoreactivity outlined the vascular structures with clear localisation in the basement membrane of the papillae. Fibronectin immunostaining in medulloblastoma yielded an easily visible network confirming the presence of capillary vessels.

In the group of mesenchymal tumours only a weak reaction for FN was found over the surface of meningioma cells. A similar distribution was found in the haemangioblastoma, however it was more limited to

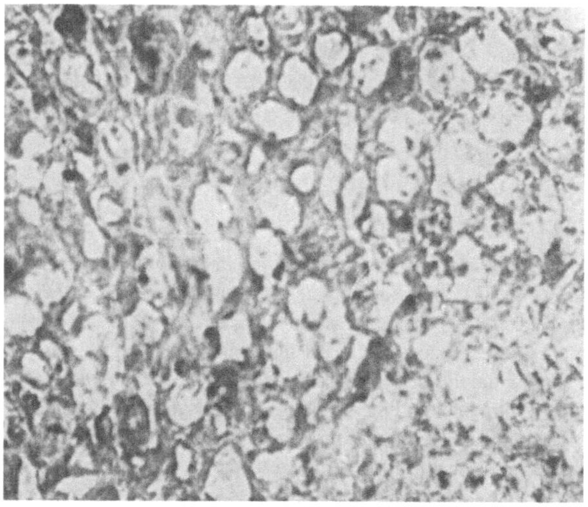

Fig. 2. Astrocytoma, case 2. Cytoplasmatic deposits of PGE_2. Frozen section. Immunoperoxidase × 320

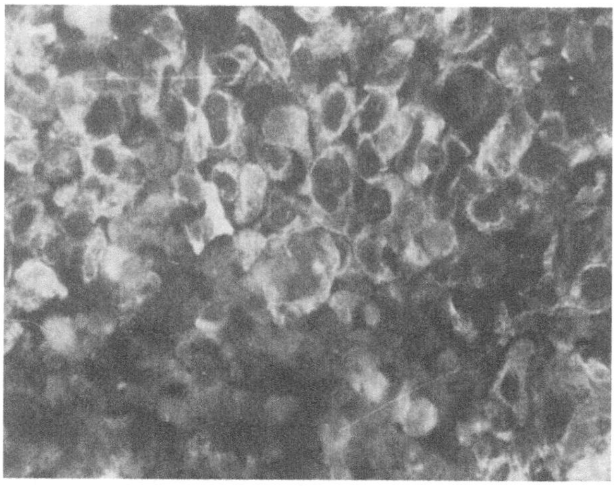

Fig. 3. Adenocarcinoma, case 14. Cytoplasmatic fluorescence of neoplastic cells. Rabbit anti-PGE_2 and goat anti-rabbit FITC × 320

Table 1. *Results of Immunohistochemical Studies in Brain Tumours*

No.	Age	Sex	Histology	Location	Immunohistochemical findings					
					FN	PGE 2	CEA	Lymphocytes		
								Pan-T	B	NK
			Neuroepithelial tumours							
1.	53	m	Astrocytoma III	supratentorial	+++	+	—	—	—	—
2.	36	m	Astrocytoma III	supratentorial	++	+	—	—	±	—
3.	58	m	Astro-oligoden droglioma II	supratentorial	+	—	—	—	—	—
4.	12	m	Ependymoma II	infratentorial	+	+	—	—	—	—
5.	22	f	Plexuspapilloma	supratentorial	+++	++	—	—	—	—
6.	64	f	Glioblastoma multiforme	supratentorial	++	+	—	—	—	—
7.	12	m	Medulloblastoma	infratentorial	+++	±	—	—	+	—
			Mesenchymal tumours							
8.	62	f	Fibromatous meningioma	supratentorial	+++	+	—	—	—	—
9.	53	m	Psammomatous meningioma	supratentorial	++	—	—	—	—	—
10.	3	m	Haemangioblastoma-tous meningioma	supratentorial	++	—	—	—	—	—
11.	23	f	Haemangioblastoma	supratentorial	+++	+	—	±	±	±
			Metastatic tumours							
12.	68	f	Macrocellular carcinoma	supratentorial	+++	+	±	±	—	—
13.	47	m	Planocellular carcinoma	supratentorial	+	+	±	±	—	—
14.	40	m	Adenocarcinoma	infratentorial	+++	+	+	±	±	±
15.	38	m	Malignant lymphoma	supratentorial	+++	—	—	—	±	—
16.	32	m	Malignant melanoma	infratentorial	+	—	—	—	—	—
			Controls							
17.	27	m	Normal	temporal lobe	+[x]	—	—	—	—	—
18.	16	f	Normal	temporal lobe	+[x]	—	—	—	—	—

Legend:
— negative result,
± trace,
+, ++, +++ various degrees of positive reactions,
[x] FN positive only in normal vessels.

the vessel walls. In the last group, comprising metastatic tumours, FN was present not only in the vessel walls but mainly around the neoplastic foci in the form of FN-positive fibrillary network.

Prostaglandin E 2 was detected only in some tissue specimens; in tumours originating from the astrocytic line PGE 2 was found in the form of cytoplasmatic deposits.

Immunoreactive PGE 2 was also disclosed in a case of plexus-papilloma, fibromatous meningioma and haemangioblastoma. In specimens of metastatic carcinomas localisation of PGE 2 was also cytoplasmatic.

Only metastatic carcinomas were positive for CEA. Monoclonal antibody used for detection of lymphocytes revealed immunoreactivity of single cells in only a few specimens.

The results of immunostaining of all the brain tumours studied are summarised in Table 1.

Discussion

Fibronectin is a substance active in mediation of cell adhesion, participating in production of intracellular substance and tissue organisation, regulation of cell movements and mediation of macrophage interaction with collagen and fibrin (Mosesson and Amrani 1980). Being produced by fibroblasts but also by astroglial cells (Vaheri *et al.* 1976) and endothelial cells it may be manufactured by various epithelial cells including those deriving from neoplastic clones. The ability of macrophages to synthesise and excrete fibronectin has been demonstrated (Bevilaqua *et al.* 1980). It is believed that the presence of cellular fibronectin is associated with cell differentiation and organogenesis (Pearlstein *et al.* 1980). FN is widely found in different tissues but primarily in the matrix and in the loose connective tissue as well as in various basement membranes.

The results we obtained with FN immunostaining in the tested tumours provided useful information on the relationship between neoplastic cells and their surrounding – extracellular matrix. In the normal brain tissue we found only scanty FN positive vessel walls, whereas in neoplasms an intricate fibronectin network was disclosed inside and around the neoplastic foci. Moreover, a well developed FN positive fibrillary network confirms the presence of angiogenesis in the residual tissue surrounding the centre of neoplastic growth. FN immunostaining seems to be very useful in visualising the abnormality of the vessel and its focal disruption.

Prostaglandins are a group of substances secreted by most animal tissues and exhibiting a bewildering array of actions. They have been detected in several human and experimental tumours (Humes and Strausser 1974). Besides their wide range of various activity they are engaged in inhibition of proliferation of malignant tumour cells such as melanoma (Simmet and Jaffe 1983), human leukaemia (Fukishima *et al.* 1982), and different other cell lines (Sakai and Yamaguchi 1984). Westphal *et al.* (1986) in their studies on glial tissue culture found that PGD 2 has suppressive action on DNA synthesis in human gliomas. This could be achieved only to some extent by PGE 2. The role of the latter seems to be different. Żółtowska (1987) in her series of 25 tumours found PGE 2 widely in the cells of all of 7 breast cancers, in 3 out of 5 malignant melanomas and in 1 out of 4 bronchial carcinomas. Earlier reports about increase of PGE 2 in human tumours were based only on tissue homogenate studies (Attalah *et al.* 1984).

The technique applied in present report may support the suggestion of other authors that tumour cells are capable of PGE 2 production. Since prostaglandins inhibit lymphocytic cytotoxicity against tumour cells in vitro (Oven *et al.* 1980) it seems that PGE 2 production may represent a tumour-versus-host reaction by subverting the cellular immune response elicited against lymphocytes. It is suggested that PGE 2 as one of those active substances produced by neoplastic cells may be involved in tumour defence mechanisms and angiogenesis (Attalah *et al.* 1984). The present study indicates that PGE 2 may be produced also by brain tumour cells. The role of prostaglandins in brain tumour pathology requires further clinical and immunomorphological studies. It is suggested that prostaglandins may play a role in tumour spread and growth in bones and that drugs which inhibit their synthesis may be of therapeutic value in bone metastising tumours (Robertson *et al.* 1976, Seyberth *et al.* 1975).

Carcinoembryonic antigen is one of best known tumour markers. Distinct positive reactions has been found in various types of carcinoma. Bręborowicz *et al.* (1985) in a series of 44 lung cancer patients detected CEA by direct immunoperoxidase technique in 38 (86%). Miyake *et al.* (1979) investigated plasma and CSF for CEA in 64 patients with brain tumours. Among 14 patients with elevated titres a significantly increased level was found in cases of metastatic brain tumours. Their findings suggest that CEA levels may be of value in the differential diagnosis of primary and metastatic brain tumours and useful in assessment of these patients after treatment. Utilisation of CEA in assessment of patients with colonic and rectal carcinoma has been also reported (Brooks *et al.* 1975, Wilson *et al.* 1974). In our investigations we were not able to confirm the presence of CEA in primary brain tumours by immunohistochemical method. Detection of CEA in plasma and CSF in patients with brain tumours (Miyake *et al.* 1979) results from the application of more sensitive techniques for this purpose, which is radio-immune-assay.

The disappointing result concerning detection of *subpopulations of lymphocytes* by monoclonal antibodies may be explained by the small size of specimens used for investigations in frozen sections. In routine paraffin technique with the more sizable blocks generally in use, regional accumulation of lymphocytes in some of the tumours has been visualised.

On the other hand, recent studies suggest that the brain tumour, like other tumours, is capable of eliciting the depression of lymphocyte activity. It is known that the immune defence system required the participation of macrophages which are involved in production of interleukinin 1. Has this lack of cooperation contributed to the failure in revealing lymphocyte markers by monoclonal antibodies in our studies? This requires further investigations.

Attempts to find tumour markers have been widely undertaken but the specificity of the reactions is often disputable. At the present stage some of them may be of great value in differentiating microscopic appearances of tumours. Further advances in immunology may result in aids to diagnosing the occurrence of malignancy and even monitoring the rate of growth, but this will require continuous clinical and laboratory research.

References

1. Attalah A, Lee JB, Ambrus JL, Ambrus CM, Karakousis C, Takita H (1984) Prostaglandin E 2 production by human tumours. Defense mechanism against the host? A preliminary report. Res Com Chem Path Pharm 43: 195–201
2. Bevilaqua M, Mosesson MW, Bianco C (1980) Fibronectin receptors on mononuclear phagocytes. Clin Res 28: 340 A–340 A
3. Bręborowicz D, Gładysz K, Słowik-Gabryelska A, Zengteler G, Bręborowicz J (1985) Carcinoembryonic antigen in lung cancer – II Immunohistochemical study (Polish) Immun Pol 10: 345–354
4. Brooks WH, Netsky MG, Levine JE (1975) Immunity and tumours of the nervous system. Surg Neurol 3: 184–186
5. Fukushima M, Kato T, Ueda R, Ota K, Narumiya S, Hayashi O (1982) Prostaglandin D 2 a potential antineoplastic agent. Biochem Biophys Res Commun 105: 956–965
6. Humes TL, Strausser HR (1974) Prostaglandins and cyclic nucleotides in Moloney sarcoma tumours. Prostaglandins 5: 183–196
7. Imieliński BL (1980) Intracranial tumours of childhood. Etiopathogenic, diagnostic and therapeutic considerations; perspectives of non-conventional treatment. Garyounis MJ 3: 187–198
8. Miyake A, Yamashita M, Kitamura K, Ishigami F (1979) Carcinoembryonic antigen (CEA) levels in patients with brain tumours. Acta Neurochir (Wien) 46: 53–57
9. Mosesson MW, Amrani DL (1980) The structure and biologic activities of plasma fibronectin. Blood 56: 145–158
10. Pearlstein E, Gold LJ, Garcia-Pardo A (1980) Fibronectin: a review of its structure and biological activity. Mol Cell Biochem 29: 103–128
11. Robertson RP, Baylink DJ, Metz SA, Cumonings KB (1976) Plasma prostaglandin E in patients with cancer with and without hypercalcaemia. J Clin Endocrinol Metab 43: 1330–1335
12. Sakai T, Yamaguchi N (1984) Prostaglandin D 2 inhibits the proliferation of human malignant tumour cells. Prostaglandins 27: 17–27
13. Schnegg JF, Diserens AC, Carrel S, Accola RS, De Tribolet N (1981) Human glioma associated antigens detected by monoclonal antibodies. Cancer Res 41: 1209–1213
14. Seyberth HW, Segre GV, Morgan JL, Sweetman BJ, Potts JT Jr, Oates JA (1975) Prostaglandins as mediators of hypercalcaemia associated with certain types of cancer. N Engl J Med 293: 1278–1283
15. Simmet T, Jaffe BM (1983) Inhibition of B 16 melanoma growth in vitro by prostaglandin D 2. Prostaglandins 25: 47–54
16. Taylor CR, Burnes J (1974) The demonstration of plasma cells and other immunoglobulin containing cells in formalin fixed paraffin embedded tissues using peroxidase-labelled antibody. J Clin Path 27: 14–20
17. Vaheri A, Ruoslahti E, Westermark B, Ponten J (1976) A common cell-type specific surface antigen in cultured human glial cells and fibroblasts: loss in malignant cells. J Exp Med 143: 64–72
18. Westphal M, Neuss M, Herrmann HD (1986) Prostaglandins: anti-proliferative effect of PGD 2 on cultured human glioma. Acta Neurochir (Wien) 83: 56–61
19. Wilson C, Normansell DE, Netsky MG (1974) Antigenicity of foetal and adult cerebral tissue of man. Neurology 24: 761–766
20. Żółtowska A (1987) Human tumours and prostaglandin E 2 (in press)

Address for correspondence: Dr. Brunon L. Imieliński, Department of Neurosurgery, Medical Academy Gdańsk – School of Medicine, Debinki 7, 80-211 Gdańsk, Poland.

Acta Neurochirurgica, Suppl. 43, 140–144 (1988)

The Investigation of Experimental Brain Tumours Using ^{31}P-MRS and ^{1}H-MRI

K. Hirakawa, Sh. Naruse, T. Higuchi, Y. Horikawa, Ch. Tanaka, and **T. Ebisu**

Department of Neurosurgery, Kyoto Prefectural University of Medicine, Kyoto, Japan

Summary

In vivo ^{31}P-magnetic resonance spectra (MRS) were obtained by the surface coil method from rat glioma, human glioblastoma, and human neuroblastoma inoculated subcutaneously in CD Fisher rats and hamsters, and the effects of chemotherapy, photoradiation therapy, and radiofrequency hyperthermia as well as ^{60}Co-irradiation were evaluated by sequentially observing spectral changes. In the ^{31}P-spectra of tumour tissue, the nucleoside triphosphate (NTP), phosphomonoesters (PME) peaks were high and the phosphocreatine peak was low compared to those of normal brain. When the antitumour agents were given and were effective, NTP peaks decreased, and inorganic phosphate increased remarkably within several hours after the treatment. ^{1}H-magnetic resonance imaging (MRI) were also obtained in some cases. Necrotic regions was detected by the ^{1}H-MRI as image changes which appeared later than those detected by MRS. It proved practical to monitor the effect of therapy by employing either ^{31}P-MRS or ^{1}H-MRI. However, the image changes which demonstrated the effect of the therapy used closely resembled those changes which occurred with the onset of necrosis in tumour tissue during tumour growth. Several problems for future application of these techniques to human brain tumours are also mentioned.

Keywords: ^{31}P-MRS; ^{1}H-MRI; experimental brain tumour; antitumour therapy.

Introduction

For the last several years, we have been investigating the energy metabolism of brain tumours in living rats and hamsters by using ^{31}P-magnetic resonance spectroscopy (MRS). However, concomitant morphological changes have been obtained histopathologically or at most by ^{1}H-magnetic resonance imaging (MRI) that was performed separately from ^{31}P-MRS. Modern technological advances have made it possible to detect the morphological changes by ^{1}H-MRI in the same animal at same time as ^{31}P-MRS measurement with the same spectrometer. In this article, we present some of the results of combined ^{31}P-MRS and ^{1}H-MRI measurements and summarize the results of various kinds of ^{31}P-MRS measurements which we have made in the past to evaluate the effects of therapies on malignant brain tumours. Some technical problems in MRS cases are also discussed.

Materials and Methods

Experimental Animals

A total of 150 CD Fisher rats and 50 hamsters was used for the experiments. A suspension of 10^6–10^7 cultured cells of rat glioma (EA 285) in the rats and human glioblastoma (KHG 2) or human neuroblastoma in the hamsters with pre- and posttreatment of antilymphocyte serum was inoculated subcutaneously in the lumbar region of the animals. When the tumours grew to over 1.5 cm in diameter, the animals were used in the experiments, which were divided into four groups: chemotherapy, photoradiation, hyperthermia, and ^{60}Co-irradiation. The animals in each group were divided into appropriate subgroups of about 10 rats or hamsters, including controls, according to the experimental plans and the treatment dose. In each group in vivo ^{31}P-MRS of the tumours were measured serially before and after the treatments using the surface coil method described below. In some animals in each group, ^{1}H-MRI were also obtained before and after the treatments using the same apparatus. Tumour masses were removed at appropriate intervals for histological examination.

Therapy Schemes

Since the details of our treatment for each of the four group have already appeared in separate publications[4, 5, 6], only an extremely short description is appropriate. In the chemotherapy group, cyclophosphamide (30 mg/kg, 300 mg/kg body weight), vincristine (0.5 mg/kg, 2 mg/kg body weight), or methotrexate (15 mg/kg, 100 mg/kg body weight) was used to determine and compare the effect of each drug with large and low doses[5]. In the photoradiation group, the effect of haematoporphyrine derivatives (HpD) was measured according to the dose[6]. Radiofrequency hyperthermia can be generated by using the same surface coil in the same spectrometer. A radiofrequency pulse with the same resonant frequency of phosphorus nucleus was applied continuously to the tumour at various power levels (3–5 Watts) for 60 minutes. Tumour temperature at a depth of 3 mm under the skin was 43.5 °C. In this group, variations in the effect of therapy obtained by using these different levels of radiofrequency power were compared[4]. In the ^{60}Co-irradiation group, a single 50 Gy dose was employed.

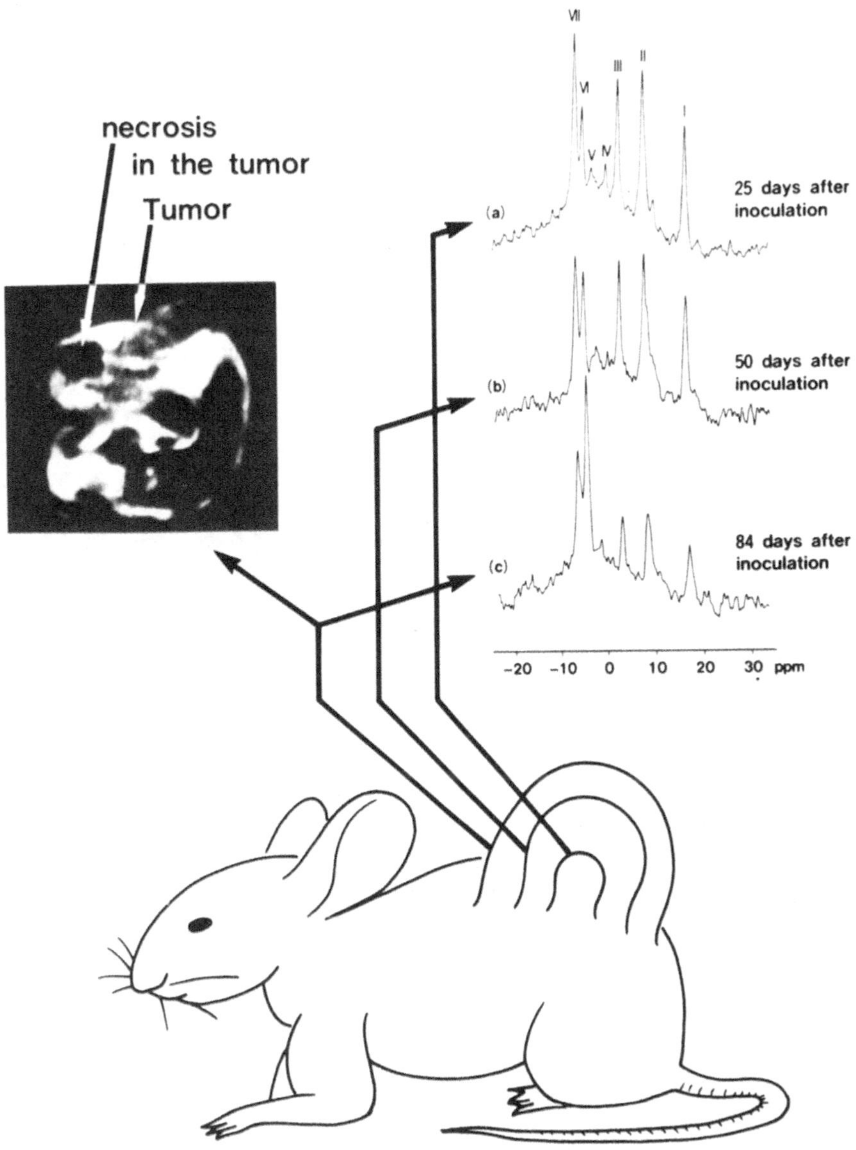

Fig. 1. ^{31}P-MRS and ^{1}H-MRI of rat glioma in three different stages of tumour growth. Sequential changes in ^{31}P-MRS during the natural growth of the implanted tumour. A gradual increase in the Pi peak and a gradual decrease in the NTP peaks were demonstrated. Necrotic lesions in the large sized tumours were clearly shown by ^{1}H-MRI

NMR Measurements

In vivo ^{31}P-MRS measurements were performed using the surface coil method usually with a SCM-200 spectrometer (JEOL, Japan: 9 cm bore, 4.7 Tesla magnet, 80.75 MHz for ^{31}P) and in some cases, for chemotherapy, with a TMR-32 spectrometer (Oxford Research System, United Kingdom: 30 cm bore, 1.8 Tesla magnet, 32.5 MHz for ^{31}P) as we have previously published. After sodium pentobarbital anaesthesia, the subject animal was put into the MR apparatus so that the tumour mass was located on a four-turn surface coil of 1.1 cm in diameter, and the resonant signal of ^{31}P was obtained from the tumour. Each of the ^{31}P-MRS measurements took about 15 minutes. From the ^{31}P-MRS obtained, tissue pH was calculated from the chemical shift of the inorganic phosphorus peak.

^{1}H-MRS were obtained with the SCM-200 spectrometer. Spin echo (SE) images with short (36 msec) and long (96 msec) echo time (Te) and inversion recovery (IR) images with a 300 msec interpulse interval (Td) were obtained by the ordinary two-dimensional Fourier transformation (2DFT) method. The data matrix numbered 256 × 256. Each image measurement took a total of about 10 minutes.

MR measurements were continued intermittently for about 7 days.

Results

^{31}P-MRS of the Tumour

Regardless of the origin of the tumour or of the histological type, the pattern of ^{31}P-MRS of the tumours was basically the same. The typical spectrum is shown in Fig. 1, where several peaks were well identified. They are, in the order of high resonant frequency: β-phosphate of nucleoside triphosphate (NTP), α-phosphate of NTP and nucleoside diphosphate (NDP) (including nicotinamide adenine dinucleotide NAD and NADH),

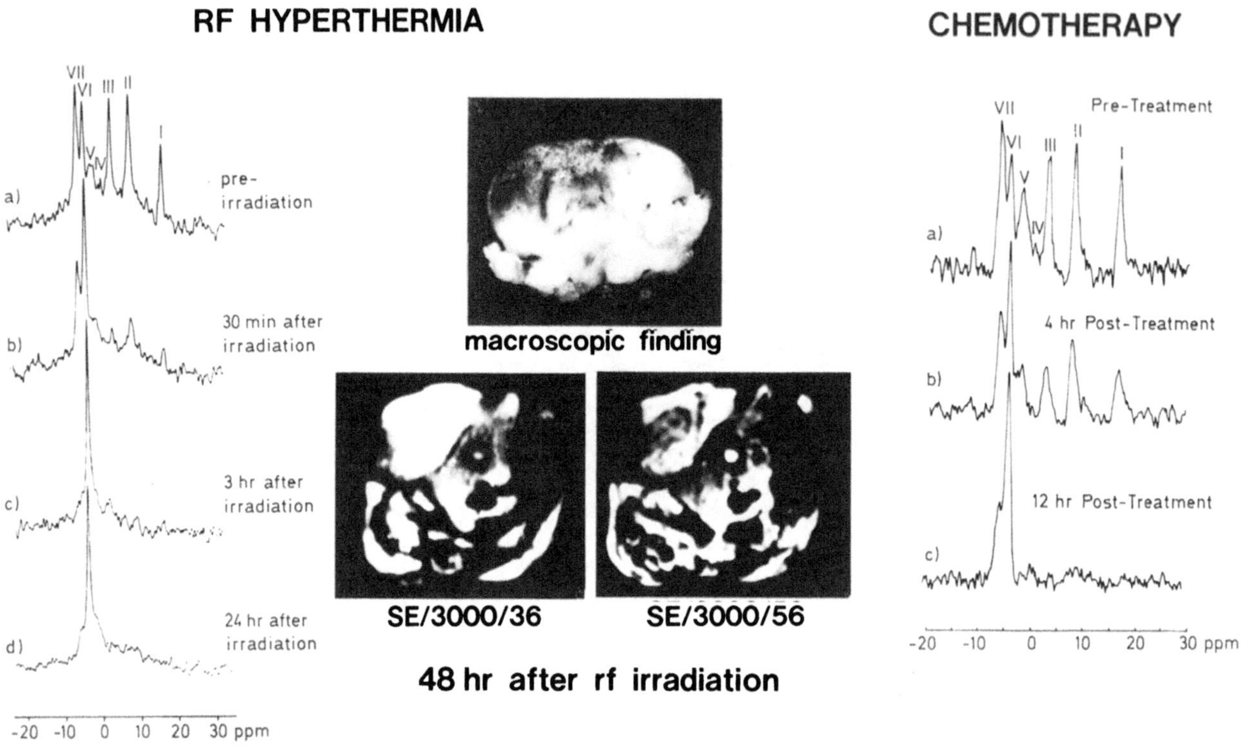

Fig. 2. Changes in ^{31}P-MRS and ^{1}H-MRI induced by the rf hyperthermia and chemotherapy. Left side: RF hyperthermia. Changes in the ^{31}P-MRS after rf irradiation at a power of 5 W. The spectrum changed to a Pi dominant pattern immediately. Macroscopic findings showed necrosis in the region from the surface to a depth of 5 mm in the tumour 48 hours after rf irradiation. This region was clearly demonstrated on ^{1}H-MRI. Right side: Chemotherapy. Changes in the ^{31}P-MRS after the administration of a high dose of vincristine. The spectrum changed to a Pi dominant pattern several hours after administration

γ-phosphate of NTP and β-phosphate of NDP, phosphocreatine (PCr), phosphodiesters (PDE), inorganic phosphate (Pi), and phosphomonoesters (PME). The spectrum pattern for tumours differed from that of the normal brain tissue. In the tumour, NTPs and PME peaks were characteristically large and Pi and PCr peaks could hardly be detected. The tissue pH was around 7.15.

^{31}P-MRS and ^{1}H-MRI During the Tumour Growth

These patterns underwent a change as the tumour growth advanced. NTP peaks gradually decreased with a concomitant increase of Pi, finally resulting in a Pi-dominant pattern (Fig. 1). Tissue pH at this stage was 7.07. A typical tumour pattern with large NTPs and a PME was seen until about 4 weeks after the tumour inoculation for the rat glioma and until about 2 weeks for human glioblastoma and neuroblastoma. A Pi dominant pattern became prominent about 12 weeks after inoculation in rat glioma cases. Histologically, these changes coincided with tissue necrosis in the center of the tumour seen in later stages of tumour growth.

When ^{1}H-MRI were applied to tumours with advanced necrosis, the necrotic portions were well identified and increased in size as time progressed (Fig. 1).

^{31}P-MRS and ^{1}H-MRI Changes Caused by Antitumour Agents

When any of the antitumour agents listed above were applied to the animals and was effective, the ^{31}P-MRS pattern changed (Fig. 2): characteristically, the NTP peaks decreased and the Pi peak increased. The PME peak also decreased. These changes did not occur in normal organs such as the muscle and brain.

When chemotherapeutic agents were applied, spectra changes occurred in cases with high doses, but not with low doses. At low doses, however, after repeated injections of the drug, spectrum changes occurred just as in high-dose administration. These changes occurred within 6–10 hours after the treatments and lasted at least for 2–6 days. Tissue pH was 7.09. When dramatic spectra changes occurred, tissue necrosis was histologically obvious.

When photoradiation was applied, the spectrum began to change within 15 minutes and showed a Pi dominant pattern with a complete loss of NTPs within 60 minutes. This effect was recognized with the injection of HpD at a dose of 10 mg/kg, but not of 5 mg/kg. Histological examination 3 days after the photoradiation showed extensive necrosis in the tumour tissue. However, the spectrum pattern reverted to a characteristic tumour pattern within 10–12 days with the regrowth of the rest of the tumour.

The spectra changes which accompanied hyperthermia occurred within 30 minutes almost the same as those mentioned above when the radiofrequency power was more than 5 W. A Pi dominant pattern became prominent 3 hours later with a tissue pH of 6.96 and lasted for at least 7 days. There was no remarkable change in the ^{1}H-MRI of the tumour up to 2–3 hours after the induction of hyperthermia even though ^{31}P-MRS had already began to demonstrate the Pi dominant pattern. However, tissue change became apparent 24–48 hours later as a high-intensity area in long SE and IR images.

As for the ^{60}Co-irradiation, no definite conclusion has been reached as yet, but it can be said that the effect of irradiation varied from tumour to tumour depending on the histological type. When the irradiation was very effective, a Pi dominant pattern appeared within 1 or 2 days, and in some cases within several hours. An increase in the height of the PCr peak was detected temporarily within 1 hour in some cases. These experiments have not yet been completed and are still in progress.

Discussion

Noticeable common feature were found in the ^{31}P-MRS in all of the tumour tissues in human and rat brain tumours: large NTPs, large PME and extremely small PCr peaks. When the tumour growth advanced to a stage where necrosis developed in the tumour, the spectrum pattern changed. NTP and PME peaks decreased with a concomitant increase in the Pi peak, and finally a Pi dominant pattern appeared. These changes may occur spontaneously in the natural course of tumour growth, but at the same time, they may occur as a response to the antitumour agents. Therefore, in planning experiments, one should be very carefuly not to misinterpret the two.

Through these experiments, we found it possible to detect spectrum changes which accompanied various kinds of therapies. However, it is necessary to determine the meaning of these phenomena. NTP peaks are the composites of ATP, UTP, GTP, and CTP[2]. They not only represent the energy metabolism of the tumour tissue, but also have something to do with the cell growth mechanism. For instance, GTP comprised 13% of the total NTP and correlates to *ras* oncogene[7], but it plays a relatively minor role in the cell growth of brain tumour. It is not yet known how N-*myc* participates. In the same way, PME represents the phospholipide of cell membranes[2]. But we know nothing about it in connection with cell growth. Therefore, it is still difficult to interpret spectra change in detail.

PH in the tumour tissue was shown to be acidic.

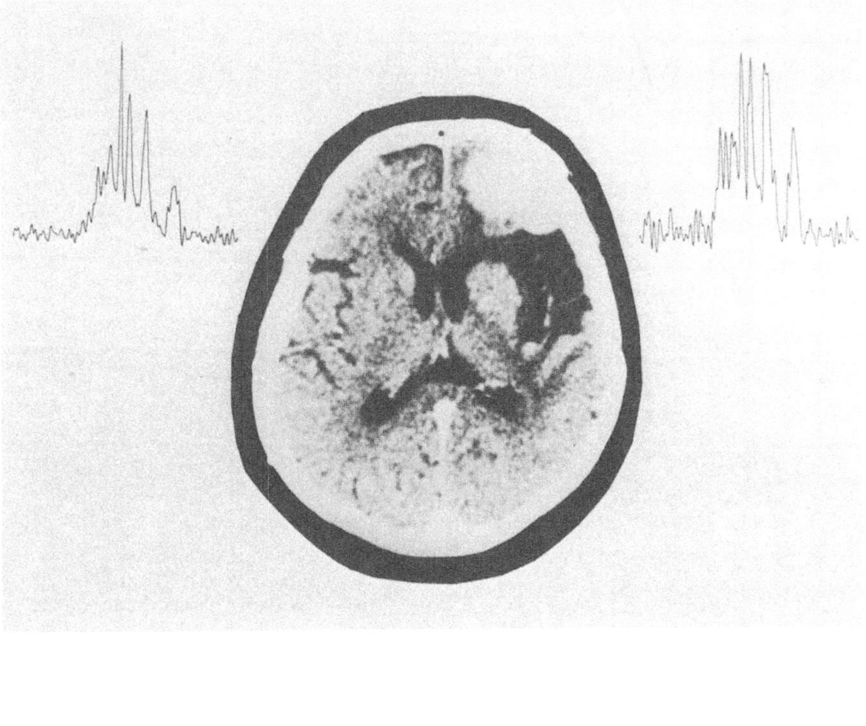

Fig. 3. ^{31}P-MRS obtained from a human brain tumour (meningioma) in vivo using a surface coil. The spectrum obtained from the tumour side showed a high PME peak compared with the spectrum from the normal side

However, recent positron emission tomography studies favoured evidence showing pH to be alkalotic. This is also one of the points to be solved in connection with tumour metabolism.

There has been a recent tendancy to use ^{1}H-MRI together with MRS to detect morphological changes. Although MRI clearly showed the presence and the extent of tissue changes, it took a couple of days to detect the phenomena, while with MRS it was possible to detect them even minutes or hours after the insult. In that sense, MRS is more sensitive and can be valuable in detecting the phenomenon in the tissue earlier than possible with MRI.

Thus, ^{31}P-MRS measurements can be applied clinically for foresee non-invasively whether or not a therapy will be effective with a particular brain tumour. Recently, we have obtained ^{31}P-MRS from human brain tumours using the surface coil method (Fig. 3). This method has its merits and demerits. Signals can only be obtained from fairly close to the surface and the proper volume measured is uncertain. To overcome this problem, again new techniques arose such as Dress[1], Spars[3] and Isis[8] method, which made it possible to obtain ^{31}P-MRS signals from greater depth in the brain. Hopefully, in the near future ^{31}P-MRS will be obtained from human brain tumours as easily as it is now for animals.

References

1. Bottomley PA, Foster TB, Darrow RD (1984) Depth-resolved surface-coil spectroscopy (DRESS) for in vivo ^{1}H, ^{31}P, and ^{13}C NMR. J Magn Reson 59: 338–342
2. Evanochko WT, Ng TC, Glickson JD (1984) Application of in vivo NMR spectroscopy to cancer. Magn Reson Med 1: 508–534
3. Luyten PR, Marien JH, Sijtsma B, Hollander JA (1986) Solvent-suppressed spatially resolved spectroscopy. An approach to high-resolution NMR on a whole-body MR system. J Magn Reson 67: 148–155
4. Naruse S, Higuchi T, Horikawa Y, Tanaka C, Nakamura K, Hirakawa K (1986) Radiofrequency hyperthermia with successive monitoring of its effects on tumours using NMR spectroscopy. Proc Natl Acad Sci USA 83: 8343–8347
5. Naruse S, Hirakawa K, Horikawa Y, Tanaka C, Higuchi T, Ueda S, Nishikawa H, Watari H (1985) Measurements of in vivo ^{31}P nuclear magnetic resonance spectra in neuroectodermal tumours for the evaluation of the effects of chemotherapy. Cancer Res 45: 2429–2433
6. Naruse S, Horikawa Y, Tanaka C, Higuchi T, Sekimoto T, Ueda S, Hirakawa K (1986) Evaluation of the effects of photoradiation therapy on brain tumours with in vivo p-31 MR spectroscopy. Radiology 160: 827–830
7. Newbold R (1984) Mutant ras proteins and cell transformation. Nature 312: 628–629
8. Ordidge RJ, Connelly A, Lohman JAB (1986) Image-selected in vivo spectroscopy (ISIS). A new technique for spatially selective NMR spectroscopy. J Magn Reson 66: 283–294
9. Rottenberg DA, Ginos JZ, Kearfott KJ, Junck L, Dhawan V, Jardan JO (1985) In vivo measurement of brain tumour pH using [11 c] DMO and positron emission tomography. Ann Neurol 17: 70–79

Address for correspondence: Kimiyoshi Hirakawa, M.D., Department of Neurosurgery, Kyoto Prefectural University of Medicine, Kawaramachi-Hirokoji, Kamigyo-ku, Kyoto 602, Japan.

Acta Neurochirurgica, Suppl. 43, 145–148 (1988)

Glucocorticosteroid Treatment of Vasogenic Oedema

Th. Rommel[1] and **W. Bodsch**[2]

[1] Neurochirurgische Klinik des Städtischen Krankenhauses Köln-Merheim
[2] Max Planck Institut für Neurologische Forschung, Abteilung für Experimentelle Neurologie, Köln-Merheim, Federal Republic of Germany

Summary

The components of vasogenic oedema associated with brain tumours were investigated in human biopsy material sampled from tumour and peritumoural tissue intraoperatively. Sixty patients with glioblastomas, gliomas, meningiomas and metastases, which had been treated with dexamethasone prior to surgery were employed for tissue measurements of water, electrolyte, haemoglobin, serum protein and dexamethasone concentrations. The quantification of serum proteins was achieved with the method described by Bodsch *et al.* Accordingly, serum proteins in the brain tissue and the blood were determined with 125 J labelled antihuman antibody. Taking into account brain haematocrit and blood-volume, quantitative measurements of the so-called oedema proteins as a measure of tumour oedema were performed. With the exception of metastases positive correlations were obtained between water and both serum proteins and sodium contents in tumours and peritumoural tissue. The serum protein content varied considerably being high in glioblastomas and low in peritumoural tissue surrounding metastases. However water and serum protein contents decreased with increasing dexamethasone concentrations in glioblastomas, while this effect was virtually absent in gliomas and meningiomas. Our results suggest a previously unknown selectivity among tumour types for the reduction of water content and serum proteins in corticosteroid treated oedematous tissue.

Keywords: Brain tumour; oedema; dexamethasone; oedema proteins.

Introduction

Several lines of evidence suggest that the extravasation of serum-derived plasma proteins into cerebral tissue concomittantly determines the distribution of water and electrolytes between extra- and intracellular sites following leakage of the blood brain barrier. This vasogenic variety of cerebral oedema is frequently associated with brain tumours and trauma in patients suffering from an increased intracranial pressure. However, for several years it has remained a matter of controversy as to whether extravascular built-up of pressure is predominantly the result of water concentrations with an insignificant contribution from plasma proteins or whether it is facilitated by the influence of proteins. Closely associated with these conjectures is the uncertainty about the mechanism of action of the adrenal glucocorticosteroids frequently used to relieve the severity of oedema and its resultant symptoms in human brain tumours. Recent studies in our laboratory of biopsy material from patients with glioblastomas, meningiomas, and metastases suggest that treatment with dexamethasone has an unequal but consistent influence on the constituents of vasogenic oedema fluid both in tumour as well as in peritumoural tissue (Bodsch *et al.* 1987). We now advance the hypothesis, that dexamethasone may be effective in reducing or retarding vasogenic oedema only in glioblastomas while exerting beneficial effects through an unknown – not oedema related – in mechanism considerably more types of human brain tumours. If this is accepted we have to consider possible detrimental or even deleterious effects of corticosteroids on oedematous and normal brain tissue. The presence of specific, high affinity binding sites for steroid hormones such as corticosterone or dexamethasone by interaction with cytosolic receptor proteins has been demonsteated in experimental and human studies (McEwen 1982, Yu *et al.* 1981). However, glucocorticosteroids appear to induce metabolic vulnerability non-specifically, enhancing the toxicity of a variety of subsequent metabolic insults (Sapolsky 1985). The aim of this report is to outline briefly the possible implications of steroid therapy to brain oedema. Cytotoxic oedema, often associated with ischaemic infarction, may also occur in restricted areas surrounding necrotic tissue of brain tumours. Moreover, it is possible that peritumoural tissue with low metabolic consumption becomes more vulnerable following high doses of steroids so that the chances of such tissue remaining viable are diminished.

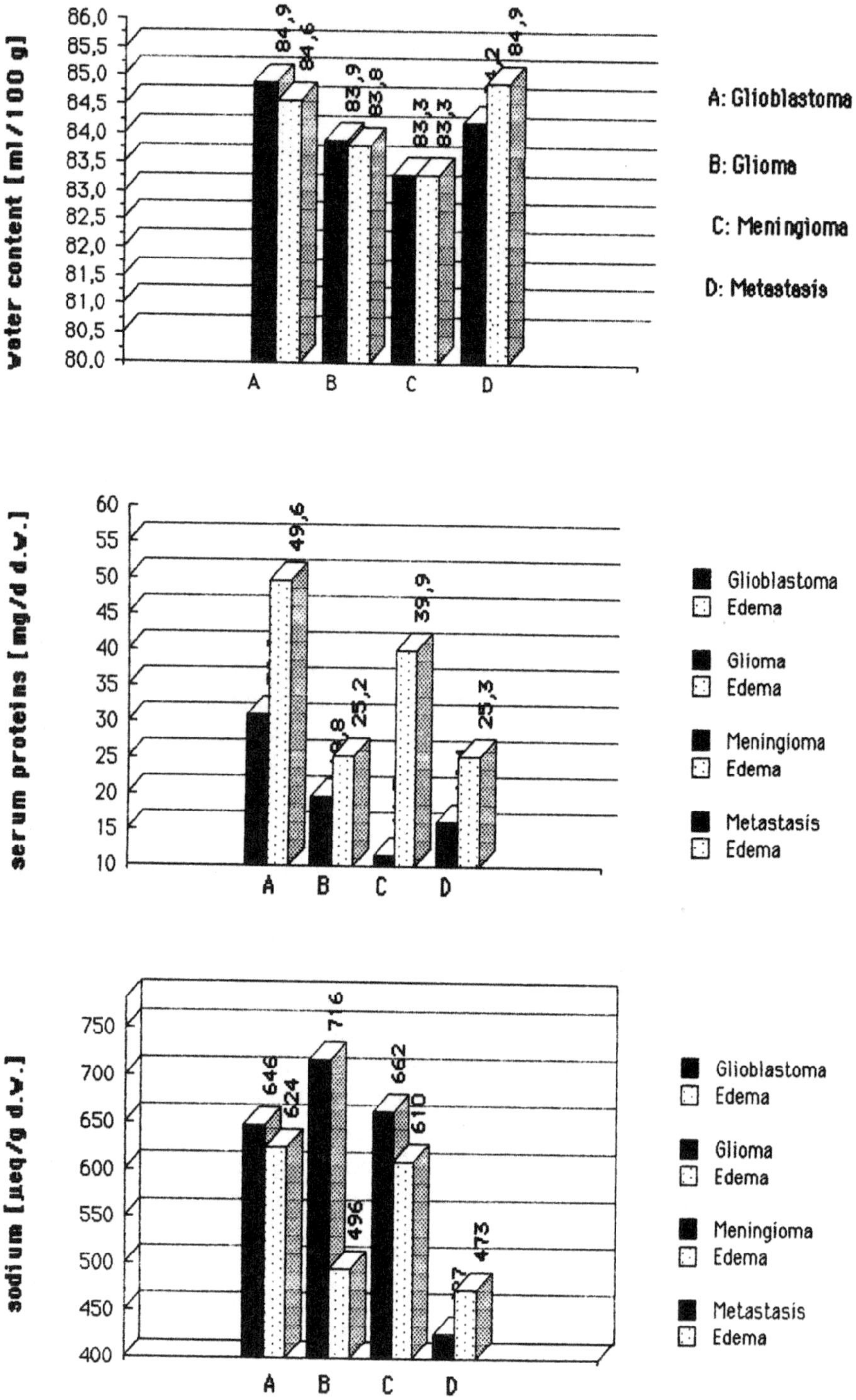

Fig. 1. Comparison between different tumour types

Comparison of Oedema Constituents Between Different Tumour Types

Analytical data for water, serum protein and sodium contents of tumour and peritumoural (oedema) tissue are illustrated in Fig. 1.

In all groups the mean water content was identical in the tumour and the peritumoural tissue except metastases, but the content of serum proteins was lower while that of sodium was higher in the tumour. Among these four groups, extravasated serum proteins were highest in regions of brain tissue close to glioblastomas, followed by meningiomas, metastases and gliomas. The rank order of sodium concentrations in peritumoural tissues was glioblastomas, meningiomas, gliomas and metastases. In all tumours cerebral blood volume was about 2.5 ml/100 g w.w and in peritumoural brain tissue 1.6–2 ml/100 g w.w. No major difference in potassium content were observed. This seems to be in line with the previous notion that peritumoural oedema does not disturb extra-intracellular ion homeostasis, although knowledge of total tissue potassium concentrations does not allow conclusions to be drawn about cell membrane ion fluxes. The relationship between the accumulation of water and the contents of sodium and extravasated proteins turned out to be more obvious when individual measurements were correlated (Bodsch *et al.* 1987). With the single exception of metastases, positive correlations were obtained between water and both serum proteins and sodium contents in tumours and peritumoural oedema. A determination of the slope of these correlations allowed the estimation of the average protein and sodium content of oedema fluid. The serum protein contents of oedema fluid varied between 4 and 16 mg/ml, with high values being observed in glioblastomas but low values being present in peritumoural oedema surrounding metastases. The sodium contents ranged from 130 to 170 μeq/ml in tumours and 130–150 μeq/ml in peritumoural tissue. Interestingly, glioblastomas exhibited a maximal protein concentration of 60–70 mg/g d.w. which did not increase further while water continued to rise. Similar high levels were not observed in other tumours, indicating that the water content did not increase as high as in glioblastomas.

Corticosteroid Treatment

Most patients that were included in the present study had been treated with dexamethasone prior to surgery. In order to determine a possible relationship between corticosteroid concentration and oedema formation, dexamethasone was measured in the tumour and the peritumoural tissue of 56 patients suffering from glioblastomas, gliomas and meningiomas (Bodsch *et al.* 1987). Although a large variability of individual tissue samples was obtained, mean values of dexamethasone concentrations were almost identical in tumour as well as peritumoural brain tissue, and also in different tumour groups. A suggested correlation of dexamethasone with water and extravasated serum proteins revealed a non-homogeneous pattern. In glioblastomas both water and serum proteins decreased in a dose-dependent way, starting from a threshold of about 500 ng/g w.w. Steroid concentrations of meningiomas and gliomas did not show correlations with either water or serum proteins. Also the peritumoural oedema responded to dexamethasone only in the glioblastoma group but not in other gliomas and meningiomas. Interstingly, the therapeutic threshold of dexamethasone was reached in gliomas and glioblastomas only when the water content was below 6 ml/g d.w., whereas in meningiomas and peritumoural oedema there was no restriction from tissue water. This suggests a passage of corticosteroids from the blood to the tissue that depends on both the tumour type and the degree of tissue hydration.

Discussion

The presence of serum proteins in vasogenic oedema may be a sensitive indicator of the severity of the blood brain barrier disturbance. Its contribution to the formation and retention of oedema, however, is still not clear. As discussed previously (Bodsch *et al.* 1982, Bodsch and Hossmann 1983), increased protein concentrations in the extracellular space od oedematous tissue completely shift the ion equilibria across the cell membrane. According to the Gibbs-Donnan law, the resting membrane potential and the ion equilibrium is maintained under physiological conditions only by the presence of polyanions (proteins) in glia as well as in neurones. A sudden cellular uptake of serum-derived proteins by nerve cells undoubtedly shifts potassium ions out of glia cells and also changes the sodium-potassium equilibrium of neurones. Resolution of oedema would then depend in the first place on the reduction of tissue electrolytes associated with the extra-intracellular passage of serum proteins. Numerous authors have confirmed that corticosteroids reduce the extent of peritumoural oedema. Possible mechanisms discussed include a reduction of the blood-tissue permeability, increased transport of sodium, potassium and water across the capillary tissue interface, stabilization of lysosomal membranes, reduced oedema spread, and an inhibition of tumour growth. Yu *et al.* (1981) identified glucocorticoid receptors in human brain tumours and related the efficiency of corticosteroid therapy to the number of such receptors. Our study confirms that meningiomas respond poorly to dexamethasone whereas a dose-dependent relationship between the drug and the reduction of oedema does exist in glioblastomas and may – to a lesser degree – also occur in other gliomas. The parallel reduction of water, sodium and proteins points to a reduction of barrier permeability as the most likely mechanism. The reason for failure of dexamethasone to improve oedema around meningiomas may have numerous reasons; vascularisation different from glioblastomas may just be one. However, meningiomas can produce oedema by mechanical pressure on the brain leading to an ischaemic type of oedema with subsequent contribution of vasogenic components. Although peritumoural serum protein concentration in meningiomas are high, the ratio of protein content between tumour and peritumoural tissue is higher than in glioblastomas, by several orders of magnitude. This indicates either a difference in origin of the underlying barrier disturbances or else a major built-up of vasogenic constituents at sites where cytotoxic oedema has already developed. This suggestion then would leave tumour tissue of meningiomas relatively unaffected by serum proteins. It is also conceivable that corticosteroids will not affect oedema spread when acting primarily on the border between tumour and peritumoural tissue. From all the data presented and discussed, it follows that different therapeutic approaches are required for each different mechanism of tumour related oedema; the presence and the relevance of ischaemic oedema must always be included in our considerations (Sapolsky and Pulsinelli 1985).

References

1. Bodsch W, Hürter Th, Hossmann K-A (1982) Immunohistochemical method for quantitative evaluation of vasogenic oedema in rat brain. Brain Res 249: 111–121
2. Bodsch W, Hossmann K-A (1983) ^{125}I-antibody autoradiography and peptide fragments of albumin in cerebral oedema. J Neurochem 41: 239–243
3. Bodsch W, Rommel Th, Grosse Ophoff B, Menzel J (1987) Factors responsible for the retention of fluid in human tumour oedema and the effect of dexamethasone. J Neurosurg 67 (in press)
4. McEwen B (1982) Glucocorticoids and hippocampus: receptors in search of a function. In: Ganten D, Pfaff D (eds) Current topics in neuroendocrinology, vol 2. Springer, Berlin Heidelberg New York, pp 23–43
5. Sapolsky RM (1985) Glucocorticoid Toxicity in the hippocampus: temporal aspects of neuronal vulnerability. Brain Res 359: 300–305
6. Sapolsky RM, Pulsinelli WA (1985) Glucocorticoids potentiate ischaemic injury to neurons: therapeutic implications. Science 229: 1397–1400
7. Yu Z-Y *et al* (1981) A study of glucocorticoid receptors in intracranial tumours. J Neurosurg 55: 757–760

Address for correspondence: Dr. Thomas Rommel, Rommel-Klinik, Bätznerstraße 96, D-7547 Wildbad, Federal Republic of Germany.

Acta Neurochirurgica, Suppl. 43, 149–153 (1988)

Transplant-induced Recovery from 6-OHDA Lesions of the Nigrostriatal Dopamineneurones in Mice

K. Shimizu, N. Tsuda, Y. Okamoto, Y. Matsui, Y. Miyao, K. Tamura, M. Yamada, S. Nakatani, T. Ikeda, and **H. Mogami**

Department of Neurosurgery, Osaka University Medical School, Osaka, Japan

Summary

Attempts to reconstruct the damaged nigrostriatal pathway in experimental models of Parkinson's disease have thus far been carried out in animals with neurotoxically induced dopamine deficiency. Our study established that unilateral 6-hydroxydopamine (6-OHDA) lesions of the nigrostriatal dopamine (DA) neurons produced a well-characterized functional asymmetry in the behaviour of the C57BL/6 (H-2^b) mice. The intraperitoneal administration of methamphetamine induced ipsilateral rotation at 7–20 turns/min 1×10^6 syngenic DA-rich cells of embryonic ventral mesencephalon were stereotaxically transplanted in the caudate-putamen. A complete recovery of methamphetamine-induced rotational response was produced around the 60th day after the syngenic cell suspension graft. And a complete compensation of the rotational response was also brought about with the DA-rich cells from embryonic ventral mesencephalon (crown-rump length; 10–13 mm) of allogenic C3H/HeN (H-2^k) mice. The FACS IV analysis revealed no H-2 (K^k and Ia^k) antigens before transplantation of these embryonic cells. Immunohistochemistry showed that the dopaminergic fibers had grown predominantly into the ipsilateral caudate-putamen. These results provide evidence of integration of syngenic and allogenic grafts and host tissue. And the immunological response in the transplanted brain are under investigation.

Keywords: Parkinson's disease; experimental model; embryonic tissue transplantation.

Introduction

At present, the most effective mode of treatment for Parksinson's disease, particularly early in its course, is the administration of both the L-isomer of 3,4-dihydroxyphenylalanine (L-DOPA), a DA precursor[8], and dopamine-receptor agonists. However, it is now clear that L-DOPA and the receptor agonists are only limited ameliorative treatments and that the course of most patients is one of inexorably progressive disability regardless of therapy[13]. On the other hand, an alternative approach in the therapeutic management of Parkinson's disease is the neurosurgical implantation of either DA neurones or adrenal chromaffin cells that have the ability of releasing DA[1, 15]. This approach has been stimulated by observing that effects of grafts of adrenal medullary cells or mesencephalic DA neurons implanted into experimental animals with DA deficiency induced by toxic compounds such as 6-hydroxydopamine (6-OHDA)[19]. Recently, Madrazo *et al.*[12] have reported on a successful transplantation of tissue from the adrenal medulla to the brain in patients with Parkinson's disease, who had a dramatic postoperative improvement in symptoms and signs.

The study is an attempt to provide a rational and effective treatment for Parksinson's disease with a firm scientific basis. An animal model of Parkinson's disease was developed in the 1970s by destroying neurones of the nigrostriatal system with a selective neurotoxin, 6-OHDA. The transplanted tissues (fetal substantia nigra, adrenal medulla) grow and reinnervate the host neostriatum, and function is restored despite the fact that the transplant is not in its normal location and does not have normal afferent connections[3, 10]. However, up to the present, such neural transplantations have scarcely been studied immunologically in graft-host interactions, as the central nervous system is considered to be an immunologically privileged organ[2, 14]. Brundin *et al.* reported that immunosuppression is necessary for optimum survival of mouse DA neurones grafted to the rat brain[5]. On the clinical side, it is particularly important that it is definitely known whether immunosuppression is necessary for neural transplantation in the brain of Parksinson's disease patients, before thoughtless clinical trials.

The present study demonstrates that (1) both syngenic and allogenic grafted DA neurones survive in the host brain, and that (2) the dopaminergic fibers grow predominantly into the ipsilateral striatum, and that (3) these grafted embryonic cells do not express the H-2 antigens before transplantation. These findings indicate that immunosuppression may not be necessary for optimum survival of allogenic mouse embryonic DA cells grafted to the mouse brain.

Materials and Methods

1. Lesion Surgery and Behavioral Testing

Female C57BL/6 mice (8–12 week at age, H-2^b) were given local unilateral stereotaxic injections of 2.5 µl of 6-OHDA (3 µg/µl in 0.2 mg/ml ascorbate saline) into the right nigrostriatal DA pathway. Two to three weeks after the lesion the mice were given 5 mg/kg of methamphetamine i.p. and their ipsiversive rotational behavior was counted for 1 min every 5 min in 25 cm diameter bowls. The rotational behavior was measured for the next 40 min. Mice that exhibited a mean of at least 5 full body turns per minute ipsilateral to the lesion were selected for neural transplantation. All mice with complete DA denervation were tested again several times for amphetamine-induced rotation at various time periods after transplantation.

2. Donor Tissue and Implantation Procedure

Donors; syngenic C57BL/6 mouse (H-2^b) embryos, or allogenic C3H/HeN mouse (H-2^k) embryos. Pairs of normal female and male mice were placed in the same cage for a 24-hr mating period. In the normal mouse, all of the neurones in the substantia nigra originate by embryonic day 12[17]. Pregnant dams were selected 12–14 days after separation from the males (*i.e.*, on gestational day 13–15) and anaesthetized with sodium pentobarbital (50 mg/kg i.p.). A caesarean section was performed and the embryos (C57BL/6, and C3H/HeN) removed from the uterine horns one at a time. The crown-rump length (CRL) of the embryos was 10–13 mm. The brains were removed from the cranium intact. The region of the ventral mesencephalon containing DA neurons was dissected under a microscope, and the donor cells were prepared according to the cell suspension method[4]. Briefly, several pieces of the ventral mesencephalon were washed twice with warm Hanks' balanced salt solution (HBSS), were then cut into six to eight smaller solid pieces and left gently in 0.1% Trypsin-containing 0.01 M phosphate buffered saline (PBS) for 20 min at 37 °C. Thereafter, the several pieces were washed 3 times with 0.6% glucose-containing saline, and were shaken gently in 50–100 µl of the same solution with a Pasteur pipette. The donor tissue was finally dissociated in a 40–80 µl volume.

Recipients were anaesthetized with sodium pentobarbital (50 mg/kg i.p.) and placed in a David Kopf stereotaxic apparatus. 5–10 µl of the cell suspension (containing 10^6 cells) was slowly injected into each host through a 10 µl Hamilton syringe stereotaxically. All graft injections were made into the head of the caudate-putamen; they were 2.5 mm mediolaterally, + 3.5 mm anteroposteriorly with respect to bregma, and 2.5 mm dorsoventrally with respect to the dural surface. After the operation, the overlying skin was sutured and the animals were allowed to recover.

3. Immunocytochemistry

At postoperative times of 50–75 days, mice were anaesthetized with sodium pentobarbital (50 mg/kg i.p.), and perfused transcardially with 60 ml of saline and an equal volume of 4% (wt/vol) paraformaldehyde in 0.1 M phosphate buffered solution adjusted to pH 7.4. Brains were dissected from the cranium, postfixed in the same fixative overnight, and were dehydrated in 10%, 20% and 25% sucrose-containing 0.1 M PBS (for 2 hr each). Sections of 4–10 µm thick were prepared on a cryostat (– 18 °C), and mounted on several slides. After redehydration at room temperature, these slides were washed 3 times in 0.01 M cold PBS for 15 min each. They were incubated with monoclonal antibody to tyrosine hydroxylase (TH)[11], which was diluted 1 : 200 with 0.1% Triton X-100 in 0.01 M PBS, for 16–24 hr at 4 °C. They were washed 3 times with 0.01 M cold PBS for 15 min each, and were incubated with fluorescein isothiocyanate (FITC) conjugated F $(ab')_2$ gout anti-mouse (GAM) immunoglobulin (Ig) G, which was diluted 1 : 200 with 0.1% Triton X-100 in 0.01 M PBS, for 16–24 hr at 4 °C. Finally, the slides were washed three times for 15 min each with 0.01 M cold PBS, and coverslipped.

4. Analysis of H-2 Antigens on Mouse Embryonic Cells by Flow Cytometry

Live 10^6 dissociated cells of the ventral mesencephalon of C3H/HeN (H-2^k) mouse embryos (on gestational day 13–15) were put in 1 ml centrifugal tubes, washed 3 times with 1 ml of 0.01 M cold PBS, suspended in 50 µl of 0.01 M PBS, and incubated with 50 µl of mouse monoclonal antibody to H-$2K^k$ antigens or Ia^k, which was diluted 1 : 100 in 0.01 M cold PBS, for 30 min at 4 °C. These cells were centrifuged (1,000 × g, 30 sec) with the Micro-Centrifuge (model 59 A, Fisher), washed 2 times with PBS, resuspended in 50 µl of 0.01 M cold PBS, and incubated with 50 µl of FITC-conjugated $GAMIgG_{2a}$ or FITC-conjugated $GAMIgG_{2b}$ diluted 1 : 50 in 0.01 M PBS for 30 min at 4 °C. They were centrifuged (1,000 × g, 30 sec), resuspended in 1 ml of 0.01 M cold PBS containing 20 µg/ml of propidium iodide, and were analyzed on the Fluorescence Activated Cell Sorter (FACS-IV, Becton Dickinson, CA, USA).

Results

1. Motor Asymmetry

All mice with DA pathway destroyed by injecting 6-OHDA were followed to test the amphetamine-induced rotational behavior until 90–110 days after 6-OHDA lesion. Non-grafted mice, which were lesioned by 6-OHDA before, showed no recovery of amphetamine-induced ipsiversive rotational behavior during the course of the experiments. On the other hand, the syngenic dopaminergic cells-grafted lesioned mice showed significantly less turning than the non-grafted lesioned mice (Student's t-tests; $p < 0.05$) in the 35 day behavior test. All grafted lesioned mice ($n = 4$) exhibited complete behavioral compensation in the amphetamine-induced rotation test at 65 days after transplantation of syngenic embryonal cells (Fig. 1). Behavioral compensation in the lesioned mice (H-2^b), into which had been transplanted the DA-rich cells of the ventral mesencephalon of allogenic C3H/HeN (H-2^k) mouse embryos, developed about 60 days after transplantation without any immunosuppressive treatment.

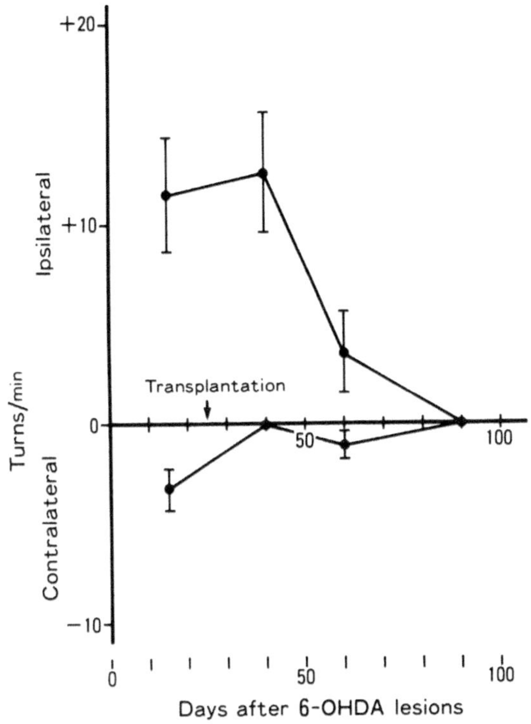

Fig. 1. Motor asymmetry (mean ± SEM of net 360° turns/min between 15 and 20 min in the direction towards the lesion side) induced by 5 mg/kg methamphetamine, before transplantation and at different timepoints after transplantation. About 10^6 dissociated DA neurones of syngeneic C57BL/6 mouse embryos (CRL/ 11.7 ± 0.3 mm) were transplanted to the 6-OHDA lesioned C57BL/ 6 mice. Values indicate that the grafted group is different from the control (non-grafted) group (Student's t-test)

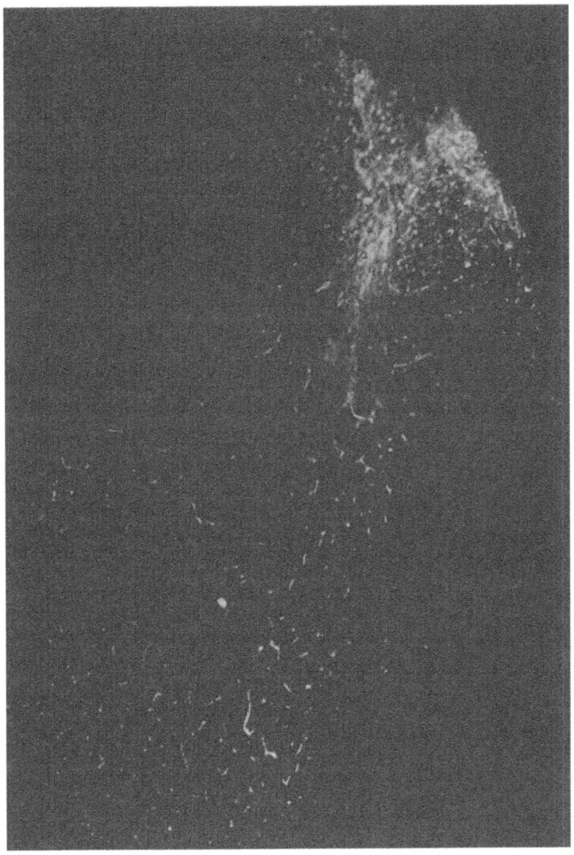

Fig. 2. TH immunoreactivity in the syngeneic embryonal cell grafts, showing immunostained neurones and processes (× 100)

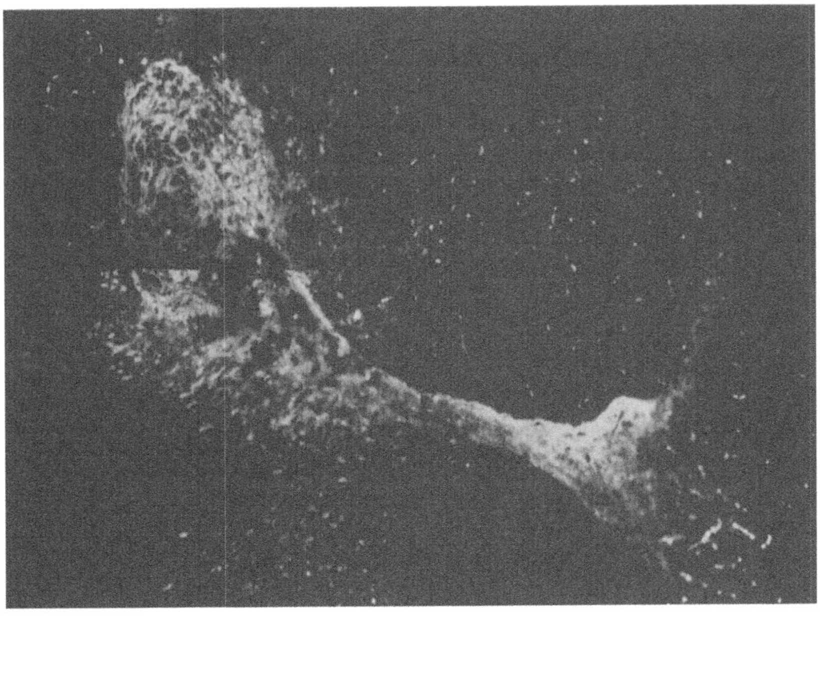

Fig. 3. TH immunoreactivity in the allogeneic embryonal cell grafts, showing embedded immunoreactive cells in the ipsilateral striatum and a dense network of neuronal processes (× 200)

2. *Morphological Findings*

4–5 months after the lesions, the mice were sacrificed by transcardial perfusion with 4% paraformaldehyde in 0.1 M PBS. Microscopic examination of the sections demonstrated that the presence of TH-immunoreactive cell clusters in the parenchyma around the needle tract, transplanted stereotaxically, in both syngenic and allogeneic embryonal cells-grafted mice showing functional recovery (Figs. 2 and 3). Fine TH-immunoreactive fibers were detectable within the host ipsilateral striatum beneath the needle tract, while they were not detectable in the contralateral striatum (Fig. 2). In one of allogenic embryonal cells-grafted mice exhibiting functional recovery, TH-immunoreactive cells were found embedded deeply in the ipsilateral striatum and a dense network of neuronal processes was seen in the same section (Fig. 3). There was a good correlation between functional recovery and the establishment of grafted TH-immunoreactive neurones, without regard to syngenic embryonal cells or allogenic cells, without any immunosuppressive treatment.

The FACS analysis revealed that the DA-rich cells of the ventral mesencephalon of C3H/HeN embryos had neither $H\text{-}2K^k$ nor Ia^k antigens.

Discussion

The present results demonstrate that the experimental C57BL/6 mice ($H\text{-}2^b$) with DA deficiency induced by 6-OHDA exhibited the same ipsiversive rotational behavior as the 6-OHDA rat models in the amphetamine-induced turning tests and that the 6-OHDA lesioned mice which received transplants of dissociated cells of the ventral mesencephalon of either syngenic C57BL/6 ($H\text{-}2^b$) or allogenic C3H/HeN ($H\text{-}2^k$) mouse embryos showed complete behavioral compensation without any immunosuppressive treatment. They confirm that the grafts of embryonal ventral mesencephalon can survive and grow, express TH-immunoreactivity, innervate the target nuclei without regard to syngenic or allogenic cells, but that have no H-2 antigens on their cells.

After non-grafted 6-OHDA lesioned mice were given the various doses (1–20 mg/kg, i.p.) of methamphetamine, amphetamine-induced ipsiversive rotational behavior was measured three times weekly. The lesioned mice exhibited the best frequency of rotations, when they were given 5 mg/kg of methamphetamine. Then the rotational behavior lasted for 35 min.

In the mice which were successfully grafted with syngenic embryonal cells, the first effects in the rotation test appeared after 5 weeks and complete behavioral compensation was exhibited by 10 weeks. This is a little later than with similar transplants utilizing rodent donors where functional effects can be observed already by 2–4 weeks after grafting[4, 5]. But this is much faster than with similar human-to-rat grafts where functional effects can be first observed at 12.5 weeks[6]. The behavioral recovery of the mice which were successfully transplanted with allogenic embryonal cells developed as fully as in the mice grafted with syngenic embryonal cells, without any immunosuppressive treatment. In other xenografted experiments, it is reported that immunosuppression is necessary for optimum survival of the xenografts[5, 6].

The grafted mouse DA neurones, without regard to syngenic or allogenic DA cells, gave rise to a terminal network in the host striatum at least as extensive as that observed for DA neuron grafts transplanted to rats with 6-OHDA lesions. By the FACS analysis, the grafted neurones of the ventral mesencephalon of mouse embryos had neither H-2 nor I_a antigens.

These findings indicate that immunosuppression is not necessary for grafting embryonal neurones. Brundin's report (1986)[6] that grafted human fetal dopamine neurones give rise to axonal processes that form a new DA terminal network in large parts of the rat striatum, and that they can produce behavioral effects in a rat model of Parkison's disease, definitely suggests that human fetal DA neurones are more suitable for transplantation in Parkinson's disease patients.

Acknowledgements

We thank Dr. Hiroshi Hatanaka for his valuable comments. Monoclonal antibodies against TH were kindly donated by him.

References

1. Backlund E-O, Granberg P-O *et al* (1985) Transplantation of adrenal medullary tissue to striatum in Parksonism. J Neurosurg 62: 169–173
2. Barker CF, Billingham RE (1977) Immunologically privileged sites. Adv Immunol 25: 1–54
3. Björklund A, Dunnet SB *et al* (1980) Reinnervation of the denervated striatum by substantia nigra transplants: functional consequences as revealed by pharmacological and sensorimotor testing. Brain Res 199: 307–333
4. Brundin P, Isacson O *et al* (1985) Monitoring of cell viability in suspensions of embryonic CNS tissue and its use as a criterion for intracerebral graft survival. Brain Res 331: 251–259
5. Brundin P, Nilsson OG *et al* (1985) Cyclosporin A increases survival of cross-species intrastriatal grafts of embryonic dopamine-containing neurons. Exp Brain Res 60: 204–208
6. Brundin P, Nilsson OG *et al* (1986) Behavioural effects of human fetal dopamine neurons grafted in a rat model of Parkinson's disease. Exp Brain Res 65: 235–240

7. Calne DB, Duvoisin RC *et al* (1984) Speculations on the aetiology of Parkinson's disease. Adv Neurol 40: 353–360
8. Cotzias GC, Papavasiliou PS *et al* (1969) Modification of Parkinsonism – Chronic treatment with L-DOPA. N Engl J Med 280: 337–345
9. Ehringer H, Hornykiewicz O (1960) Verteilung von Noradrenalin und Dopamin (3-Hydroxytyramin) im Gehirn des Menschen und ihr Verhalten bei Erkrankungen des Extrapyramidalen Systems. Klin Wochenschr 38: 1236 – 1239
10. Freed WJ, Hoffer BJ *et al* (1984) Transplantation of catecholamine-containing tissues to restore the functional capacity of the damaged nigrostriatal system. Neural transplants: development and function. Plenum Press, New York, pp 373–406
11. Hatanaka H, Arimatsu Y (1984) Monoclonal antibodies to tyrosine hydroxylase from rat pheochromocytoma PC12h cells with special reference to nerve growth factor-mediated increase of the immunoprecipitable enzymes. Neurosci Res 1: 253–263
12. Madrazo I, Drucker-Colin R *et al* (1987) Open microsurgical autograft of adrenal medulla to the right caudate nucleus in two patients with intractable Parkinson's disease. N Engl J Med 316: 831–834
13. Markham CH, Diamond SG *et al* (1986) Modification of Parkinson's disease by long-term levodopa treatment. Arch Neurol 43: 405–407
14. Mason DW, Charlton HM *et al* (1985) Immunology of allograft rejection in mammalia. Neural grafting in mammalian CNS. Elsevier, Amsterdam, pp 91–98
15. Olson L (1985) On the use of transplants to counteract the symptoms of Parkinson's disease; Background, experimental models, and possible clinical applications. Synaptic plasticity. Guilford, New York, pp 485–505
16. Parkinson J (1817) An essay on the shaking palsy. Sherwood, Neely, and Jones, London
17. Taber-Pierce E (1973) Time of origin of neurons in the brain stem of the mouse. Prog Brain Res 40: 53–65
18. Trétiakoff C (1919) Contribution a l'etude de l'anatomie pathologique du Locus Niger de Soemmering. Thèse, Université de Paris, Sorbonne
19. Ungerstedt U (1968) 6-Hydroxy-dopamine induced degeneration of central monoamine neurons. Eur J Pharmacol 5: 107–110
20. Yokochi M, Narabayashi H *et al* (1984) Juvenile parkinsonism – some clinical, pharmacological, and neuropathological aspects. Adv Neurol 40: 407–413

Address for correspondence: K. Shimizu, M.D., Department of Neurosurgery, Osaka University Medical School, 1-1-50, Fukushima, Fukushimaku, Osaka 553, Japan.

Acta Neurochirurgica, Suppl. 43, 154–158 (1988)

Possible Synapse Formation by Embryonic Cerebellar Tissue Grafted into the Cerebellum of the Weaver Mutant Mouse

H. Takayama, Sh. Toya, T. Shinozaki, H. Inoue, M. Otani, Sh. Kohsaka[1], and **Y. Tsukada**[1]

Department of Neurosurgery, and [1] Department of Physiology, School of Medicine, Keio University, Tokyo, Japan

Summary

The weaver mutant mouse is characterized by degeneration of cerebellar granule cells in early post-natal stage. In the present study, the possibility of synapse formation in the weaver mouse cerebellum by implanted granule cells was analyzed immunohistochemically with antiserum against synaptic vesicle protein, Synapsin I. Normal cerebellar tissue, obtained from 15-day-old CBA/JNCij mouse embryos, was transplanted into the cerebellum of 4-week-old weaver mice. 6 weeks after the transplantation, the grafted tissue was clearly detected in the host cerebellum, developing a trilaminar organization. A number of granule-like cells were observed in the folia of the host cerebellum, suggesting that the implanted granule cells may have migrated from the grafted tissue into the host cerebellum. Some areas in the host cerebellum as well as in the grafted tissue were intensely stained by anti-Synapsin I serum, indicating that the implanted granule cells make synaptic contacts with the neuronal cells.

Keywords: Synapsin I; weaver mutant mouse; transplantation; cerebellum.

Introduction

Transplantation of neural tissue has been used extensively as a viable approach to studying the development of the neural cells in the central nervous system[4–7]. To investigate the behaviour of the transplanted neuronal cells in the host brain, it is of advantage to use a mutant mouse which is deficient in a single class of neuronal cells. The weaver mutant mouse is characterized by degeneration of cerebellar granule cells in early post-natal development[8–10]. Electron microscopic studies demonstrated that the synaptic contacts between the Purkinje cell dendrites and the granule cells are extremely poor in the cerebellum of this mutant[11]. In the present study, cerebellar tissue taken from 15-day-old normal mouse embryos (CBA/JNCij) was transplanted into the cerebellar vermis of 4-week-old weaver mutant mice, and we have investigated immunohistochemically whether the implanted granule cells can survive and develop synaptic contact in the weaver mouse cerebellum. Immunohistochemical examinations were conducted by using antiserum to Synapsin I, which is a phosphoprotein specific to the membranes of synaptic vesicle[1, 2, 13].

Materials and Methods

B6CBA/F 1 (wv/+) mice were purchased from the Jackson Laboratory (Bar Harbor, Me., U.S.A.). The homozygous weaver mutant mice were obtained by mating known heterozygotes for the weaver mutation and were identified by their abnormal gait, and by their smaller size compared with their littermates. Timed pregnant CBA/JNCij mice were purchased from Nippon Bio-supply Center (Tokyo, Japan).

The cerebellar primordia were dissected from 15-day-old CBA/JNCij mouse embryos (E 15). Four-week-old weaver mice were anaesthetized with ether and ketamine hydrochloride and held in a stereotaxic apparatus with ear bars. Under aseptic conditions, a small craniotomy was performed in the middle of the occipital bone to expose the surface of the cerebellar vermis and a syringe containing small pieces of embryonic cerebellar tissue was lowered to a depth of 1.0 mm from the cerebellar surface. The grafted tissue was stereotaxically injected to the host cerebellum by depressing the plunger.

The mice were sacrificed 6 weeks after the transplantation, and perfused through the left cardiac ventricle with 10% formaldehyde in phosphate buffered saline (PBS). The fixed brains were then dehydrated, embedded in paraffin and cut sagittally into 4-μm thick sections. Ungrafted cerebella from 10-week-old weaver mutant and normal CBA/JNCij mice were also prepared for immunostaining in the same manner. The deparaffinated sections were immunohistochemically stained with anti-Synapsin I serum by the PAP method[12].

Results and Discussion

Synapsin I is a phosphoprotein localized at the nerve terminal region of the great majority of neurones of the central and peripheral nervous system[1, 2, 13]. As shown in Fig. 1 a, the molecular layer and the glomeruli in the granule cell layer were positively stained with anti-Synapsin I serum in the normal control mouse. The positive staining in the molecular layer is suggestive

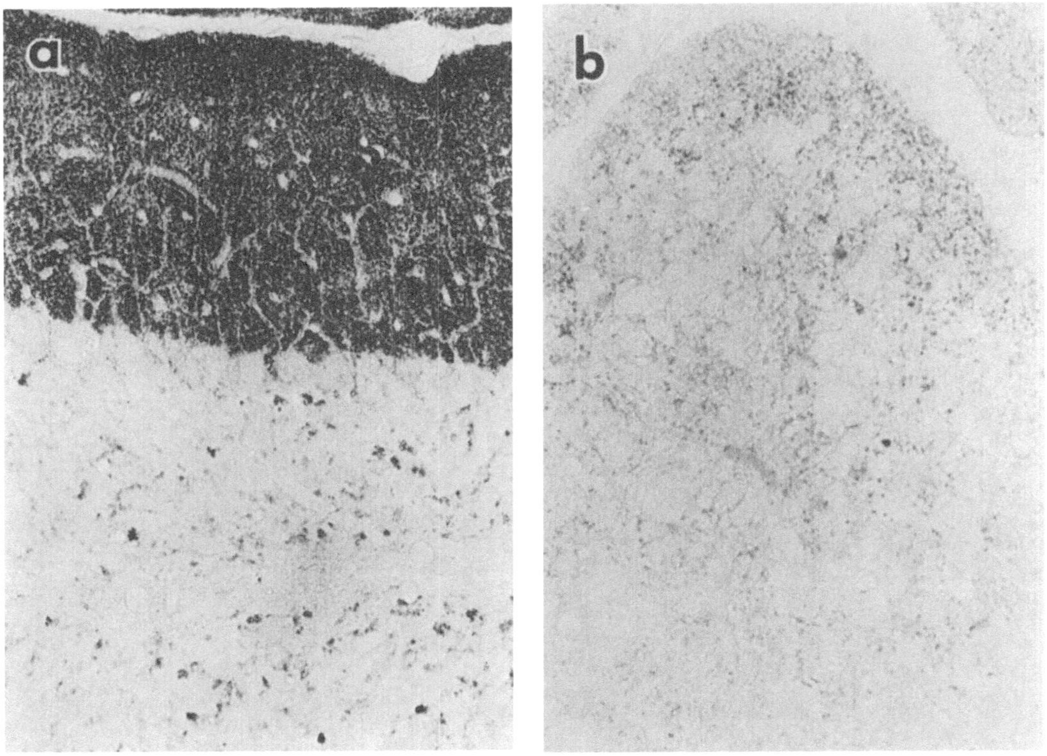

Fig. 1. Sections from the cerebella of 10-week-old control (a) and weaver (b) mice, immunohistochemically stained with antiserum to Synapsin I. × 100

of the existence of numerous synapses between the parallel fibres and the Purkinje cell dendrites. In the weaver mutant mouse, however, Synapsin I immunoreactivity in the molecular layer was greatly reduced, and only sparsely distributed dots of immunoreactivity were detected in the whole cerebellum (Fig. 1 b). Dolphin and Greengard[3] also reported that the amount of Synapsin I is markedly reduced in the weaver mouse cerebellum by 66% compared to the control. These results agree with the reduction of synapses in the weaver mutant cerebellum due to degeneration of granule cells. Consequently, if intense immunoreactivity with Synapsin I is observed in the weaver mutant mouse with grafted tissue, it is likely that synaptic contact will be developed by the implanted neuronal cells.

The survival and development of the grafted tissue was examined 6 weeks after the transplantation by haematoxylin-eosin staining. A large mass of grafted tissue was detected in the host cerebellum of the recipient weaver mutant mouse (Fig. 2 a). It appeared to be anatomically integrated with the host brain, and had developed a trilaminar organization. Many granule-like cells and Purkinje cells were visible, forming the granule cell, Purkinje cell and molecular layers. Most interestingly, a number of granule-like cells were detected in the folia of the host cerebellum (Fig. 2 b). This finding indicates that the implanted granule cells may have migrated from the grafted tissue into the host cerebellum.

To study the possibility of synapse formation by the implanted granule cells, these sections were processed for Synapsin I immunohistochemistry. Fig. 3 a shows the immunoreactivity in the grafted tissues. The area surrounded by the granule cell layer was positively stained by the antiserum, which indicates that synapse formation between implanted granule-like cells and Purkinje cells may have developed in the grafted tissue. Moreover, some areas in the host cerebellum, where the implanted granule-like cells had migrated, as indicated in Fig. 2 b, were also positively stained with anti-Synapsin I serum (Fig. 3 b). This result raises the possibility that implanted granule cells extend fibres and make synaptic contacts with the host neuronal cells.

In the present study, it was clearly demonstrated that implanted granule cells integrate themselves with the host weaver mouse cerebellum as well as with the grafted tissue. The distribution of Synapsin I immunoreactivity also suggests that implanted granule cells make synaptic contacts with the neuronal cells.

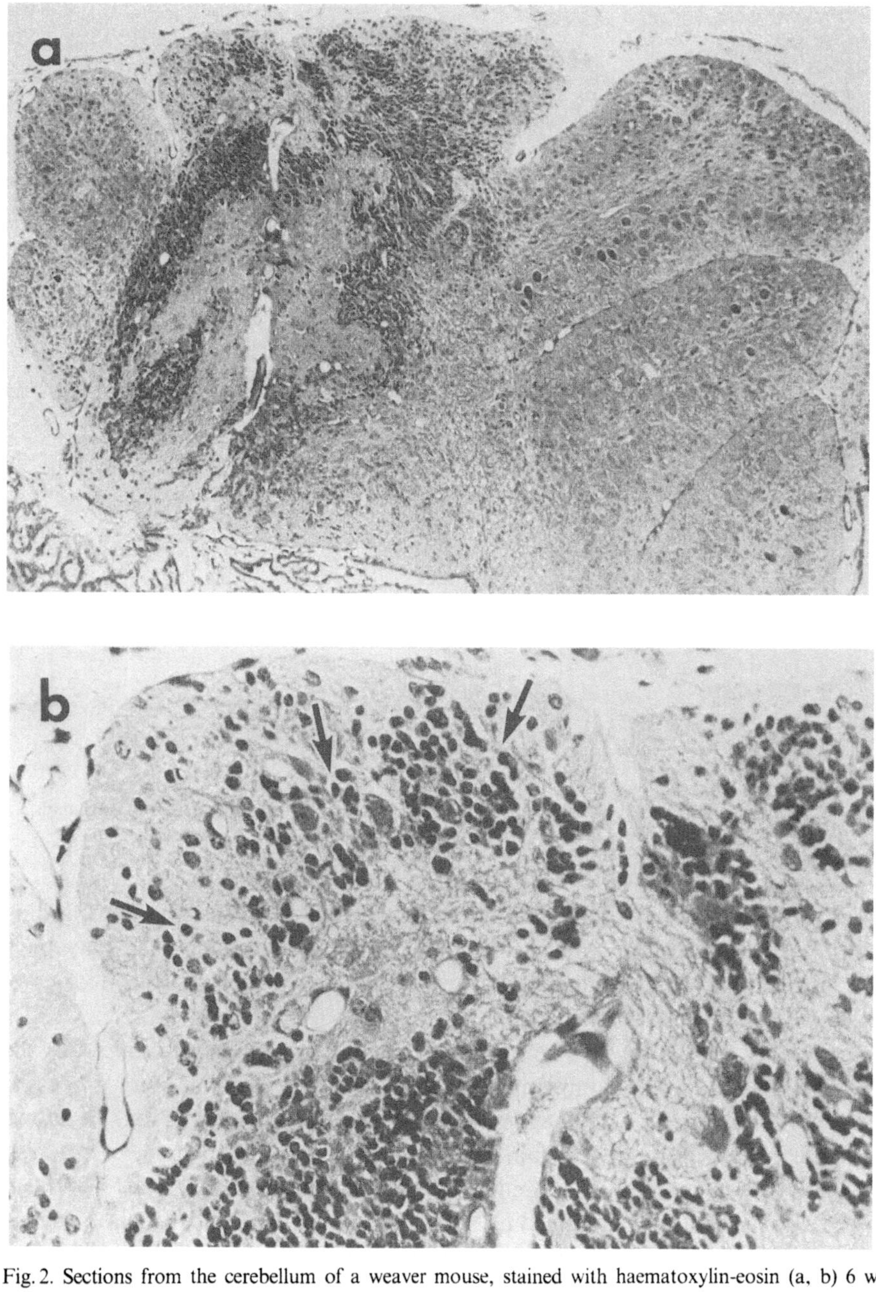

Fig. 2. Sections from the cerebellum of a weaver mouse, stained with haematoxylin-eosin (a, b) 6 weeks after the transplantation. (a) A large mass of grafted tissue was observed. Granule-like cells, Purkinje cells and the molecular layers were developed. × 25. (b) High magnification of the host cerebellum where the implanted granule-like cells were detected as indicated by arrows. × 100

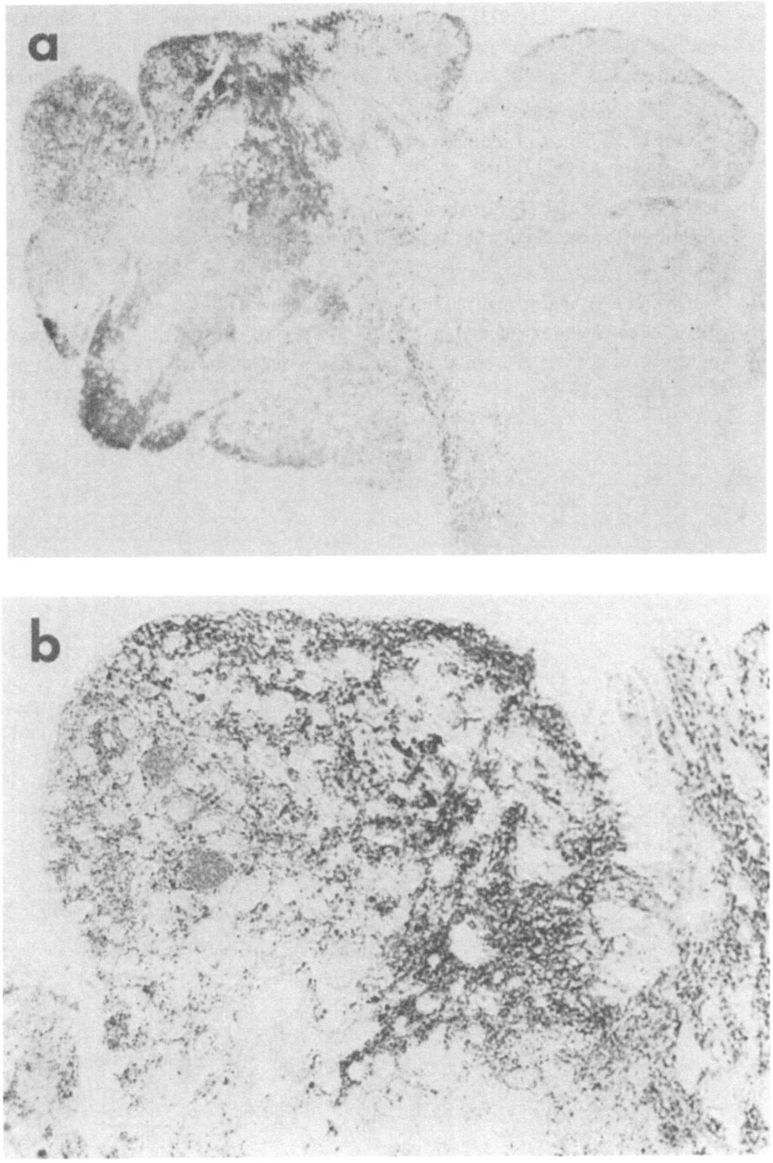

Fig. 3. Sections from the cerebellum of a weaver mouse, stained with anti-Synapsin I serum (a, b) 6 weeks after the transplantation. (a) Synapsin I staining of the section adjacent to that in Fig. 2a. × 25. (b) Synapsin I staining of the section adjacent to that in Fig. 2b. × 100

Acknowledgement

Authors are grateful to Dr. T. Ueda, University of Michigan, for providing us an anti-Synapsin I serum.

References

1. Bloom FE, Ueda T, Battenberg E, Greengard P (1979) Immunocytochemical localization, in synapses, of protein I. An endogeneous substrate for protein kinases in mammalian brain. Proc Natl Acad Sci USA 76: 5982–5986
2. De Camilli P, Harris SM, Huttner WB, Greengard P (1983) Synapsin I (protein I), a nerve terminal-specific phosphoprotein. II. Its specific association with synaptic vesicles demonstrated by immunocytochemistry in agarose-embedded synaptosomes. J Cell Biol 96: 1355–1373
3. Dolphin AC, Greengard P (1981) Presence of protein I, a phosphoprotein associated with synaptic vesicles, in cerebellar granule cells. J Neurochem 36: 1627–1631
4. Inoue H, Kohsaka S, Otani M, Toya S, Tsukuda Y (1986) The effect of arcuate nucleus transplantation on the development of the anterior pituitary in monosodium glutamate-treated rats. Neurosci Res 3: 555–567
5. Inoue H, Kohsaka S, Yoshida K, Otani M, Toya S, Tsukada Y (1985) Cyclosporin A enhances the survivability of mouse cerebral cortex grafted into the third ventricle of rat brain. Neurosci Lett 54: 85–90
6. Inoue H, Kohsaka S, Yoshida K, Otani M, Toya S, Tsukada Y (1985) Immunohistochemical studies on mouse cerebral cortex grafted into the third ventricle of rats treated with cyclosporin A. Neurosci Lett 57: 289–294
7. Kohsaka S, Yoshida K, Inoue Y, Shinozaki T, Takayama H, Inoue H, Mikoshiba K, Takamatsu K, Otani M, Toya S, Tsukada Y (1986) Transplantation of bulk-separated oligodendrocytes into the brains of shiverer mutant mice: immunohistochemical and electron microscopic studies on the myelination. Brain Res 372: 137–142

8. Rakic P, Sidman RL (1973) Sequence of developmental abnormalities leading to granule cell deficit in cerebellar cortex of weaver mutant mice. J Comp Neurol 152: 103–132

9. Rakic P, Sidman RL (1973) Organization of cerebellar cortex secondary to deficit of granule cells in weaver mutant mice. J Comp Neurol 152: 133–162

10. Rezai Z, Yoon CH (1972) Abnormal rate of granule cell migration in the cerebellum of "weaver" mutant mice. Dev Biol 29: 17–26

11. Sotelo C (1975) Anatomical, physiological and biochemical studies of the cerebellum from mutant mice. 2. Morphological study of cerebellar cortical neurons and circuits in the weaver mouse. Brain Res 94: 19–44

12. Sternberger LA, Hardy PH, Cuculis JJ, Meyer HG (1970) The unlabelled antibody-enzyme method of immunohistochemistry. Preparation and properties of soluble antigen-antibody complexes (horseradish-antihorseradish peroxidase) and its use in identification of spirochetes. J Histochem Cytochem 18: 315–333

13. Ueda T, Maeno H, Greengard P (1973) Regulation of endogeneous phosphorylation of specific proteins in synaptic membrane fractions from rat brain by adenosine 3' : 5'-monophosphate. J Biol Chem 248: 8295–8305

Address for correspondence: Hideichi Takayama, M.D., Department of Neurosurgery, School of Medicine, Keio University, 35 Shinanomachi, Shinjuku-ku, Tokyo 160, Japan.

Acta Neurochirurgica, Suppl. 43, 159–161 (1988)

Intraventricular Transplantation of Omentum for Treatment of Hydrocephalus. An Experimental Study in Dogs

J. M. González-Darder[1], J. L. Barcia-Salorio[2], J. Barberá[1], and J. Broseta[3]

[1] Departamento de Neurocirugía, Hospital Mora, Cádiz, [2] Departamento de Neurocirugía, Hospital Clínico Universitario, Valencia, [3] Departamento de Neurocirugía, Hospital Clínico Universitario, Salamanca, Spain

Summary

In 17 mongrel dogs an experimental non-communicating hydrocephalus was developed by intracisternal injection of kaolin solution. In 10 of them a portion of omentum was isolated and after suturing end-to-end and the vascular pedicle to the temporal artery and vein the graft was introduced transcortically into the lateral ventricle. Metrizamide ventriculography, intraventricular basal and pulse pressure amplitude, compliance and resistance to CSF drainage were performed to evaluate related changes after surgery. Brain and omentum specimens were submitted for pathological studies. In summary, intraventricular omentum transplantation reduced intracranial pressure, compliance and resistance in experimental hydrocephalus whereas ventricular size remained unchanged. Pathological studies showed graft survival in most animals.

Keywords: Omentum transplantation; intracranial pressure; cerebrospinal fluid; hydrocephalus.

Introduction

Capability of CSF absorption in transplanted or transposed greater omentum has been shown in several experimental studies[4, 5, 6]. Although intracranial transplantation of greater omentum is progressively being used for brain revascularization it has been seldom applied in treating communicating hydrocephalus[4].

The aim of this study was to determine the role of the intraventricular transplantation of the omentum in the management of experimental non-communicating hydrocephalus.

Material and Methods

Experimental hydrocephalus was produced in 17 mongrel dogs by intracisternal injection of sterile kaolin solution. In 5 control animals saline was injected after the cisternal puncture. In 10 acute hydrocephalic animals the portion of omentum to be transplantated was isolated together with the gastroepiploic artery and vein as vascular pedicle. A right temporoparietal craniectomy was done and the superficial temporal artery and vein prepared. A microvascular end-to-end arterial and venous suture was then carried out between gastroepiploic and temporal vessels and finally the transplanted omentum was introduced transcortically into the lateral ventricle and fixed to the dura mater.

Metrizamide ventriculography, measure of intraventricular basal (VP) and pulse pressure amplitude (PA), compliance determination through a bolus injection test (C) and resistance to CSF drainage (R) were performed in order to evaluate the progression of the hydrocephalus and related changes after the surgical procedure. Brain, omentum graft, and ventricular wall were submitted for pathological studies.

Results

Table 1 summarizes the mean values of experimental parameters. Hydrocephalic animals were studied a

Table 1. *Parameters in Experimental Groups*

Parameter	Normal	Hydrocephalus	Omentum
Ventricular index	0.66 ± 0.02	0.71 ± 0.03	0.68 ± 0.02*
VP (mm Hg)	6.4 ± 1.6	22.8 ± 9.2*	9.7 ± 6.4
PA (mm Hg)	1 ± 0.5	4.9 ± 3.4	5.3 ± 2.1*
C (ml/mm Hg) · (10^{-3})	17.5 ± 5.6	6.7 ± 2*	16.1 ± 8.5
R (mm Hg/ml/min)	22.8 ± 9.1	78.8 ± 37.7**	13.6 ± 10.3

VP, basal ventricular pressure; PA, pulse amplitude; C, compliance; R, resistance. Values are means ± standard error. * $p < 0.05$, ** $p < 0.01$.

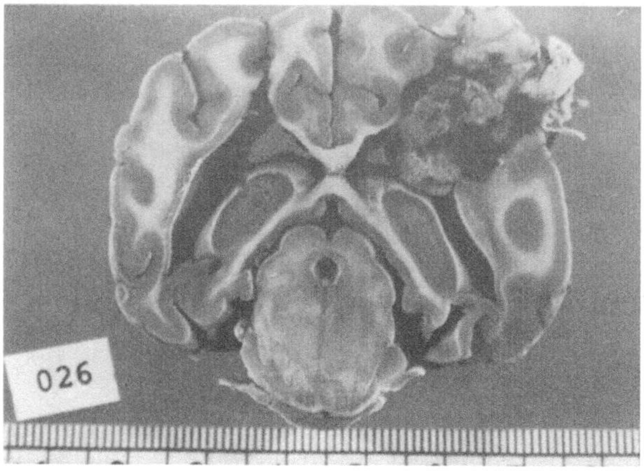

Fig. 1. Transverse brain section showing the intraventricular location of the omentum

mean of 17 days after kaolin injection when hydrocephalus was in acute stage. In the serie, 4 out of 10 transplanted animal died during postoperative period. The others had a follow-up free of complications, being evaluated between 2 and 3 weeks after surgery. The patency rate for the anastomosis at autopsy was 83%. In all cases the omentum was properly placed in the lateral ventricle (Fig. 1). Pathological studies showed small areas of necrosis and fatty degeneration but the graft survived showing wide vascular connections with the brain, dura mater and muscle (Fig. 2).

Discussion

Our experimental study shows that the intraventricular omentum reduced the intracranial pressure, compliance and resistance in kaolin induced non-communicating hydrocephalus whereas ventricular size remained unchanged. These findings suggested the omentum draining ability by reduction of resistance to CSF outflow. However, since though it could not modify other factors involved in chronic hydrocephalus development, consequently there was no ventricular reduction[3]. This could be related to some pathological changes observed in the ependymal wall during hydrocephalus development[3]. However we believed that pulse pressure generated in the choroid plexuses could be a major factor in hydrocephalus pathogenesis[1]. Pulse pressure seemed undamped in kaolin induced hydrocephalus due to the obliteration of subarachnoid space[3], and in transplanted animals the intraventricular omentum would behave as an extra pulsating plexus increasing therefore pulse pressure amplitude. Thus a high pulse wave would produce a mechanical effect creating and maintaining the ventricular enlargement[2].

In conclusion, intraventricular transplantation of omentum is a new option for treating non-communicating hydrocephalus that needs further experimental studies.

References

1. Dirocco C, Pettorossi V, Calderelli M, Mancinelli R, Velardi F (1977) Experimental hydrocephalus following mechanical increment of intraventricular pulse pressure. Experientia 33: 1470–1472
2. Foltz EL (1984) Hydrocephalus and CSF pulsality: clinical and laboratory studies. In: Shapiro K, Marmarou A, Portnoy H (eds) Hydrocephalus. Raven Press, New York, pp 337–362

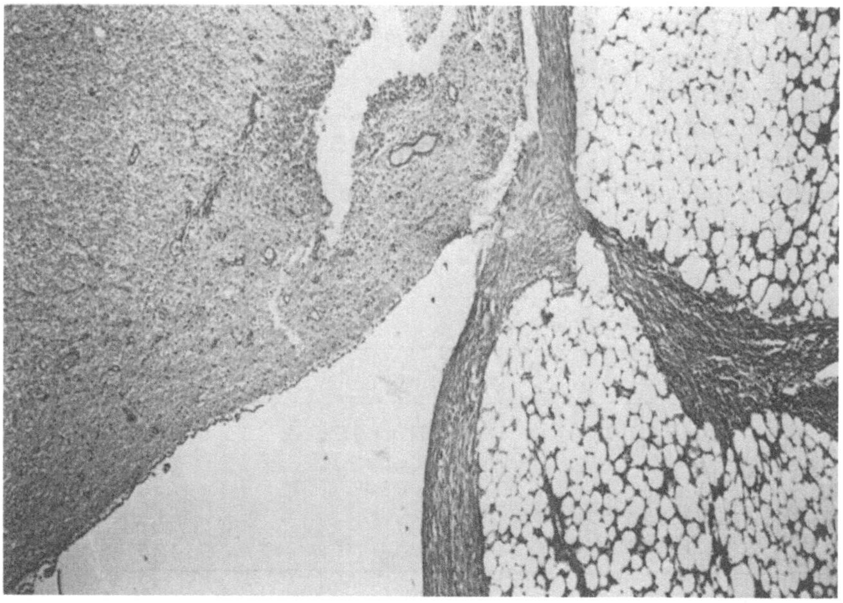

Fig. 2. Photomicrograph showing the normal appearance of the brain and omentum closed to the lateral ventricle

3. González-Darder JM, Barberá J, Cerdá-Nicolás M, Segura D, Broseta J, Barcia-Salorio JL (1984) Sequential morphological and functional changes in kaolin-induced hydrocephalus. J Neurosurg 61: 918–924
4. Levander B, Grandberg PO, Hindmarsh T (1978) Lumbo-omental shunt for drainage of cerebrospinal fluid in hydrocephalus. Acta Neurochir (Wien) 44: 1–11
5. Yaşargil Y, Yoneskawa Y, Denton I, Pirot H, Benes I (1974) Experimental intracranial transplantation of autrogenic omentum majus. J Neurosurg 39: 213–217
6. Yonekawa Y (1978) Experimental intracranial transplantation of omentum majus in dogs: a tentative new treatment for hydrocephalus and cerebral ischaemia. Arch Jap Chir 47: 3–17

Address for correspondence: J. González-Darder, M.D., Departamento de Neurocirugía, Hospital Mora, E-11003 Cádiz, Spain.

Acta Neurochirurgica, Suppl. 43, 162–167 (1988)

Proton Magnetic Resonance Imaging in Experimental Cerebral Ischaemia*

P. C. van Rijen, A. Verheem, and C. A. F. Tulleken

Department of Neurosurgery, University Hospital, Utrecht, The Netherlands

Summary

MRI was performed on 11 primates, 7 with a proximal and 4 with a distal MCA occlusion. Chronic implanted electrodes created only minor image disturbances. The development of oedema formation was visualised in repetitive imaging. The site of the MCA occlusion determined the infarct-size.

Keywords: Cerebral ischaemia; oedema; image artifacts; magnetic resonance imaging.

Introduction

In our research on acute and chronic focal ischaemia, we use the primate model with middle cerebral artery (MCA) occlusion. For measurements of local cerebral blood flow (CBF), somatosensory evoked potentials (SSEP) and EEG, small electrodes are introduced in the cortex and the deep structures. To assess the extension of infarction we first used X-ray CT, but the electrodes and also the high density bone of the skull disturb the image in the region of interest. In our experiments pathologic changes on CT were seen 24 to 36 hours after MCA occlusion.

Proton magnetic resonance imaging (MRI), sensitive to changes in content and physiochemical properties of tissue water, can be used to detect cerebral oedema produced under ischaemic conditions[4, 15].

In this study the development of cerebral infarction in the acute and chronic phase was followed. The influence of changing the site of MCA occlusion on the degree of infarction was determined. It was tested whether it was possible to perform MRI without image disturbance by the intracerebral electrodes.

Method and Material

Imaging was performed on 11 adult monkeys of the Macaca Iris species, with a weight between 2–4.5 kg. The mean cerebral volume was 60 cc. Experimental focal ischaemia was achieved by occluding the right MCA in the awake state.

Approximately one week before occlusion, under halothane anaesthesia, a ligature was guided around the MCA using the transorbital approach[11]. The ligature was pulled through a silicone catheter and fixed under the skin of the skull. The orbit was closed and the animal was allowed to recover. To achieve occlusion, a small skin incision was made under local anaesthesia, the ligature identified, tightened and secured.

– In 7 animals (group 1) the ligature was placed just proximal to the branching of the lenticulostriate arteries (LSA) which supply the basal ganglia[14]. Four weeks before occlusion an array of 26 intracerebral (in cortex and thalamus) and 4 epidural electrodes were placed on stereotactically defined positions. The intracerebral electrodes were made of a platinum wire of 0.2 mm diameter with a copper connecting wire. Epidural electrodes were made of 0.8 or 0.6 mm diameter silver wire and copper connection. All used materials were non-ferromagnetic. A connector was fixated on the skull with teflon screws. Electrodes and connector were sealed with dental acrylic.

In this group of seven animals, the infarctsize and -location was determined and the disturbance of the electrodes on image quality was tested.

– In 4 animals (group 2) the ligature was situated distally to the branching of the LSA and just proximally to the bifurcation of the MCA in order to decrease the volume of infarction.

In three of these animals no electrodes were introduced and only a dummy connector was placed on the skull. In this group, series of imaging were performed to detect the development of infarction in time, and relaxation times were measured.

By using a cradle on which the connector could be tightly screwed, the head was immobilized and the monkey was restrained in a comfortable way. During imaging the electrodes were not connected to any measurement equipment.

MRI scans were performed using a Philips Gyroscan whole body imager with a magnetic field of 0.5 Tesla (monkeys 1–5) and 1.5 Tesla in the remaining experiments. A 14 cm diameter surface coil placed in the coronal plane was used for signal detection. Because T 2 changes are the most sensitive parameters in detecting early infarction[5, 7, 10, 13] a multislice spinecho sequence was used with repetition time TR of 2,100 mS and two echotimes TE 1, TE 2 of 50 and 100 mS. The slice thickness was 7 mm with an interslice gap of 0.7 mm. Two acquisitions were made in a 256 × 256 matrix, resulting in a scan duration of 18 minutes.

* This work was funded through a grant from the Dutch Heart Foundation.

Coronal slices were made perpendicular to the cantomeatal line. When time series were made, care was taken that the slice position remained in the same plane.

Infarctsize was determined by subjective measurement of the hyperintense area in the 100 mS echo image. Using a region of interest (ROI) function in each slice, and multiplying the measured hyperintense area with the total slice thickness, a volume was calculated.

The transversal relaxation time T 2 was calculated using the formula $T2 = (TE2 - TE1)/Ln(I1/I2)$ in which I is the signal intensity in first and second echo image. In the slice where the insula was visualised the signal intensity of the infarction over a ROI of 1 cm^2 was measured. A control value was measured on the same location in the opposite hemisphere. These regions included white and grey matter.

In four experiments longitudinal realaxation time T 1 was measured using a mixed sequence. This was a combination of inversion recovery pulses (TR 2240 mS, inversion time 360 mS) and spin echo pulses (TR 760 mS, TE 50 mS 4 echoes), in single slice mode. T 1 was calculated using the formula $I = Pd.exp(-TE/T2) \cdot (1 - exp(-TR/T1))$ where Pd is the protodensity.

When the monkey survived after occlusion, the experiment was terminated after 38–42 days by administration of 14-C labeled deoxyglucose (50 μCi/kg) followed by the preparation of autoradiograms of coronal brain slices[12]. The deoxyglucose uptake is an indication for metabolic activity, on the autoradiograms the necrotic areas are clearly seen.

Results

Group 1, Proximal Occlusion, 7 animals

Proximal occlusion in the Macaca Iris monkey results within a few minutes in contralateral hemiparesis, facial palsy, hemianopia and a decreased level of consciousnes. CBF (hydrogen clearance method) in the MCA territory and SSEP amplitude after median nerve stimulation decreases within minutes.

MRI performed on the first or second day after occlusion always showed a large hyperintense area primary located over the insula and parts of the temporal-, frontal- and parietal cortex near the lateral sulcus. Also the subcortical white matter, head of the caudate nucleus, internal capsule and lenticular nucleus are involved. The volume of the hyperintense area was calculated, see Table 1.

In each experiment there was a marked compression of the lateral ventricle on the ischaemic side and a shift of the longitudinal cerebral sulcus to the nonischaemic side. 6 out of 7 monkeys died in the acute phase under signs of high intracranial pressure. Because of the rapid progression of this process, in three monkeys imaging could only be performed postmortem. Except for the absence of a flow phenomenon in the coronally cut anterior cerebral artery, pre- as well as postmortem images showed sharp boundaries of the hyperintense area and a good soft tissue contrast, even at 20 hours postmortem (monkey 3).

On coronal slices some disturbance of the image was seen (Fig. 1).

By taking a transversal image plane it could be dem-

Table 1

Monkey number	MRI at hours postocclusion	Infarct in cc	Survival in hours	Comment
1	56	6.7	68	severe midline shift
2	56 (postmortem)	11.8	54.5	severe midline shift
3	36 (postmortem)	12.1	17	severe midline shift
4	33	8.7	66	moderate midline shift
5	31	10.1	36	moderate midline shift
6	29 (postmortem)	12.5	19	severe midline shift

Table 2

Monkey 7 Time	Infarct in cc	T 2 right mS	T 2 left mS	T 1 right mS	T 1 left mS
preoccl.	—	74	75	840	832
29 hours	6.7	100	69	1178	924
78 hours	11.2	125	68	1367	909
6 days	12.2	133	65	1361	888
10 days	10.2	125	68	1332	836
21 days	8.9	119	69	1338	844
37 days	7.9	110	67	—	—

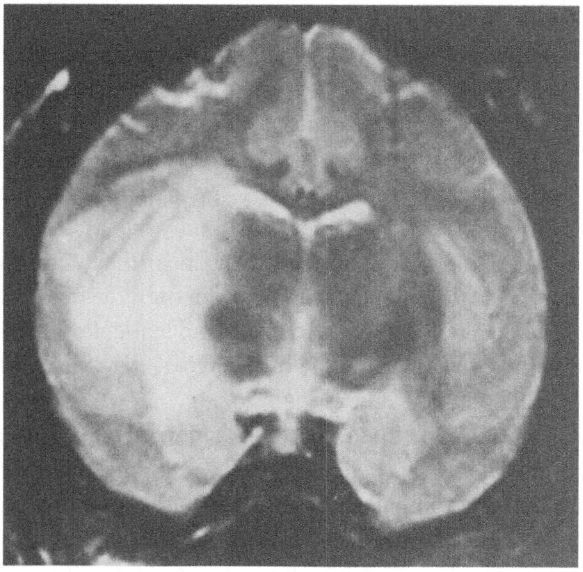

Fig. 1. Monkey 7, 29 hours after proximal MCA occlusion SE 2,100/100

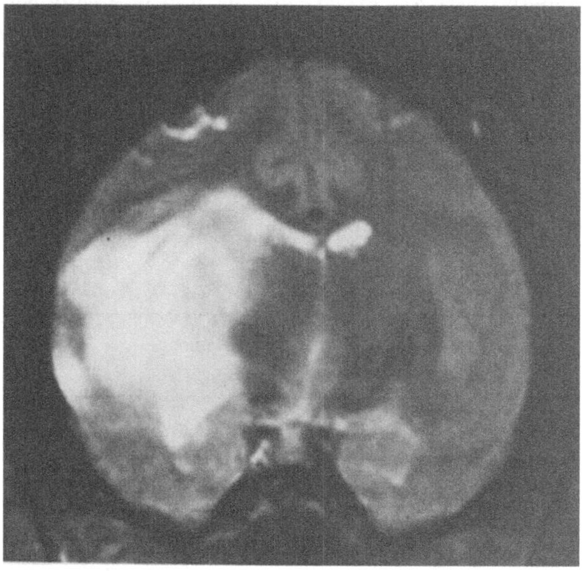

Fig. 2. Monkey 7, 78 hours after occlusion SE 2,100/100

onstrated that this disturbance was caused by the 0.8 mm diameter silver epidural electrodes. By using 0.6 mm electrodes in following experiments this problem clearly diminished. All intracortical electrodes showed as small black spots because of the absence of signal producing tissue.

Only one monkey survived the acute phase and was monitored up to 37 days after occlusion, see Table 2. In this monkey 29 hours after right MCA occlusion a stratified hyperintense area is seen in the right hemisphere (Fig. 1). There is a slight midline shift and no compression of the lateral ventricle. In Fig. 1 the track of a platinum depth electrode in the left hemisphere is seen. At 78 hours postocclusion there is a midline shift and a compression of the lateral ventricle. The stratification is now more pronounced. Hyperintense show the cortex of the insula and the grey matter of caudate and lenticular nucleus, separated by the less intense white matter of external and internal capsule (Fig. 2). 6 days postocclusion the midline shift is increased and the boundaries of the hyperintense areas are confluent. 10 days p.o. the midline shift is decreased as well as the infarct size. At 21 days p.o. there is only a minimal shift and the small hyperintense area has a patchy appearance. At 37 days p.o. the monkey has a slight paresis of the left extremities, a hemianopia and a decreased sensibility in left extremities and face. On MRI there remains a hyperintense area over the striate body and caudate nucleus and surrounding the insula (Fig. 3). This corresponds extremely well with the necrotic areas found in deoxyglucose preparation on the same level (Fig. 4). In this figure the decreased metabolic activity of the intact right cortex is seen, possibly evoked because of deafferentiation by the infarction.

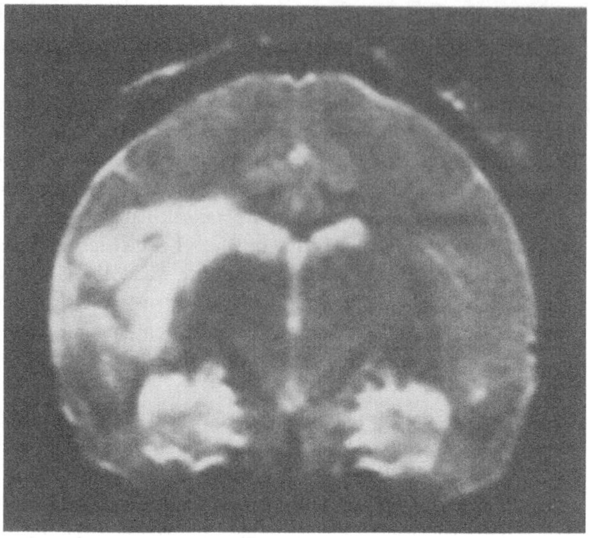

Fig. 3. Monkey 7, 37 days after occlusion SE 2,100/100

Group 2, Distal Occlusion, 3 animals without electrodes

In these 3 experiments series of images were made immediately after occlusion to detect the early changes (see Table 3).

Only one (number 8) out of these 3 monkeys survived the distal occlusion. On MRI there was only a moderate hyperintense area, first seen after 24 minutes, over the right insula region, with a maximum volume

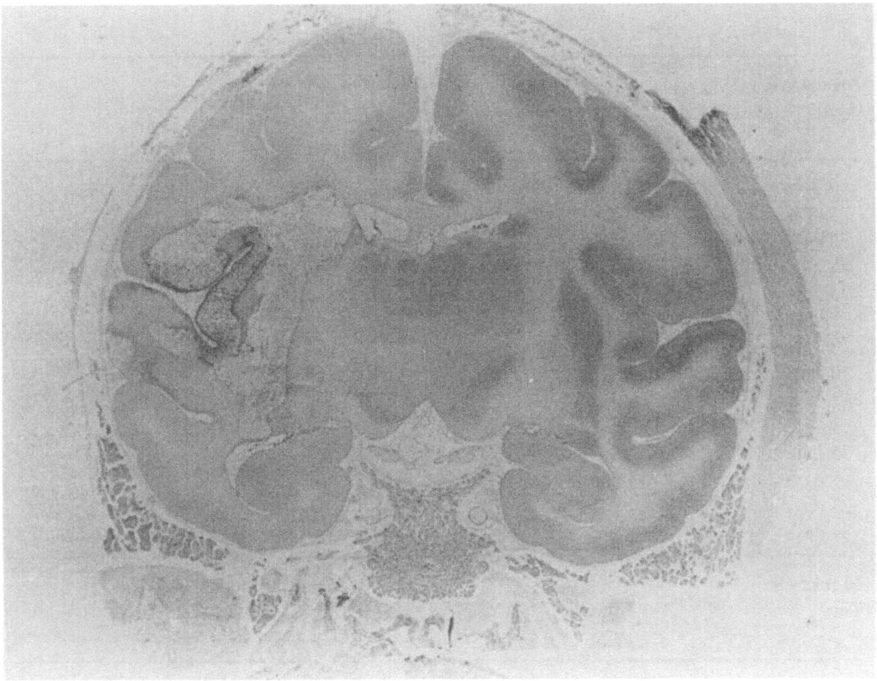

Fig. 4. Monkey 7, autoradiogram in plane of Fig. 3

measured at 5 days p.o. At that moment the insula, head of the caudate nucleus and a small lateral part of striate body are involved. There was no midline shift, and no compression of the lateral ventricle. Deoxyglucose autoradiograms showed necrotic areas consistent with the hyperintense areas on MRI performed 37 days p.o.

In monkey 9 there was a rapid progression in infarct volume. After 8 hours there is a compression of the lateral ventricle, and after 25 hours a midline shift is present. The hyperintense area is strictly bound to the gray matter of the insula. The caudate nucleus and striate body are not involved. In 53 hours the monkey died under signs of high intracranial pressure.

In monkey 10 there also is a rapid progression. The hyperintense area appears after 29 minutes over the insula, and spreads out over the temporal lobe and parts of the parietal and frontal cortex. After 5 hours a compression of the lateral ventricle exists and after 24 hours a midline shift is present. The basal nuclei are not involved. At 48 hours p.o. a treatment is started with mannitol i.v. (3 g/kg/24 h) and furosemide i.v. (3 mg/kg/24 h). After 6 hours there is a marked improvement of consciousness and activity which remains stable for the next three days. MRI 56 hours p.o. shows only a slight increase in infarct volume. 104 hours p.o. a decrease in infarct volume is visualised, and the relaxation times T 2 and T 1 are shortened. The monkey deceased after 7 days because of electrolyte derangement.

In monkey 11, with implanted electrodes, in the first three hours after occlusion no neurologic deficit, CBF or SSEP changes could be detected. MRI 8 hours p.o. showed two small hyperintense areas in the temporal and parietal gray and white matter near the insula. This became more clearly visualised after 56 hours in a T 2 reconstructed image in which T 2 values are represented by pixel intensity (Fig. 5). No midline shift or compression of the ventricles occurred. The size of the hyper-

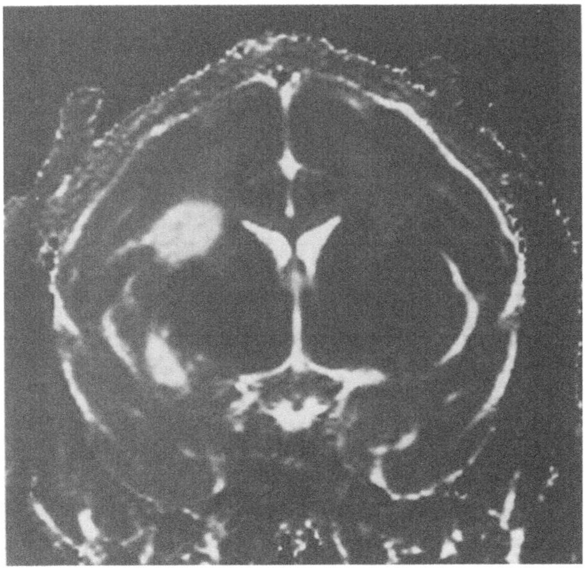

Fig. 5. Monkey 11, 56 hours after distal MCA occlusion T 2 reconstructed image

Table 3

Monkey 8 Time	Infarct in cc	T 2 right mS	T 2 left mS	Comment
0.24 h.	1.1	80	90	hyperintense area over insula
1.42 h.	2.1	85	83	
3.09 h.	2.9	83	86	
4.21 h.	3.0	86	85	
5.52 h.	3.3	88	81	
7.05 h.	3.7	86	82	
8.20 h.	3.5	91	85	
10.29 h.	3.9	97	84	
11.34 h.	4.3	97	86	
24.46 h.	5.2	112	83	
35.21 h.	5.2	117	78	
5 days	6.9	118	81	
23 days	3.7	118	86	
37 days	2.1	115	78	

Monkey 9 Time	Infarct in cc	T 2 right mS	T 2 left mS	Time	T 1 right mS	T 1 left mS
preoccl.	—	70	70	preoccl.	960	913
0.54 h.	hyperint.	78	70	1.55 h.	1054	981
2.59 h.	3.5	81	68	3.57 h.	1095	974
5.30 h.	5.7	79	72	7.03 h.	1151	966
8.18 h.	7.3	82	70	25.55 h.	1261	913
25.39 h.	10.1	102	68			
survival time 53 hours						

Monkey 10 Time	Infarct in cc	T 2 right mS	T 2 left mS	Time	T 1 right mS	T 1 left mS
preoccl.	—	69	69	3.42 h.	1097	967
0.29 h.	0.5	74	73	9.18 h.	1163	969
1.08 h.	1.4	75	68	26.57 h.	1391	931
1.50 h.	1.5	78	69	56.35 h.	1424	934
2.59 h.	1.9	84	73	105.08 h.	1191	957
5.01 h.	2.3	97	69			
6.51 h.	3.8	82	73			
8.39 h.	4.3	87	67			
23.43 h.	8.6	111	65			
28.25 h.	10.2	123	70			
55.45 h.	10.9	127	68			
104.25 h.	9.4	102	63			
survival time 7 days						

Table 4

Monkey 11 Time	Infarct in cc	T 2 right mS	T 2 left mS	T 1 right mS	T 1 left mS
preoccl.	—	77	76	668	678
8 hours	1.14	82	70	—	—
56 hours	1.16	109	73	884	688
5 days	1.23	118	73	988	697
10 days	0.92	115	71	945	734
21 days	0.91	112	72	950	769
41 days	0.31	95	72	889	705

intense areas diminished in time as did the relaxation time T 2 (see Table 4). Autoradiograms again showed a perfect correlation with the MRI findings.

Discussion

Oedema formation under ischaemic conditions can be interpreted as an accumulation of "free" water. Loss of energy rich phosphates leads to a disturbance of the Na/K homeostasis over the cell membrane and intracellular fluid accumulation begins. In time also leakage of fluid from capillary to extracellular space exists[6, 9]. In ischaemic areas, cellular degradation products attract fluid by osmotic force. For an increase in the volume of oedema under ischaemic conditions, it is essential that there is a collateral blood supply present[1].

In NMR physics the increased free water content implicates a prolongation of longitudinal (T 1) and transversal (T 2) relaxation times[3, 8]. In T 2 weighted spinecho images this prolongation is represented as an increased signal intensity.

In all distal occlusion experiments it was demonstrated that there was a rapid increase of water content in the gray matter of the insula, which could be detected even within 30 minutes. When the monkey survived, it was observed that oedema formation increased until the 6th day postocclusion. After this period the volume of the hyperintense area decreased as did the relaxation times. This indicates a decrease in tissue water content. Imaging after 37 days revealed a small hyperintense area which was conform to the necrotic areas in the autoradiogram. So hyperintensity in the acute and subacute phase does not always implicate an irreversible lesion.

The infarct localisation in the distal occlusion group is clearly outside the region of the basal ganglia, as was expected on the basis of the vascularisation pattern. Despite this distal occlusion, 2 monkeys developed a massive cortical and white matter oedema. It is illustrative that dehydrating therapy in one of these monkeys decreased the size of the hyperintense area and shortened the relaxation times T 1 and T 2.

In the proximal occlusion group, the infarction also comprised the basal ganglia as was expected. There was a massive oedema formation in every experiment. This indicates that although a large drop in perfusion of the MCA territory was present, there was still sufficient collateral circulation to supply the oedema fluid.

From the introduced electrodes, with a different magnetic susceptibility relative to brain tissue, only the large epidural electrodes produced localised magnetic field changes[2]. In the image this is presented as a black area surrounded by a hyperintense rim. The smaller intracortical electrodes did not influence the magnetic field, so it was possible to perform qualitative good imaging with electrodes of limited dimensions.

Images made with 0.5 Tesla magnetic field strength (monkeys 1–5) showed a lower signal to noise ratio compared to 1.5 Tesla images. Disturbance of the image by the epidural electrodes increased with higher magnetic field strength.

It is demonstrated that MRI is a very sensitive tool in observing the developing ischaemia. It proved to be possible to combine this method with functional observations by means of chronic implanted electrodes.

References

1. Bell BA (1985) CBF and time thresholds for the formation of ischaemic cerebral oedema, and effect of reperfusion in baboons. J Neurosurg 62: 31–41
2. Bellon EM (1986) MR artifacts: a review. AJR 147: 1271–1281
3. Brant-Zawadzki M (1984) NMR in experimental cerebral oedema: value of T 1 and T 2 calculations. AJNR 5: 125–129
4. Brant-Zawadzki M (1986) MR imaging of acute experimental ischaemia in cats. AJNR 7: 7–11
5. Bryan RN (1983) Nuclear magnetic resonance evaluation of stroke. Radiology 149: 189–192
6. Go KG (1984) Pathophysiological aspects of brain oedema. Clin Neurol Neurosurg 86: 77–80
7. Kato H (1985) Correlations between proton nuclear magnetic resonance imaging and retrospective histochemical images in experimental cerebral infarction. J Cereb Blood Flow Metab 5: 267–274
8. Kato H (1986) Characterisation of experimental ischaemic brain oedema utilizing proton nuclear magnetic resonance imaging. J Cereb Blood Flow Metab 6: 212–221
9. Kucharczyk W (1987) Magnetic resonance imaging of cerebral ischaemia and infarction. In: Kressel HY (ed) Magnetic resonance annual 1987. Raven Press, New York, pp 49–69
10. Mano I (1983) Proton nuclear magnetic resonance imaging of acute experimental cerebral ischaemia. Invest Radiol 17: 345–351
11. O'Brien MD (1973) Transorbital approach for occluding the middle cerebral artery without craniectomy. Stroke 4: 201–206
12. Sokoloff L (1977) The 14-C deoxyglucose method for the measurement of lokal cerebral glucose utilisation. J Neurochem 28: 897–916
13. Unger EC (1987) Acute cerebral infarction in monkeys an experimental study using MR imaging. Radiology 162: 789–795
14. Yonas H (1981) Selective lenticulostriate occlusion in the primate. Stroke 12: 567–572
15. Zabramski JM (1985) Magnetic resonance imaging: comparative study of radiofrequency pulse techniques in the evaluation of focal cerebral ischaemia. Neurosurgery 16: 502–510

Address for correspondence: P. C. van Rijen, Department of Neurosurgery, University Hospital, P.O. Box 16250, NL-3500 CG Utrecht, The Netherlands.

Acta Neurochirurgica, Suppl. 43, 168–171 (1988)

Intracellular Recordings from Neurones in Rat Cerebral Cortex During Hypoxia

I. A. Langmoen and **J. Berg-Johnsen**

Department of Neurosurgery, Rikshospitalet, Institute for Experimental Medical Research, University of Oslo and Department of Neurosurgery, Ullevål Hospital, Oslo, Norway

Summary

Neurones in rat hippocampal cortex were exposed to hypoxia while their membrane properties and responses to different kinds of stimuli were recorded with an intracellular electrode. The initial changes consisted of a small depolarization followed by a hyperpolarization. Following these early events the neurones lost their membrane potential through a large depolarization. Similar changes were observed in neurones where the Na/K-ATPase was blocked by ouabain. Responses to direct application of the transmitters GABA and glutamate, which was lost at this point, were restored by passive reestablishment of the membrane potential with current through the intracellular electrode.

Introduction

Cerebral ischaemia is characterized by loss of EEG activity and consciousness within seconds[4, 17]. Somewhat later there is a block of evoked responses[1, 23], a strong depolarization of neuronal membranes[5, 12] and marked changes in extracellular ion concentrations[8, 9]. As the mechanisms underlying these events have not been fully explained, we have: 1. studied changes in membrane potential and electrical resistance of neurones during hypoxia and have compared these changes to those occuring when energy dependant Na^+/K^+ exchange is blocked, and 2. tested responses to the putative transmitters glutamate and GABA during hypoxia.

Methods

The hippocampal slice technique has been described in detail previously[13] and will therefore be discussed only briefly. Young Wistar rats (150–200 g) were decapitated and the brain removed. Thin slices (400–600 μm) were cut by hand from one hippocampus and transferred to an incubation chamber of the Haas type[6]. The slices were superfused by a gas mixture (95% O_2, 5% CO_2) and bathed in artificial cerebrospinal fluid[13]. The temperature was kept at 32–35 °C in most experiments. Hypoxia was produced by changing to a gas mixture containing 95% N_2/5% CO_2.

Recording electrodes were pulled from thin borosilicate glass tubing and filled with 3 M potassium acetate. The electrodes were placed under visual control using a stereomicroscope. The potential across the cell membrane was monitored by a conventional DC amplifier with an active bridge circuit for passing current through the recording electrode. Simultaneous print-out was provided by an Apple II microcomputer or a chart recorder (Gould Brush 220). Data were further collected by photographs from a digital storage oscilloscope (Hitachi VC-6041). Electrodes for iontophoretic application of glutamate and GABA (tip diameter 1–2 μm) were made as described in earlier publications[2, 7].

Results

Stable intracellular recordings were obtained from 45 neurones penetrated in the pyramidal cell layer of the CA 1 region showing electrophysiological properties typical of pyramidal cells[13, 19, 20]. All cells showed stable recordings for at least 10 minutes before hypoxia, had overshooting action potential (amplitude exceeding 70 mV) and resting membrane potential of at least − 55 mV.

When the gas superfusing the slices was changed from 95% O_2/5% CO_2 to 95% N_2/5% CO_2 (Figs. 1 and 2 A), a large depolarization of the membrane was

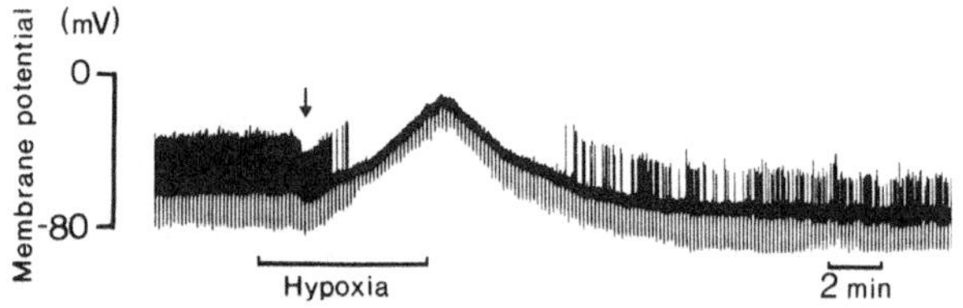

Fig. 1. Continuous recording of membrane potential from a neurone in the hippocampal CA 1 region. Downward vertical deflections represent membrane potential deflections to a current pulse of constant strength and thus are a measure of the input resistance of the neurone. Upward deflections are spontaneous action potentials which are shortened due to the limited speed of the chart recorder response. Note early hyperpolarization (arrow)

observed, usually starting after about 2 minutes. Smaller changes in the membrane potential were often seen prior to this. These consisted of a slight depolarization followed by a small hyperpolarization. The latter (Fig. 1, arrow) was associated with a decrease in input resistance. Within 2–4 minutes from the start of the large depolarization most neurones lost their membrane potential. In many neurones the depolarization progressed with a constant rate ($\delta v/\delta t$) until the membrane potential was close to 0 mV. The membrane potential then remained at or close to 0 mV ("plateau") until the preparation was reoxygenated. During the "plateau" the input resistance of these neurones was markedly reduced, often by 75–90%. On reoxygenation the membrane potential was restored with a variable time-course. In some neurones the repolarization was much faster than the depolarization, whereas in others it took up to 10 min. If hypoxia had lasted for 10 min or more (at temperatures close to normal body temperature) the membrane potential was generally not restored on reoxygenation.

Other neurones showed a very fast termination of the large depolarization; the membrane potential falling from $-40/-30$ to 0 mV within a few seconds (Fig. 2 A). These neurones generally lost their membrane resistance completely, as if the electrode had slipped out of the cell. The input resistance was, however, not restored by small upward and downward movements of the electrode tip and occasionally membrane potential and input resistance were re-established in these neurones upon reoxygenatation. It is not possible to conclude on whether these neurones suffered from a complete loss of electrical isolation from their surroundings due to the membrane being riddled during hypoxia or whether the electrode had just drifted out of the cell due to tissue movements produced by oedema.

There was often a difference between the membrane potential before and after hypoxia. Some neurones showed a posthypoxic hyperpolarization with reduced excitability. This phase was usually transient and was often followed by a slightly hyperexcitable state with moderate spontaneous activity. Occasionally, the neurones were left in a markedly hyperexcitable phase following hypoxia, with intense activity superimposed on a depolarized membrane potential.

If reduction of input resistance and loss of membrane potential during hypoxia is caused by energy dependant failure of sodium pumps (Na/K-ATPase), one would expect ouabain, which is known to block this pump, to produce similar changes. Accordingly,

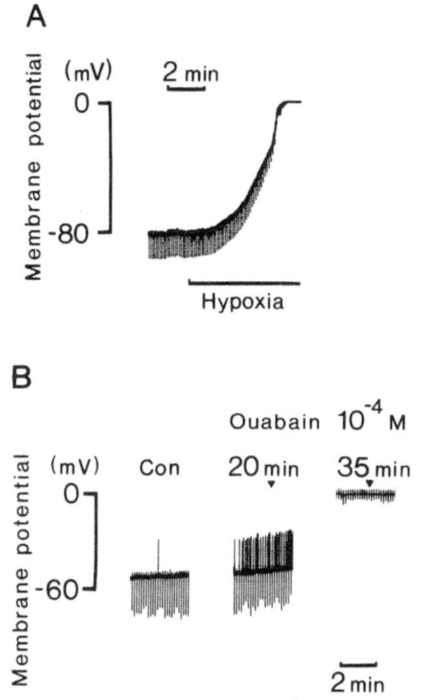

Fig. 2. Continuous recording of membrane potential and input resistance (given by the amplitude of the downward vertical deflections) from two different neurones in the hippocampal CA 1 region. A) During hypoxia. B) During administration of 10^{-4} M ouabain. Upward deflections 20 minutes after ouabain represent spontaneous action potentials which are shortened due to the limited speed of the chart recorder response

experiments were performed during administration of 10^{-4} M ouabain (mixed in the perfusate). As can be seen from Fig. 2 B, pump failure resulted in an early depolarization with hyperexcitability (20 min) and later in a loss of membrane potential with a pronounced fall in input resistance (35 min). The time course of the large depolarization was similar to that of hypoxia.

Glutamate and GABA are two of several probable neuro-transmitters between the hippocampal-pyramidal cell and its afferents. When topically applied to the dendrites both glutamate[14, 21] and GABA[2, 15] cause depolarization of the cell membrane. With the large depolarization produced by hypoxia, both the glutamate response and the depolarizing GABA response were gradually reduced, together with activity evoked by synaptic stimulation and by depolarizing pulses through the intracellular electrode. In neurones where the electrical resistance of the membrane was not lost during the "plateau", the membrane potential could be restored passively by passing current through the intracellular recording electrode. Following passive re-establishment of the membrane potential during hypoxia, responses to the different stimuli where tested again. As expected activity evoked by synaptic stimulation was still lost. Both the glutamate response and the depolarizing GABA response, however, were restored. Action potentials could still be evoked by depolarizing pulses early in the plateau phase, albeit threshold was increased.

Discussion

The initial changes of neurones exposed to hypoxia consisted of a small depolarization followed by a hyperpolarization of a few mV. This is in accordance with earlier reports[5, 11, 22]. Ouabain also tended to produce an initial small depolarization with hyperexcitability (c.f. Schiff and Somjen[18]), but no hyperpolarization.

Following these initial changes a large depolarization developed together with a pronounced reduction of the electrical resistance of the membrane. Both the depolarization and the fall in resistance could be reproduced by ouabain. Accordingly, loss of membrane potential and reduction in electrical resistance during hypoxia may well be explained by failure of energy dependant Na^+/K^+ exchange (c.f. Lipton and Whittingham[16]). The drop in input resistance may, however, also be a consequence of the depolarization as these neurones usually lose input resistance when they are depolarized (c.f. Hablitz and Langmoen[7]). As hypoxia causes a release of glutamate[3], and glutamate causes a depolarization of these neurones with a reversal potential close to 0 mV due to opening of Na^+/K^+ channels[7], release of transmitter glutamate would produce responses similar to what we have observed during hypoxia. At present it seems more likely, however, that the depolarization is caused by pump failure and that glutamate is released as a consequence of the depolarization, although other mechanisms still must be considered.

In a series of publications Hansen (c.f. ref.[10]) has described the time course of changes in extracellular ion activities during hypoxia. Usually there is a slow increase in $[K^+]$ up to 10–12 mM from where a fast rise up to about 60 mM occurs. Most of the delay (usually 2 minutes) which occured before the start of the depolarization in the present experiments was due to the volume of the input tubes carrying the gas mixture. When comparing our results with those of Hansen it seems to us that the membrane potential is lost before the large ion displacements of his "phase 2" occurs. This is supported by our observation that action potentials could be triggered in some cases (after passive repolarization) during the early part of the "plateau". This finding is difficult to explain if largely normal ion gradients across the membrane are still preserved at this point.

During hypoxia both the glutamate response and the depolarizing GABA response were lost. Both reappeared, however, following passive re-establishment of the membrane potential, suggesting that loss of synaptic transmission during hypoxia is not due to changes in the postsynaptic receptor.

Acknowledgements

This study was supported by Dr. Dedichens Foundation, The Norwegian Research Council for Science and the Humanities and Anders Jahres Foundation.

References

1. Andersen P (1960) Interhippocampal impulses. II. Apical dendritic activation of CA 1 neurons. Acta Physiol Scand 48: 178–208
2. Andersen P, Dingledine R, Gjerstad L, Langmoen IA, Mosfeldt Laursen A (1980) Two different responses of hippocampal pyramidal cells to application of gamma-amino-butyric acid (GABA). J Physiol 305: 279–296
3. Bosley TM, Woodhams PL, Gordon RD, Balazs R (1983) Effects of anoxia on the stimulated release of amino acid neurotransmitters in the cerebellum in vitro. J Neurochem 40: 189–201
4. Dennis C, Kabat H (1939) Behaviour of dogs after complete temporary arrest of the cephalic circulation. Proc Soc Exp Biol Med 40: 559–581
5. Grossman RG, Williams VF (1971) Electrical activity and ultrastructure of cortical neurons and synapses in ischaemia. In: Brierly JB, Meldrum BS (eds) Brain hypoxia. Heineman, London, pp 61–75
6. Haas HL, Schaerer B, Vosmansky M (1979) A simple perfusion chamber for the study of nervous tissue slices in vitro. J Neurosci Methods 1: 323–325
7. Hablitz JJ, Langmoen IA (1982) Excitation of hippocampal pyramidal cells by glutamate in the guinea-pig and rat. J Physiol 325: 317–331
8. Hansen AJ (1977) Extracellular potassium concentration in juvenile and adult rat brain cortex during anoxia. Acta Physiol Scand 99: 412–420
9. Hansen AJ (1978) The extracellular potassium concentration in brain cortex following ischaemia in hypo- and hyperglycemic rats. Acta Physiol Scand 102: 324–329
10. Hansen AJ (1985) Effect of anoxia on ion distribution in the brain. Physiol Rev 65: 101–148
11. Hansen AJ, Hounsgaard J, Jahnsen H (1982) Anoxia increases potassium conductance in hippocampal nerve cells. Acta Physiol Scand 115: 301–310
12. Kolmodin GM, Skoglund CR (1959) Influence of asphyxia on membrane potential level and action potentials of spinal moto- and interneurons. Acta Physiol Scand 45: 1–18
13. Langmoen IA, Andersen P (1981) The hippocampal slice in vitro. A description of the technique and some examples of the opportunities it offers. In: Kerkut GA, Wheal HV (eds) Electrophysiology of isolated mammalian CNS preparations. Academic Press, London, pp 55–105

14. Langmoen IA, Hablitz JJ (1981) Reversal potential for glutamate responses in hippocampal pyramidal cells. Neurosci Letters 23: 61–65
15. Langmoen IA, Andersen P, Gjerstad L, Mosfeldt Laursen A, Ganes T (1978) Two separate effects of GABA on hippocampal pyramidal cells in vitro. Acta Physiol Scand 102: 28–29 A
16. Lipton P, Whittingham TS (1979) The effect of hypoxia on evoked potentials in the in vitro hippocampus. J Physiol 287: 427–438
17. Rossen R, Kabat H, Anderson JP (1943) Acute arrest of cerebral circulation in man. Arch Neurol Psychiat (Chic) 50: 510–528
18. Schiff SJ, Somjen GG (1985) Hyperexcitability following moderate hypoxia in hippocampal tissue slices. Brain Res 337: 337–340
19. Schwartzkroin PA (1975) Characteristics of CA 1 neurons recorded intracellularly in the hippocampal in vitro slice preparation. Brain Res 85: 423–436
20. Schwartzkroin PA (1977) Further characteristics of hippocampal CA 1 cells in vitro. Brain Res 128: 53–68
21. Schwartzkroin PA, Andersen P (1975) Glutamic acid sensitivity of dendrites in hippocampal slices in vitro. In: Kreutzberg GW (ed) Advances in neurology, vol 12. Raven Press, New York, pp 45–51
22. Speckmann E-J, Caspers H, Sokolov W (1970) Aktivitätsänderung spinaler Neurone während und nach einer Asphyxia. Pflügers Arch Ges Physiol 319: 122 – 138
23. Williams V, Grossman RG (1970) Ultrastructure of cortical synapses after the failure of presynaptic activity in ischaemia. Anat Rec 166: 131–142

Address for correspondence: Iver A. Langmoen, M.D., PhD., Department of Neurosurgery, National Hospital (Rikshospitalet), N-0027 Oslo 1, Norway.

Acta Neurochirurgica, Suppl. 43, 172–176 (1988)

Pathophysiological Investigation of Experimental Cerebral Ischaemia Using in vivo ^{31}P-NMR Spectroscopy and ^{1}H-MRI

T. Higuchi, S. Naruse, Y. Horikawa, C. Tanaka, T. Ebisu, K. Yamamoto, and **K. Hirakawa**

Department of Neurosurgery, Kyoto Prefectural University of Medicine, Kyoto, Japan

Summary

The cerebral energy metabolism and brain oedema were investigated in three experimental cerebral ischaemia models using ^{31}P-NMR spectroscopy (MRS) and ^{1}H-NMR imaging (MRI) in the same subject animal. These measurements were performed also in experimental brain oedema models and the findings were compared with each other. ^{31}P-MRS showed an ischaemic pattern in all of the cerebral ischaemia models, that is, ATP and PCr peaks decreased, and the Pi peak increased and shifted to a higher resonant frequency. However, ^{31}P-MRS did not show any remarkable change in the brain oedema models. On the other hand, ^{1}H-MRI clearly demonstrated brain oedema in the brain oedema model. In the cerebral ischaemia models, ^{1}H-MRI findings differed depending upon the type of model, namely the most marked brain oedema was detected in the unilateral middle cerebral arterial occlusion model and no marked change was detected in the temporary four vessel occlusion model. It was thought that this difference depended on the severity of the ischaemic insult. Accordingly, the fundamental pathophysiological problem of cerebral ischaemia was the energy metabolism disturbance with the brain oedema being associated with this disturbance but occuring secondarily. However, in the brain oedema model the main pathological change was the increase in tissue water.

Keywords: Cerebral ischaemia; brain oedema; ^{31}P-MRS (NMR spectroscopy); ^{1}H-MRI (NMR-Imaging).

Introduction

Since the introduction of in vivo NMR many studies have been conducted on pathological brain applying ^{31}P-NMR to the study of the energy metabolism and ^{1}H-NMR to brain oedema. We reported that sequential changes in ^{31}P-NMR spectra (^{31}P-MRS) following brain ischaemia indicate abrupt disturbances in the energy production of the brain and that the prolongation of proton relaxation times (both the longitudinal relaxation time (T 1) and the transverse relaxation time (T 2)), in ischaemic brain tissue indicates an increase in the free water content and changes in the state of the tissue water[2, 3, 5, 7–9, 11, 13]. Moreover, ^{1}H-NMR images (^{1}H-MRI) clearly demonstrate the spatial distribution of these changes in live animals. However, the previously reported studies have been performed independently using only ^{31}P-MRS or ^{1}H-NMR. In this study we tried to investigate both the brain energy metabolism and ischaemic brain oedema in the same subject animal using a single NMR apparatus. Several types of experimental cerebral ischaemia models were examined and characteristic findings were detected which differed markedly from the results obtained from the investigation of experimental brain oedema models.

Materials and Methods

We used 48 Wistar rats, weighing about 200 g each, and 16 mongolian gerbils, weighing about 80 g each. Experimental cerebral ischaemia was induced in Wistar rats by 1) temporary 30 min four vessel occlusion, by a modification of the method reported by Pulsinelli and Brierley[14], 2) unilateral middle cerebral arterial occlusion[16], and 3) in mongolian gerbils by unilateral carotid arterial occlusion.

Experimental vasogenic and cytotoxic brain oedema were induced by cold injury and by triethyl tin (TET) intoxication respectively in Wistar rats with our previously reported procedure[7]. All of the experimental models were prepared under sodium pentobarbital anaesthesia (35 mg/kg i.p.) which was maintained by 0.5% halothane inhalation during the NMR measurements. ^{31}P-MRS and ^{1}H-MRI measurements were performed sequentally up to a maximum of 7 days after the induction of the lesions.

Both ^{31}P-MRS and ^{1}H-MRI were performed with the same SCM 200 spectrometer (JEOL, Tokyo, Japan) which has a superconducting magnet with a 9-cm bore and is operated at 4.7 Tesla. The ^{31}P-MRS were measured by using a surface coil, 1.1 cm in diameter, with a pulse width of 14 μsec. Scanning was performed repeatedly 400 times at 1.9-second intervals. A 30-Hz noise filter was applied to the free-induction decay signal prior to Fourier transformation. Each ^{31}P-MRS measurement took about 15 minutes. The ^{1}H-MRI were obtained by using a 7-cm, saddle-type coil. Spin echo (SE) images with short (36 msec) and long (96 msec) echo times (Te) and Inversion recovery (IR) images with a 300 msec interpulse interval (Td) were obtained by the ordinary two-dimensional Fourier

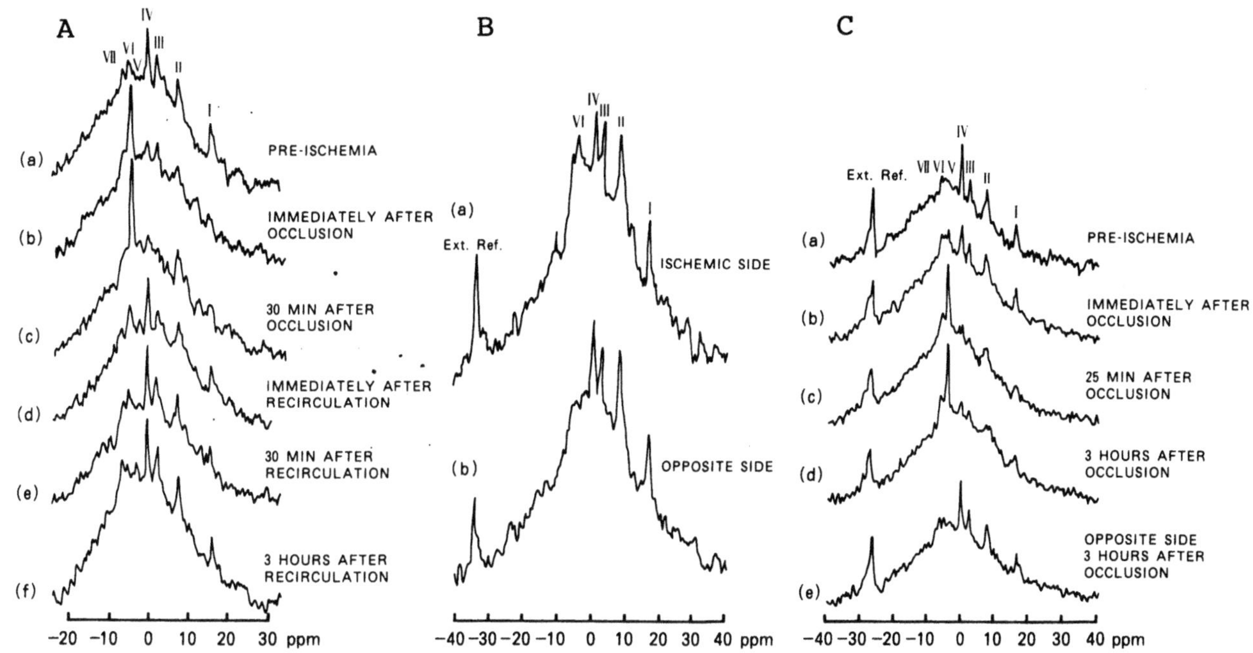

Fig. 1. Sequential ^{31}P-MRS changes in three experimental cerebral ischaemia models. A) temporary 30 min four vessel occlusion model in rat, B) unilateral middle cerebral arterial occlusion model in rat, C) unilateral carotid arterial occlusion model in mongolian gerbil

transformation (2 DFT) method with a 3.0 sec repetition time each. The data matrix pixels numbered 256 × 256. Each image measurement took a total of about 10 minutes.

Results

A typical ^{31}P-MRS of a normal brain demonstrated seven sharp peaks resulting from the highly mobile phosphorus in the tissue. These peaks were identified as follows, in order from the higher resonant frequency: (I) β-ATP, (II) α-ATP and α-ADP with contribution from NAD^+/NADH, (III) γ-ATP and β-ADP, (IV) phosphocreatine (PCr), (V) phosphodiesters (PDE), (VI) inorganic phosphate (Pi), (VII) phosphomonoesters (PME) (Fig. 1). ^{1}H-MRI of a normal brain clearly demonstrated the cerebral cortex, white matter, hippocampus, caudate nucleus, thalamus and cerebrospinal fluid (Fig. 2).

1. Temporary Four Vessel Occlusion Model in Rats

Immediately after the interruption of cerebral circulation resonant ATP and PCr peaks decreased, while the Pi peak increased. During the 30 min ischaemia the chemical shift of the increased Pi peak decreased compared to the chemical shift of the Pi peak in a normal state. Tissue pH calculated from this chemical shift decreased to 6.52 from the normal value of 7.27 during the ischaemic period. ATP and PCr peaks recovered and the Pi peak decreased rapidly when the cerebral circulation was restored. The Pi peak also began to shift back to the normal position. The ^{31}P-MRS showed an almost normal pattern within 30 minutes after the recirculation (Fig. 1). On the other hand, ischaemic brain oedema was not detected by ^{1}H-MRI, even though measurements were continued sequentially for 48 hours after the ischaemia was initiated (Fig. 2).

2. Unilateral Middle Cerebral Arterial (MCA) Occlusion Model in Rats

In this model the ^{31}P-MRS measurements were quite difficult, because of the existence of temporal muscle and subcutaneous effusion caused by the MCA occlusion procedure. However, the ^{31}P-MRS obtained after the resection of temporal muscle and scalp and drying of the surface of the skull showed slightly decreased ATP and PCr peaks and a slightly increased Pi peak (Fig. 1). It was supposed that the reason why the changes in the ^{31}P-MRS in this model were not as marked as in the temporary four vessel occlusion model was that the area measured contained some normal brain tissue. ^{1}H-MRI clearly demonstrated the ischaemic brain oedema extension as a marked high intensity signal area in the parietal cortex in all images even with varying pulse sequences (two SE images and one IR image) (Fig. 2).

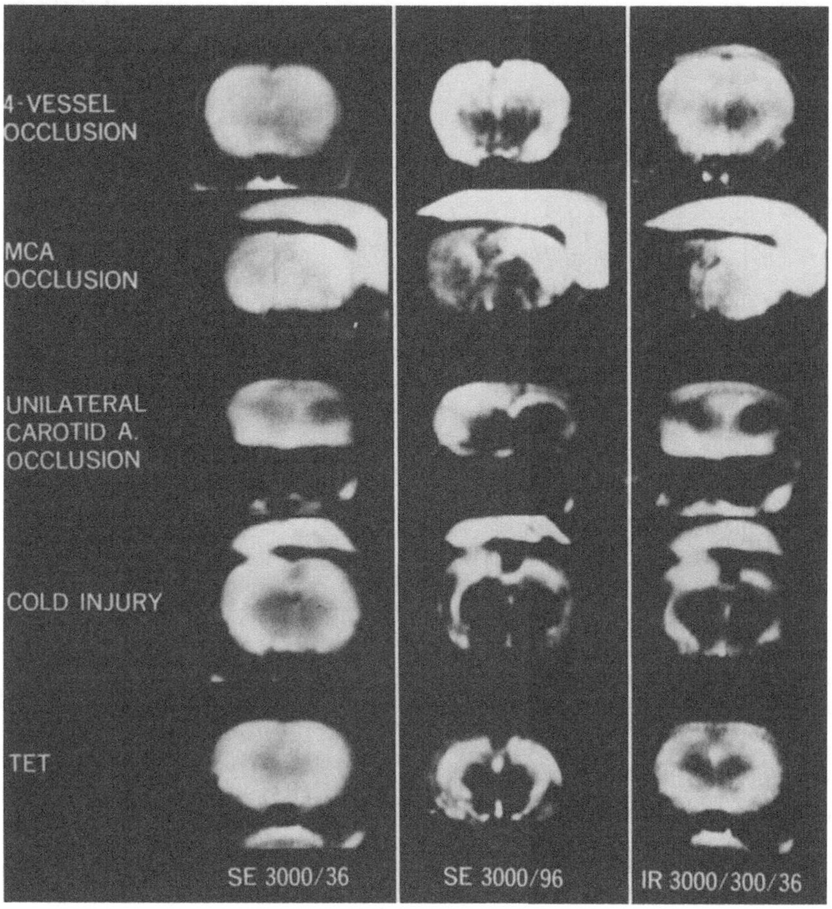

Fig. 2. A typical ^{1}H-MRI of each experimental model

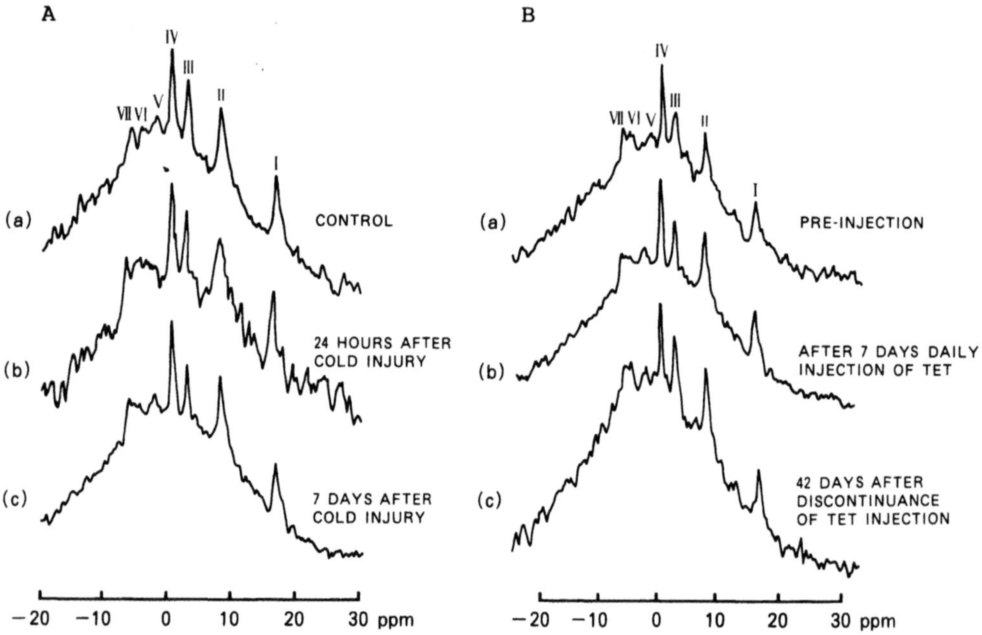

Fig. 3. Sequential ^{31}P-MRS changes in two experimental brain oedema models. A) Cold injury brain oedema model, B) TET intoxication induced brain oedema model

3. Unilateral Carotid Arterial Occlusion Model in Mongolian Gerbils

In this model the unilateral carotid artery was occluded but cerebral ischaemia did not occur in all of the animals[2]. The incidence of cerebral ischaemia was about 30%. In the gerbils in which cerebral ischaemia was induced, ^{31}P-MRS obtained from the ligated hemisphere showed an ischaemic pattern similar to that measured during the ischaemic period in the temporary four vessel occlusion model in rats. However, the ^{31}P-MRS obtained from the contralateral hemisphere showed a normal pattern (Fig. 1). ^{1}H-MRI demonstrated ischaemic brain oedema as a high intensity signal area over the whole hemisphere on the ligated side in SE images with a long Te. However, the signal intensity of the brain oedema region in this model was weaker than that demonstrated in the MCA occlusion model with rats (Fig. 2).

4. Experimental Brain Oedema Model in Rats

^{31}P-MRS showed an almost normal pattern in both models (Fig. 3). On the other hand, ^{1}H-MRI clearly demonstrated the brain oedema extension as a marked high intensity signal area in all images regardless of the pulse sequence. In the cold injury induced brain oedema model, the lesion occurred in the parietal cortex and extended to the ipsilateral white matter and even to the contralateral white matter through the corpus callosum. In the TET intoxication induced brain oedema model, the lesion was confirmed bilaterally throughout the white matter (Fig. 2).

Discussion

MRI (^{1}H-NMR imaging) has become a practical clinical tool in the diagnosis of neurological disease. Especially in the diagnosis of ischaemic brain disease, MRI is thought to be able to demonstrate the region of ischaemic brain oedema in the earliest stage, because it can detect minimal changes in tissue water content. In this study, ^{1}H-MRI demonstrated clearly the extension and the degree of brain oedema in experimental animals (Fig. 2). As a new biochemical technique ^{31}P-MRS has come to be commonly used to study energy metabolism in live animals without excision of tissue or extraction of compounds[1]. We have reported energy metabolism changes in experimental cerebral ischaemia and brain tumour and confirmed the usefulness of this technique for the evaluation of several different treatments in prior papers[2–12]. Recently, it has become possible to obtain ^{31}P-MRS of human organs using a MRI system operated with more than 1.5 Tesla. By using both ^{31}P-MRS and ^{1}H-MRI the pathophysiology of the disease can be analyzed from many different points of view. Given such a new set of circumstances, it has become important to investigate experimental pathological models using both ^{31}P-MRS and ^{1}H-MRI in order to gain the necessary basic data for clinical use of these techniques.

In this study, ^{31}P-MRS showed drastic changes in the experimental cerebral ischaemia models (Fig. 1), that is, the immediate consumption of high energy phosphorus compounds (ATP and PCr), the accumulation of inorganic phosphate, and intracellular acidosis. On the contrary, in the experimental brain oedema models, ^{31}P-MRS did not show any marked changes (Fig. 3) although ^{1}H-MRI clearly demonstrated an increase in both water content and T 1 and T 2 values in that area of the brain with oedema. The reason for this difference was thought to be the pathophysiological differences between the two models. Initially, in the experimental cerebral ischaemia model, the disturbance of energy production occurred immediately after the interruption of cerebral circulation. Brain oedema was associated with this disturbance occuring secondarily. On the other hand, in the experimental brain oedema model, energy metabolism disturbances were not detectable by ^{31}P-MRS measurements and the main pathological change was the increase in tissue water.

The extent and the degree of the brain oedema associated with the brain ischaemia differ depending upon the severity of the induced ischaemic insult, which depends on the duration of the cerebral ischaemia and the location of the cerebral arterial occlusion. Even if the ischaemia is made to affect the whole brain and the cerebral energy metabolism is disturbed severely, if the duration of the ischaemic insult is short enough as in the temporary four vessel occlusion model in rats, it is supposed that the cerebral energy metabolism disturbance is reversible and brain oedema does not occur so severely (not clearly detected by ^{1}H-MRI). As for the water content in brain tissue after a 30 minute four vessel occlusion, it has been previously reported that there was an increase of less than 2%[15]. When the cerebral artery is occluded permanently as in the MCA occlusion model in rats and the unilateral carotid arterial occlusion model in gerbils, the severity of the brain oedema is thought to depend upon the quantity of the collateral circulation after the arterial occlusion. In general, when the cerebral artery is occluded peripherally, as in the MCA occlusion model in rats, the collateral circulation supplies less blood to the affected area. Accordingly, the severity of the ischaemic brain oedema should depend on the peripheral location of the occlusion. In our study, the brain oedema in the MCA occlusion model in rats was clearly demonstrated as a high signal intensity area but was more spatially limited than it was in the unilateral carotid arterial occlusion model in gerbils. The increase in cerebral water content, measured in each hemisphere after the ischaemic insult, was about 3% in both models. However, because the extension of brain oedema in the gerbil model was larger than it was in the MCA occlusion model in rats, the increase in the water content in the brain oedema area in the latter was supposed to be larger than it was in the former.

Since ^{1}H-MRI is very sensitive to increases in tissue water content and to changes in the state of tissue water it is very practical for use in the diagnosis of ischaemic brain disease. However, the fundamental pathophysiological problem of the disease is the energy metabolism disturbance which occurs immediately after the interruption of cerebral circulation and before the increase in tissue water. Accordingly, it is important to investigate the cerebral energy metabolism in order to diagnose the disease in the earliest stage. ^{31}P-MRS is very practical for use in measuring this energy metabolism and is expected to become a new diagnostic tool for ischaemic brain disease.

References

1. Ackerman JJH, Grove TH, Wong GG *et al* (1980) Mapping of metabolites in whole animals by ^{31}P-NMR using surface coils. Nature 283: 167–170
2. Higuchi T, Naruse S, Horikawa Y *et al* (1986) Energy metabolism variation in experimental cerebral ischaemia in mongolian gerbils examined by using in vivo ^{31}P-NMR. Jpn J Stroke 8: 433–439 (in Jap. with Eng. abstr)
3. Horikawa Y, Naruse S, Hirakwa Y *et al* (1985) In vivo studies of energy metabolism in experimental cerebral ischaemia using topical magnetic resonance. Changes in ^{31}P-nuclear magnetic resonance spectra compared with electroencephalograms and regional cerebral blood flow. J Cereb Blood Flow Metabol 5: 235–240
4. Naruse S, Higuchi T, Horikawa Y *et al* (1986) Radiofrequency hyperthermia with successive monitoring of its effects on tumours using NMR spectroscopy. Proc Natl Acad Sci USA 83: 8343–8347
5. Naruse S, Hirakawa K (1986) Brain oedema studied by magnetic resonance. Seminars in Neurology 6: 53–63
6. Naruse S, Hirakawa K, Horikawa Y *et al* (1985) Measurements of in vivo ^{31}P-nuclear magnetic resonance spectra in neuroectodermal tumours for the evaluation of the effects of chemotherapy. Cancer Res 45: 2429–2433
7. Naruse S, Horikawa Y, Tanaka C *et al* (1982) Proton nuclear magnetic resonance studies on brain oedema. J Neurosurg 56: 747–752
8. Naruse S, Horikawa Y, Tanaka C *et al* (1984) In vivo measurement of energy metabolism and the concomitant monitoring of electroencephalogram in experimental cerebral ischaemia. Brain Res 296: 370–372
9. Naruse S, Horikawa Y, Tanaka C *et al* (1984) Measurements of in vivo energy metabolism in experimental cerebral ischaemia using ^{31}P-NMR for the evaluation of protective effects of perfluorochemicals and glycerol. Neurol Res 6: 169–175
10. Naruse S, Horikawa Y, Tanaka C *et al* (1985) Observations of energy metabolism in neuroectodermal tumours using in vivo ^{31}P-NMR. Magnetic Resonance Imaging 3: 117–123
11. Naruse S, Horikawa Y, Tanaka C *et al* (1986) Significance of proton relaxation time measurement in brain oedema, cerebral infarction and brain tumours. Magnetic Resonance Imaging 4: 293–304
12. Naruse S, Horikawa Y, Tanaka C *et al* (1986) Evaluation of the effects of photoradiation therapy on brain tumours with in vivo P-31 MR spectroscopy. Radiology 160: 827–830
13. Naruse S, Takada S, Koizuka I *et al* (1983) In vivo ^{31}P NMR studies on experimental cerebral infarction. Jpn J Physiol 33: 19–28
14. Pulsinelli WA, Brierly JB (1978) A new model of bilateral hemispheric ischaemia in the unanesthetized rat. Stroke 10: 262–272
15. Take Y, Yamazaki N, Fukuda N *et al* (1984) Physiological and biochemical study of temporary cerebral ischaemic rats produced by bilateral vertebral and carotid artery occlusion. Folia Pharamcol Japon 84: 471–483 (in Jap. with Eng. abstr)
16. Tamura A, Graham DI, McCulloch J *et al* (1981) Focal cerebral ischaemia in the rat: 1. Description of technique and early neuropathological consequences following middle cerebral artery occlusion. J Cereb Blood Flow Metabol 1: 53–60

Address for correspondence: Toshihiro Higuchi, M.D., Department of Neurosurgery, Kyoto Prefectural University of Medicine, Kawaramachi Hirokoji, Kamigyo-ku, Kyoto 602, Japan.

Acta Neurochirurgica, Suppl. 43, 177–181 (1988)

Effect of Nimodipine on Mitochondrial Respiration in Different Rat Brain Areas After Subarachnoid Haemorrhage*

R. Rodriguez y Baena, P. Gaetani, V. Silvani, G. Spanu, and **F. Marzatico**[1]

Department of Surgery, Neurosurgical Section, Faculty of Medicine and [1] Institute of Pharmacology, Faculty of Science, University of Pavia, Italy

Summary

The mitochondrial respiration was evaluated in three different rat brain areas (cerebral cortex, hippocampus and brain stem) after experimental subarachnoid haemorrhage (SAH). The haemorrhage was induced by injecting 0.35 ml of autologous arterial blood into cisterna magna. Intravenous administration of Nimodipine (2 μg/kg/min for 30 minutes) was started immediately after the haemorrhage induction. At the set time (1 hour after SAH procedure), animals were sacrificed and non-synaptic mitochondria from the above mentioned areas were isolated. The following respiratory parameters were evaluated utilizing glutamate plus malate and succinate plus rotenone as substrates: state 3, state 4, uncoupled state, respiratory control ratio (RCR) and ADP/O ratio. SAH significantly influences respiratory parameters, mainly RCR; the cerebral cortex and brain stem seem to be more sensitive during the acute phase of vasospasm which follows SAH procedure. Nimodipine treatment significantly ameliorates mitochondrial respiratory conditions.

Keywords: Mitochondria; experimental subarachnoid haemorrhage; nimodipine; brain areas.

Introduction

The calcium-theory of neuronal cell damage has been recently adapted to subarachnoid haemorrhage (SAH)[16]. An altered homeostasis of Ca^{++} in consequence of the haemorrhage would lead to membrane dysfunction, significant brain metabolic disturbances and modifications of the membrane electrochemical gradient[27] until complete neuronal death[16]. Despite lower rates of patients which present fixed neurological deficits, attention has been drawn to the diffuse encephalopathy caused by subarachnoid deposition of blood and its effects on brain energy metabolism[10] leading to long-term cognitive disturbances[20]. Nimodipine (Bay e-9736) has been proposed for the treatment of patients affected by aneurysmal SAH to antagonize vasospasm mechanisms and to improve both cerebral blood flow and neuronal resistance against the altered homeostatic conditions[12, 26]. Clinical trials with Nimodipine demonstrated a non significant reduction of vasospasm and a marked improvement of emotional and psychological sequelae in patients operated for aneurysmal SAH[3, 4, 20]. Experimental work on monkeys and rats demonstrated that during SAH the $CMRO_2$ significantly decreases, suggesting a significant impairment of the oxidative machinery of the brain[9, 10]. The mitochondria play a pivotal role in oxidative brain metabolism, showing the sensibility of their respiratory parameters after ischaemic-anoxic injury[11, 15, 23, 24]. The purpose of this study is to verify in different rat brain areas whether Nimodipine could improve mitochondrial respiration after experimental induction of SAH.

Material and Methods

The experiments were performed on male Sprague-Dawley rats (Charles River strain, Italy), weighing 325–350 g.

Experimental SAH Induction

Anaesthesia is induced with 3% halothane (70%: 30% N_2O: O_2) and maintained with 0.75% halothane (in a gas mixture). After a scalp incision on the dorsal midline, a burr hole is performed, by refrigerated twist drill, at the interparietal/occipital suture connection, just on the midline. The bone is carefully cleaned and a small catheter (polyethylene PE-10, 0.62 mm OD) is inserted into cisterna magna. Suitable placement of the catheter is assessed testing cerebrospinal fluid (CSF) flowing throughout the catheter and observing (with magnification) the lower distal part of the catheter through the atlanto-occipital membrane.

Two–three days after the catheter implantation, rats are anaesthetized as described before. Arterial blood samples are anaerobically drawn each 10 minutes and pH, pCO_2, pO_2 and arterial blood pressure were maintained in standard conditions. Autologous

* This experimental work was supported in part by a Grant of Regione Lombardia, Milano, Italy, 1985, and by Bayer Italia S.p.A.

Table 1. *Respiratory Activities of Cerebral Cortex Non-Synaptic Mitochondria.* Effect of Nimodipine treatment (2 µg/Kg/min for 30 minutes) after SAH (see material and methods). Control: sham operated rats; SAH: hemorrhagic treatment with veicle: SAH + N: hemorrhagic treated with Nimodipine

RESPIRATORY PARAMETERS	GLUTAMATE PLUS MALATE		
	CONTROLS	SAH	SAH + N
STATE 3	105.9 +/- 12.2	88.7 +/- 9.7	90.3 +/- 5.8
UNCOUPLED STATE	91.8 +/- 7.1	79.7 +/- 7.3	93.7 +/- 7.1
STATE 4	16.7 +/- 2.4	26.3 +/- 1.5°°	18.1 +/- 1.7
R.C.R.	5.5 +/- 0.5	3.4 +/- 0.3°°°	5.0 +/- 0.5***
ADP/O	2.5 +/- 0.2	2.1 +/- 0.2	2.2 +/- 0.6
RESPIRATORY PARAMETERS	**SUCCINATE PLUS ROTENONE**		
	CONTROLS	SAH	SAH + N
STATE 3	120.3 +/- 7.9	138.4 +/- 6.9	140.1 +/- 11.9
UNCOUPLED STATE	131.2 +/- 12.3	164.9 +/- 11.1	145.2 +/- 14.7
STATE 4	37.4 +/- 2.5	44.5 +/- 2.2	42.8 +/- 3.8
R.C.R.	3.1 +/- 0.1	3.1 +/- 0.1	3.3 +/- 3.3
ADP/O	1.3 +/- 0.1	1.2 +/- 0.1	1.2 +/- 0.1

The State 3, State 4 and Uncoupled State are expressed as nat $O_2 \cdot min^{-1} \cdot mg\ Prot.^{-1}$
Statistical analysis with analysis of variance (ANOVA) and Tukey's Test.
Statistical differences: Controls vs. SAH: °°$p < 0.02$, °°°$p < 0.01$
SAH vs. SAH + Nimodipine: ***$p < 0.01$

arterial blood is collected from the femoral artery, and an aliquot of 0.35 ml are injected into the cisterna magna via the subcutaneous catheter in about 2 minutes. Before SAH induction, CSF sample of about 0.01–0.03 ml is gently drawn according to Delgado *et al.*[7]. Animals are held in a 20° head down position.

Experimental Groups

Animals are divided in the following experimental groups: (a) Control rats (submitted to surgical procedure and injected with 0.35 ml of saline solution at 37 °C); (b) SAH group (rats subjected to SAH procedure with injection of 0.35 ml of autologous arterial blood); (c) Nimodipine treated group (rats subjected to SAH procedure and subsequently treated with Nimodipine infusion: 2 µg/kg, starting immediately after SAH induction, for 30 minutes). Biochemical evaluations are performed at 1 hour after experimental SAH induction.

Mitochondrial Isolation and Biochemical Evaluations

Rats are sacrificed by decapitation. Brain is quickly removed (about 15 seconds) and immediately immersed in cold isolation medium (IM: sucrose 0.32 M plus EDTA 1 mM at pH 7.4). Dissection of the brain areas is performed in a pre-cooled (−5 °C) box[13] in about 60–90 seconds. All manipulations are than carried out at 0–4 °C. Brain areas are homogenized in IM (10 w/v) in a pre-cooled Potter Braun S Homogenizer (six strokes up and down, 600 rpm). The homogenate is than submitted to subcellular fractionation to obtain non-synaptic mitochondrial fraction utilizing a Ficoll density gradient[19].

Mitochondrial respiratory parameters (state 3, state 4, uncoupled state, respiratory control ratio (RCR), ADP/O ratio) are evaluated in presence of glutamate plus malate and succinate plus rotenone as oxidisable substrates, in isotonic medium and in presence of 5 mM of K^+ at 23 °C[6]. Protein content is evaluated[21] in non-synaptic mitochondrial fraction. Statistical analysis is performed with analysis of variance (ANOVA test and Tukey's Test).

Results

All the differences reported for the three rat brain areas were observed with glutamate plus malate as oxidable substrates, while using succinate plus rotenone, no statistical differences were found.

Cerebral Cortex

Table 1 shows the influence of SAH and nimodipine treatment on mitochondrial respiratory parameters. SAH induces a significant increase of State 4 with a consequent decrease of RCR. After nimodipine administration State 4 shows values near to control condition with a significant increase of RCR if compared to SAH animals.

Table 2. *Respiratory Activities of Non-Synaptic Mitochondria From Hippocampus*. Effect of Nimodipine treatment (2 µg/Kg/min for 30 minutes) after SAH (see material and methods). Control: sham operated rats; SAH: hemorrhagic treatment with veicle: SAH + N: hemorrhagic treated with Nimodipine

RESPIRATORY PARAMETERS	GLUTAMATE PLUS MALATE		
STATE 3	91.3 +/- 7.0	75.5 +/- 8.7	78.8 +/- 5.4
UNCOUPLED STATE	94.2 +/- 7.7	73.2 +/- 9.9	75.7 +/- 6.7
STATE 4	20.9 +/- 1.9	22.4 +/- 2.9	17.4 +/- 2.2
R.C.R.	4.4 +/- 0.3	3.5 +/- 0.4	4.8 +/- 0.5
ADP/O	2.5 +/- 0.2	1.9 +/- 0.4°°°	2.1 +/- 0.2
RESPIRATORY PARAMETERS	**SUCCINATE PLUS ROTENONE**		
	CONTROLS	SAH	SAH + N
STATE 3	107.9 +/- 10.7	121.7 +/- 9.6	107.3 +/- 6.1
UNCOUPLED STATE	126.5 +/- 2.4	124.9 +/- 12.5	121.0 +/- 13.6
STATE 4	35.7 +/- 3.7	38.7 +/- 2.9	35.5 +/- 1.9
R.C.R.	3.1 +/- 0.1	3.1 +/- 0.1	3.1 +/- 0.2
ADP/O	1.7 +/- 0.1	1.2 +/- 0.1	1.3 +/- 0.1

The State 3, State 4 and Uncoupled State are expressed as nat $O_2 \cdot min^{-1} \cdot mg$ Prot.$^{-1}$
Statistical analysis with analysis of variance (ANOVA) and Tukey's Test.
Statistical differences: Controls vs. SAH: °°°$p < 0.01$

Hippocampus

The non-synaptic mitochondria of hippocampus show only a significant decrease of ADP/O ratio after SAH induction; thereafter, no statistical differences exist between controls and SAH animals after nimodipine treatment (Table 2).

Brain Stem

Table 3 shows for non-synaptic mitochondria a pattern similar to that observed in cerebral cortex, except for significantly lower values of State 3 after nimodipine treatment.

Discussion

Modifications of brain energy metabolism have been previously underaestimated in the discussion of pathophysiological aspects of SAH. Subarachnoidal deposition of blood may be responsible for altered oxidative and energy metabolism, leading to significative impairment whether in absence of ischaemic damage or fixed neurological deficits due to vasospasm[10, 20, 25]. Experimental works have shown the existence of a significant derangement of cerebral metabolism after SAH[9, 10]. Recently, Delgado *et al.*[8] have demonstrated after experimental SAH in rats a closed correlation between modifications of regional cerebral blood flow (decrease of about 20%) and an increased consumption of glucose (of about 30%). The regional increase of glucose consumption along with the observation of a reduced $CRMO_2$[32] and the increased lactate/pyruvate ratio found in human cerebro-spinal fluid[10] show that the cerebral oxidative metabolism is compromized after SAH. The cerebral oxidative machinery is strictly dependent on mitochondrial function: this was extensively studied to check the cell damage during anoxic-ischaemic insult in the brain[11, 15, 23, 24].

In our study we have shown the existence of characteristic alterations in respiratory parameters of non-synaptic mitochondria which indicate at a biochemical level an impairment of cerebral oxidative metabolism. It is well known that the State 3 reflects oxygen uptake capacity of the mitochondria and is well correlated with the functional status of the electron transfer chain, depending on oxidisable substrates. State 4 reflects the dissipation rate of the mitochondrial electrochemical

Table 3. *Respiratory Activities of Non-Synaptic Mitochondria From Brain Stem.* Effect of Nimodipine treatment (2 μg/Kg/min for 30 minutes) after SAH (see material and methods). Control: sham operated rats; SAH: hemorrhagic treatment with veicle: SAH + N: hemorrhagic treated with Nimodipine

RESPIRATORY PARAMETERS	GLUTAMATE PLUS MALATE		
	CONTROLS	SAH	SAH + N
STATE 3	119.4 +/- 11.1	101.1 +/- 11.6	91.7 +/- 6.7*
UNCOUPLED STATE	110.1 +/- 9.3	88.7 +/- 10.8	111.1 +/- 19.5
STATE 4	24.3 +/- 3.0	32.3 +/- 5.2°°	20.7 +/- 3.3
R.C.R.	5.1 +/- 0.4	3.3 +/- 0.4°°°	4.9 +/- 0.7
ADP/O	2.3 +/- 0.1	1.7 +/- 0.2	2.1 +/- 0.2
RESPIRATORY PARAMETERS	**SUCCINATE PLUS ROTENONE**		
	CONTROLS	SAH	SAH + N
STATE 3	135.1 +/- 13.8	141.7 +/- 10.9	123.9 +/- 5.9
UNCOUPLED STATE	13.6.4 +/- 12.3	147.9 +/- 11.8	124.8 +/- 8.6
STATE 4	46.7 +/- 4.2	46.4 +/- 3.8	41.4 +/- 3.5
R.C.R.	2.9 +/- 0.1	3.1 +/- 0.1	3.1 +/- 0.2
ADP/O	1.2 +/- 0.1	1.1 +/- 0.1	1.3 +/- 0.1

The State 3, State 4 and Uncoupled State are expressed as nat $O_2 \cdot min^{-1} \cdot mg\ Prot.^{-1}$
Statistical analysis with analysis of variance (ANOVA) and Tukey's Test.
Statistical differences: Controls vs. SAH: °°$p < 0.02$, °°°$p < 0.01$
SAH vs. SAH + Nimodipine: *$p < 0.05$

gradient and indirectly give information about selective properties of the mitochondrial membrane. The significant decrease of RCR (State 3/State 4) in cerebral cortex and in brain stem depends both on a decrease of State 3 (although not statistically significant) and on the more important increase of State 4: this would indicate the role of mitochondrial membrane dysfunction in the genesis of cerebral damage after SAH.

A massive neuronal membrane dysfunction has been demonstrated as an important pattern of different pathological conditions of the central nervous system. In SAH, the diffusion of blood in the subarachnoidal space enhances movements of Ca^{++} and K^{+} between extracellular and cellular compartments and Ca^{++} has been proposed as the final common agent causing cell damage[5, 16].

The purpose of our investigation was to verify the effect of treatment with nimodipine, a slow calcium-channel blocker with specific action on cerebral vessels[17, 22, 29], in order to antagonize the effects of SAH on non-synaptic mitochondrial function. Some controversial results have been recently published about the effect of nimodipine on brain metabolic state in experimental ischaemic conditions[30, 31]. SAH is a peculiar disease with different characteristics: the sudden hypoperfusion and the subsequent phase of arterial vasospasm together with the direct effect of blood deposition in subarachnoidal space could mimic an ischaemic condition. In this respect, experimental SAH could be an useful model to test effects of dihydropyridinic calcium-antagonists on brain metabolism. In the early stages of SAH, CBF decreases in rat hemispheres to about 40%[28]: this characteristic decrease in cerebral perfusion could be responsible for an incomplete ischaemia leading to mitochondrial impairment. Nimodipine treatment significantly improves mitochondrial function. In treated animals after SAH values of RCR are similar to those of control animals; this would be related to better conditions of cerebral perfusion. This study suggests that nimodipine treatment would play a significant role in reducing or preventing mitochondrial membrane derangement after the haemorrhage.

Acknowledgement

The authors are grateful to L. Caresia, M.D., P. Salvini, M.D., G. Fenu, Ph.D., for their precious technical help.

References

1. Ackrell BAC, Kearney EB (1975) Mammalian succinate dehydrogenase. In: Fleischer S, Packer L (eds) Methods in enzymology. Academic Press Inc, New York London, pp 466–483
2. Allen GS, Banghart SB (1979) Cerebral arterial spasm: Part 9. In vitro effects of nifedipine on serotonin-, phenylephrine-, and potassium-induced contractions of canine basilar and femoral arteries. Neurosurgery 40: 433–441
3. Allen GS, Ahn HS, Preziosi TJ, Battye R *et al* (1983) Cerebral arterial spasm – a controlled trial of nimodipine in patients with subarachnoid haemorrhage. New Engl J Med 303: 619–524
4. Auer LM, Brandt L, Ebeling U, Gilsbach J, Groeger U, Harders A, Ljunggren B, Oppel F, Reulen HJ, Saeveland H (1986) Nimodipine and early operation in good condition SAH patients. Acta Neurochir (Wien) 82: 7–13
5. Barry KJ, Mikkelsen RB, Schucart W, Keough EM, Gavris V (1985) The isolation and characterization of Ca^{++}-accumulating subcellular membrane fractions from cerebral arteries. J Neurosurg 62: 729–736
6. Clark JB, Nicklas WJ (1970) The metabolism of rat brain mitochondria. Preparation and characterization. J Biol Chem 245: 4724–4731
7. Delgado TJ, Brismar J, Svengaard NA (1985) Subarachnoid haemorrhage in the rat: angiography and fluorescence microscopy of the major cerebral arteries. Stroke 16: 595–602
8. Delgado TJ, Arbab MA-R, Diemer NH, Svengaard NA (1986) Subarachnoid haemorrhage in the rat: cerebral blood flow and glucose metabolism during the late phase of cerebral vasospasm. J Cereb Blood Flow Metabol 6: 590–599
9. Fein JM (1975) Cerebral energy metabolism after subarachnoid haemorrhage. Stroke 6: 1–8
10. Fein JM (1976) Brain energetics and circulatory control after subarachnoid haemorrhage. J Neurosurg 45: 498–506
11. Ginsberg MD, Mela L, Wrobel-Kuhl K (1977) Mitochondrial metabolism following bilateral cerebral ischaemia in the gerbil. Ann Neurol 1: 519–527
12. Gioia AE, White RP, Bakthian B, Robertson JT (1985) Evaluation of efficacy on intratecal nimodipine in canine models of chronic cerebral vasospasm. J Neurosurg 62: 721–728
13. Glowinski J, Iversen LL (1966) Regional studies of catecholamines in the rat brain. J Neurochem 13: 655–669
14. Heffez DS, Passoneau JV (1985) Effect of nimodipine on cerebral metabolism during ischaemia and recirculation in the mongolian gerbil. J Cereb Blood Flow Metabol 5: 523–528
15. Hillered L, Siesjo BK, Arfors KE (1984) Mitochondrial response to transient forebrain ischaemia and recirculation in the rat. J Cereb Blood Flow Metabol 4: 438–446
16. Hubschmann OR, Nathanson DC (1985) The role of calcium and cellular membrane dysfunction in experimental trauma and subarachnoid haemorrhage. J Neurosurg 62: 698–703
17. Kazda S, Mayer D (1985) Postischaemic impaired reperfusion and tissue damage: Consequence of calcium-dependent vasospasm? In: Godfraind T, Vanhoutte PH, Govoni S, Paoletti R (eds) Calcium entry blockers and tissue protection. Raven Press, New York, pp 129–138
18. Lai JCK, Clark JB (1976) Preparation and properties of mitochondria from synaptosomes. Biochem J 154: 423–432
19. Lai JCK, Walsh JM, Dennis SC, Clark JB (1977) Synaptic and non-synaptic mitochondria from rat brain: isolation and characterization. J Neurochem 28: 625–631
20. Ljunggren B, Sonesson B, Saaveland H, Brandt L (1985) Cognitive impairment and adjustement in patients without neurological deficits after aneurysmal SAH and early operation. J Neurosurg 62: 973–979
21. Lowry OH, Rosebrough BJ, Farr AL, Randall RJ (1951) Protein measurement with the folin phenol reagent. J Biol Chem 193: 265–275
22. Mabe H, Nagai H, Takagi T, Umemura S, Ohno M (1986) Effect of nimodipine on cerebral functional and metabolic recovery following ischaemia in the rat brain. Stroke 17: 501–505
23. Ozawa K, Seta K, Araki H, Handa H (1967) The effect of ischaemia on mitochondrial metabolism. J Biochem 61: 512–515
24. Rehncrona S, Mela L, Siesjo BK (1979) Recovery of brain mitochondrial function in the rat after complete and incomplete cerebral ischaemia. Stroke 10: 437–446
25. Ropper AH, Zervas NT (1984) Outcome 1 year after SAH from cerebral aneurysm. Management morbidity, mortality, and functional status in 112 consecutive goodrisk patients. J Neurosurg 60: 909–915
26. Sahlin C, Delgado T, Owman C, Svengaard NA (1986) Changes in cerebral blood flow and metabolism following intraarterial or local administration of Nimodipine, before and after experimental subarachnoid haemorrhage in baboons. Stroke 17: 220–224
27. Schanne FAX, Kane AB, Young EE, Farber JL (1979) Calcium dependence of toxic cell death: a final common pathway. Science 206: 700–702
28. Solomon RA, Lobo Antunes J, Chen RYZ, Bland L, Chien S (1985) Decrease in cerebral blood flow in rats after experimental subarachnoid haemorrhage: a new model. Stroke 16: 58–64
29. Towart R (1981) The selective inhibition of serotonine contractions of rabbit cerebral vascular smooth muscle by calcium-antagonistic dihydropyridines. An investigation of the mechanism of action of nimodipine. Circ Res 48: 650–657
30. Van Rempts G, Haseldonckx M, Van Demen B, Wanters L, Borgers M (1986) Structural damage of ischaemic brain: involvement of calcium and effects of post-ischaemic treatment with calcium entry blockers. Drug Develop Res 8: 387–395
31. Vilbulsresth S, Dietrich DW, Busto R, Ginsberg MD (1987) Failure of nimodipine to prevent ischaemic neuronal damage in rats. Stroke 18: 210–216
32. Voldby B, Enevoldsen EM, Jensen FT (1985) Regional CBF, intraventricular pressure, and cerebral metabolism in patients with ruptured intracranial aneurysms. J Neurosurg 62: 48–58

Address for correspondence: P. Gaetani, M.D., Department of Surgery, Neurosurgical Section, University of Pavia, IRCCS Policlinico S. Matteo, I-27100 Pavia, Italy.

Acta Neurochirurgica, Suppl. 43, 182–185 (1988)

Pathomechanism of Brain Oedema in Experimental Intracerebral Mass Haemorrhage

Th. Wallenfang[1], **G. Fries**[1], **J. P. Jantzen**[2], **J. Bayer**[3], and **F. Trautmann**

[1] Neurosurgical Clinic, [2] Clinic for Anaesthesiology, [3] Clinic for Internal Diseases, University of Mainz Medical School, Mainz, Federal Republic of Germany

Summary

The prognosis of intracerebral haemorrhage is extremely poor when arterial hypertension is present. We investigated elastance of the brain tissue and brain hydraulic conductivity in normotensive (MAP ~ 110 mmHg) and hypertensive (MAP ~ 170 mmHg/angiotensin infusion) cats following a stereotactically produced intracerebral haemorrhage. For 12 hours after the onset of haemorrhage we registered the course of ICP, subsequently the water content of cortex, white matter and basal ganglia as well as the interstitial concentration of serum proteins in the corresponding regions were determined (Evans-Blue, immunofluorescence). Hypertension was associated with a slight increase in ICP and tissue water content but with a 3 fold more elevated content of interstitial serum proteins. Immunofluorescence analysis showed the oedema in the hypertensive group to be vasogenic in nature and in the normotensive cats to originate from the haematoma itself. The generally poor outcome of intracerebral mass haemorrhage in hypertensive subjects could be ascribed to the nature of the concomitant brain oedema, as described in this study.

Keywords: Intracerebral haemorrhage; brain oedema; arterial hypertension; animal experiments.

Introduction

Our own statistical evaluations[9, 10] as well as studies by others[3, 4, 8] on the clinical course of intracerebral mass haemorrhage (ICH) emphasize the importance of arterial blood pressure during the acute phase of ICH. Among 117 conservatively treated patients suffering from ICH, of whom 81 were hypertensive and 36 were normotensive, the following differences were observed: Within 5 days after ICH survival rates decreased to 70% in both groups. Only 60% of hypertensive patients had survived until the 15th day, whereas there was only a slight tendency towards deterioration among normotensive patients. In both groups no significant differences in the size of the haematomas were seen in the CT follow-up study. However, the normotensive patients had essentially more extended areas of perifocal hypodensity. To elucidate the pathomechanisms of brain oedema, intracranial pressure and hydraulic conductivity – factors that obviously determine these differences in our clinical results – we developed an experimental model of ICH.

Methods

2 ml of freshly drawn autologous arterial blood were used to produce ICH. The blood was injected stereotactically into the internal capsule of the right hemisphere of anesthetized, paralyzed and normoventilated cats (2.8–4.1 kg b.wt.). Anesthesia was maintained with chloralose (12.5 mg/kg kg^{-1} h^{-1}) and hexacarbacholinebromide (0.6 mg initially and then 0.3 mg every 30 min). Directly after the injection of the blood, Evans-Blue dye (2 ml of 2% solution) as a marker of plasma proteins was administered intravenously.

Epidural pressure monitoring (Gaeltec pressure transducer) and blood pressure monitoring were started 2 hours after the onset of ICH.

Three experimental groups were compared:

– normotensive group (n = 9) with MAP of 110 mmHg for 12 hours

– hypertensive group (n = 7) with MAP of 165 mmHg for 12 hours (AABP elevated by continuous infusion of angiotensin (Hypertensin-CIBA, 3–60 μg/h)

– control group (n = 6) with MAP of 110 mmHg for 12 hours but without ICH.

12 hours after the injection of ICH the brains of the animals were frozen in situ with liquid N_2, cut into coronal slices of 3–4 mm and investigated as follows:

The water content and the Evans-Blue content of white matter, adjacent to and remote from the lesion and of the opposite hemisphere were measured. For the determination of the Evans-Blue content we slightly modified the formamide extraction method described by Durward *et al.*[1].

The extravasated plasma proteins of brain oedema were marked using an immunofluorescence sandwich technique[11] with rabbit antibodies against cat plasma proteins and FITC-conjugated goat antibodies against rabbit (Behring; ORDF 02, OTFK 04).

Significant differences with a mean error of $\alpha = 5\%$ ($p = 0.05$)

were evaluated by the Wilcoxon-test combined with the method of Bonferroni-Holm.

Results

In both normotensive and hypertensive cats with ICH a steep increase in epidural pressure was seen with blood injection. In normotensive animals mean values of epidural pressure rose to 40 mmHg, and in hypertensive animals to 51 mmHg during the blood injection.

This increase associated with blood injection was followed by a steep fall of epidural pressure to values of 20 mmHg (normotensive) and 25 mmHg (hypertensive). These values tended to further decrease towards the end of the 12 hours survival time. Normal values for epidural pressure (less than 10 mmHg) were recorded in the control group (Fig. 1).

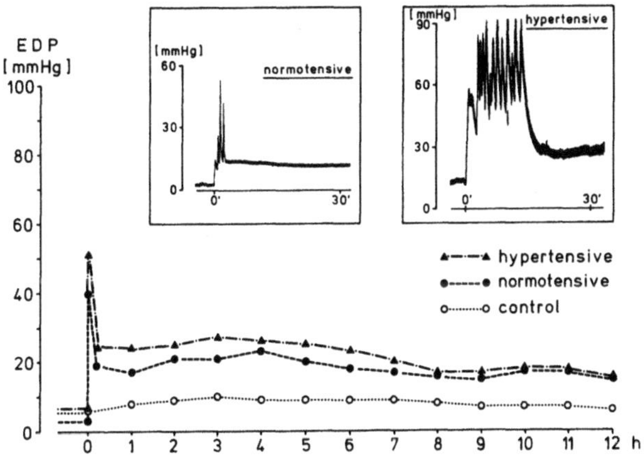

Fig. 1. Epidural pressure in experimental intracerebral haemorrhage under normotensive and hypertensive conditions (mean values). Inset: Steep increase in EDP during blood injection in both groups, markedly higher when intracranial compliance is reduced (hypertensive). After blood injection EDP ranged between 25 mmHg (hypertensive) and 20 mmHg (normotensive)

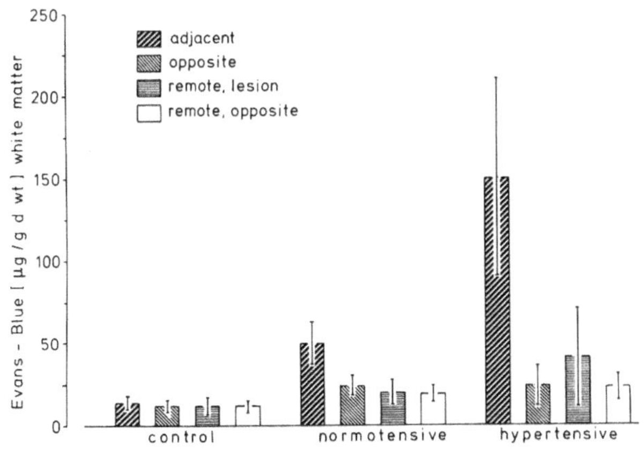

Fig. 2. Evans-Blue content in the different areas of the white matter in both experimental groups with ICH and the control group (mean values and standard deviation). Significant differences (p = 0.05) between the three experimental groups (control vs. normotensive, control vs. hypertensive, normotensive vs. hypertensive)

12 hours after the onset of ICH, water content of the white matter adjacent to the lesion was similar in both experimental groups with ICH:

Compared to normal values of 69 g/100 g w.wt. (control), water content of the perifocal area increased by 5.5 g/100 g w.wt. in the normotensive group and by 5.8 g/100 g w.wt. in the hypertensive group.

Significant differences were found with respect to the Evans-Blue content in the vicinity of the haematoma. Mean values of the Evans-Blue content in the perifocal area of ICH were 15 μg/g d.wt. in the control group, 50 μg/g d.wt. in the normotensive and 150 μg/g d.wt. in the hypertensive groups (Fig. 2).

12 hours after ICH the perifocal area of normotensive animals showed a diffuse infiltration with plasma proteins. These plasma proteins were part of the fluid component emerging almost completely from the haematoma within 12 hours after ICH. Vasogenic oedema was rarely seen in this group.

The hypertensive animals developed pronounced vasogenic oedema with perivascular accumulation of plasma proteins. In this group the fluid compound of the haematoma did not infiltrate the penumbra, but seemed to be captured in the haematoma for at least 12 hours (Fig. 3).

Discussion

Recent analysis of the data of 117 conservatively treated patients with intracerebral haemorrhage showed that the prognosis is largely established by the extent of intracerebral haemorrhage and the level of consciousness on the one hand and by the systemic blood pressure on the other hand. Thus the prognosis of the patients was extremly poor when ICH was associated with systemic hypertension and more favourable, when midline shift and perifocal hypodensity after ICH could be observed in normotensive patients[9]. These prognostic factors have been studied by a new experimental model of ICH in cats.

When intracerebral haemorrhage was caused by a stereotactic injection into the internal capsule a progressive increase in ICP was observed during the first expansion period. This lasts as long as the brain tissue provides resistance to the distention of the extracellular space. However, intracranial compartments of normotensive and hypertensive animals reacted differently in the time course of ICH. Whilst ICP and water con-

Fig. 3. a) Haematoma of a hypertensive animal (× 250) with an accumulation of plasma proteins (*P*). b) Haematoma of a normotensive animal (× 250) after drainage of plasma proteins. c) Perifocal tissue of a hypertensive animal (× 250) with marked perivascular extravasation of plasma proteins indicating vasogenic oedema (*E*). d) Perifocal tissue of a normotensive animal (× 250) with diffuse plasma infiltration (*P*) originating from the haematoma (*H*) and with sparse vasogenic oedema

tent around the haematoma in the hypertensive group were not higher than in normotensive animals, differences were found in the degree of vasogenic brain oedema.

The content of Evans-Blue as a plasma protein marker in the perifocal area demonstrated that hypertensive ICH was associated with a 3 fold greater increase of extravasated interstitial serum proteins. This agrees with findings of Ishii *et al.*[4], who described a disturbed blood-brain-barrier after 30 min of acute arterial hypertension and intracerebral haemorrhage. The immunofluorescence technique confirmed such a disturbance of blood-brain-barrier function with serum proteins infiltrating the direct perivascular regions of the affected hemisphere.

Normotensive ICH was followed by a diffuse plasma infiltration of the perifocal area with the plasma originating from the haematoma itself as indicated by immunofluorescence. The reduced perifocal Evans-Blue content indicates an only slightly compromised blood-brain-barrier. This suggests that two different pathomechanisms are involved in the formation and resolution of brain oedema after ICH:

Under normotensive conditions, ICH results in an expansion of interstitial space, subsequently draining serum from the haematoma into the CSF system. Such drainage takes place in accordance with the bulk-flow-theory, described by Reulen *et al.*[6,7]. The rapid distribution of haematoma-serum through the interstitial space along the pressure gradient away from the ICH site provides the substrate for the hypodense area, seen in the CT scan, which is more pronounced in normotensive subjects. This stimulated volume increase does not necessarily worsen prognosis, as proven by the clinical data on survival rate after the 5th day[10]. Arterial hypertension is associated with a reduced tissue compliance, as indicated by the more pronounced increase of epidural pressure during ICH in this group[9]. This results in a delayed drainage of haematoma. The effect of this delay combines with the vasogenic oedema, resulting in an increased interstitial space occupation. This provides the basis for the poor outcome in hypertensive patients, with mortality figures ranging around 48%[10].

These apparent differences among pathomechanisms are to be ascribed to different degrees of reduction of hydraulic conductivity of the brain tissue. Arterial hypertension appears to reduce the hydraulic conductivity by means of a reduced cerebrovascular compliance[5].

From these experiments it may be concluded that:

In those patients presenting with intracerebral haemorrhage that are not subject to surgical intervention according to the criteria established by Schürmann *et al.*[8], outcome is determined to some extent by the haemorrhage itself, but more so by specific accompanying pathophysiology, including altered tissue compliance and differences in hydraulic conductivity of the brain.

References

1. Durward QJ, Del Maestro RS, Amacher Al, Farrar JK (1983) The influence of systemic arterial pressure and intracranial pressure on the development of cerebral vasogenic oedema. J Neurosurg 95: 803–809
2. Enzmann DR, Britt RH, Lyona BE, Buxton JL, Wilson DA (1981) Natural history of experimental intracerebral haemorrhage: sonography, computed tomography and neuropathology. AJR 2: 517–526
3. Gänshirt H, Keuler R (1980) Intrazerebrale Blutungen. Nervenarzt 51: 201 – 206
4. Ishii S, Koike J, Hatashita S (1983) Brain oedema in hypertensive intracerebral haemorrhage. In: Mizukami M *et al* (eds) Hypertensive intracerebral haemorrhage. Raven Press, New York, pp 11–17
5. Rapoport SI (1979) Roles of cerebrovascular permeability, brain compliance, and brain hydraulic conductivity in vasogenic brain oedema. In: Popp AJ *et al* (eds) Neurol trauma. Raven Press, New York, pp 51–61
6. Reulen HJ, Graham R, Spatz M, Klatzo I (1977) Role of pressure gradients and bulk flow in dynamics of vasogenic brain oedema. J Neurosurg 46: 24–35
7. Reulen HJ, Tsuyumu M, Tack A, Fenske AR, Prioleau GR (1978) Clearance of oedema fluid into cerebrospinal fluid. J Neurosurg 48: 754–764
8. Schürmann K, Dei Anang K (1980) Indication and contraindication for surgery of spontaneous intracerebral haematomas. Neurosurg Rev 3: 17–22
9. Wallenfang Th, Fries G, Bayer J, Schürmann K (1985) Immunohistochemical investigation in experimental intracerebral haemorrhage (ICH) and its correlation with findings in clinical studies. In: Miller JD, Teasdale GM, Rowan JO, Galbraith SL, Mendelow AD (eds) Intracranial pressure VI. Springer, Berlin Heidelberg New York, pp 532–537
10. Wallenfang Th, Fries G (1986) Der Einfluß der Blutdruckverhältnisse auf das Hirnödem bei der konservativ therapierten spontanen intrazerebralen Massenblutung. Klinische und experimentelle Aspekte. In: Voth D (ed) Neurochirurgica Moguntiaca. de Gruyter, Berlin, pp 75–84
11. Wilmes F, Hossmann KA (1979) A specific immunofluorescence technique for the demonstration of vasogenic brain oedema in paraffin embedded material. Acta Neuropathol (Berl) 45: 47–51

Address for correspondence: Prof. Dr. Th. Wallenfang, Neurochirurgische Universitätsklinik, Langenbeckstrasse 1, D-6500 Mainz, Federal Republic of Germany.

Acta Neurochirurgica, Suppl. 43, 186–188 (1988)

Regulation of Brain Water and Electrolyte Contents: the Opposite Actions of Central Vasopressin and Atrial Natriuretic Factor (ANF)

T. Dóczi, F. Joó, P. Szerdahelyi, and **M. Bodosi**

Department of Neurosurgery, Central Research Laboratory, University Medical School, and Laboratory of Molecular Neurobiology, Biological Research Center, Szeged, Hungary

Summary

Considerable evidence indicates that regulation of the ionic environment of the brain is coordinated by a central neuroendocrine system capable of affecting the capillary endothelium, the choroid plexus, and the astroglia. All three cell groups are responsible for precise control of brain volume through adjustment of cell water and electrolyte content. With these considerations in mind, we have attempted to elucidate the possible involvement of the central vasopressin (AVP) and atrial natriuretic factor (ANF) systems in the regulation of the water and ion homeostasis of the brain tissue of rats:

Vasopressin-positive vascular connections, investigated by immuno-electronhistochemistry, were found in close or direct contact with brain microvessels. Central administration of AVP (125 ng) or DDAVP (0.5 µg), with or without an accompanying water load, brought about a 1–1.3% water accumulation. Brain oedema caused by experimental subarachnoid haemorrhage had a different course in Wistar and Brattleboro DI rats, the latter being unable to synthetize AVP. These findings suggest that the centrally released AVP leads to brain water accumulation by increasing the water permeability of capillaries, and may facilitate the production of brain oedema in various pathological conditions. On the other hand, central administration of synthetic rat ANF (2 µg) prevented the water accumulation elicited in rat brain by systemic hypoosmolar fluid load, and led to a significant sodium loss from the nervous tissue by altering the capillary sodium permeability. The better understanding of these hormone receptors and their manipulations have exciting clinical implications.

Keywords: Brain oedema; central vasopressin; atrial natriuretic factor; central neuroendocrine system.

Introduction

In 1981 M. Raichle put forward the hypothesis that a central neuro-endocrine system regulates the brain ion homeostasis and volume[10]. The hypothesis is that three cell groups (brain capillary endothelial cells, secretory cells of the choroid plexus, and astroglia) regulate the internal ionic environment of the brain. A unique element of this hypothesis is that the regulation of the ion and volume homeostasis of the brain is orchestrated by a central neuroendocrine system capable of affecting all three cell types. What is meant by central as opposed to peripheral as concerns the neuroendrocrine system is well demonstrated on the example of AVP. Schultz *et al.*, Brownfields and Kozlowski, and Buijs *et al.* showed that the plasma and the cerebrospinal AVP originate from different sources[2, 3, 19]. Further evidence of the existence of a separate AVP system is provided by the inability of AVP to cross the blood-brain barrier (BBB) and blood-cerebrospinal fluid (CSF) barrier[15, 16]. On the other hand, the secretion of AVP into the CSF may be influenced by the same stimuli that govern its secretion into the blood[18].

Since the discovery of potent diuretic, natriuretic and vasorelaxant activities in extracts of the rat atria, various forms of atrial natriuretic factor (ANF) have been isolated from human and rat atria[8]. The occurrence of large amounts of ANF-like immunoreactive nerve cells was recently reported in human and rat brain[17]. The presence and widespread distribution of ANF in the rat brain were demonstrated by radioimmunoassay for rat ANF, showing the prevalent location of rat ANF in the hypothalamus and septum[17]. It is unlikely that the ANF in the brain is synthetized in the heart and transported from the heart to the brain via the circulation, *i.e.* the ANF in the brain does not originate from the heart. As regulation of the ANF level in the brain differs from that in the heart, the existence of a separate central ANF system can also be hypothetized[8, 17].

AVP and ANF have been implicated in the regulation of systemic water and electrolyte balances and blood pressure[8]. All membranes known to regulate water permeability are, in fact, under the influence of circulating AVP and possibly ANF[8, 14]. The brain vasculature is not freely permeable to water and ions[9], in fact, it exhibits a permeability similar to those of secreting epithelia with tight junctions[9]. According to Rodriguez, the CSF and interstitital fluid (ISF) flows should serve as pathways in neuroendocrine integration[13]. His generally accepted concept is strongly supported by the recent findings of Rennels *et al.*[12]. They demonstrated that solutes in the CSF have access to the extracellular space throughout the brain within a matter of minutes (2–4 minutes), via fluid pathways along the intraparenchymal vasculature. Thus, the water and ion permeability of brain microvessels can be influenced in the following way: AVP or ANF released into the CSF may reach the abluminal side of the capillaries through CSF and ISF transports and change the permeability properties of the membrane either directly or by modulating the central noradrenergic system. In certain areas, on the other hand, direct vasopressinergic innervation of microvessels may operate[7].

With these considerations in mind, we have attempted to elucidate the possible involvement of the central AVP and ANF systems in the regulation of the water and ion homeostasis of the brain tissue of rats[4, 5, 6].

Material and Methods

Measurement of brain water and electrolyte contents after central administration of ANF or AVP: AVP (50 µU or 50 mU) or l-desamino-8-D-arginine-vasopressin (DDAVP) (0.5 µg), or synthetic rat ANF (syn rANF), sequence 101–126 of the precursor (0.2 or 2 µg), with or without a hypoosmolar fluid load, was administered into the right lateral ventricle under sodium pentobarbital anaesthesia by means of stereotactic puncture. The hypoosmolar fluid load (2.5% dextrose in water to a volume of 15–20% of the body weight) was administered intravenously (i.v.) at a rate of 0.5 ml/min. The rats were killed by decapitation 40 minutes after the start of the treatment. The hemispheres were investigated separately. The tissue water content and dry weight were established by weighing before and after drying at 110 °C for 40 hours. Sodium and potassium contents were determined with atomic absorption spectrophotometry. Serum osmolality was determined by the freezing point depression method. Systemic blood pressure and blood gases were monitored. The technique used to measure sodium uptake from blood to brain is a modification of the intracarotid bolus injection method devloped by Oldendorf: the test solute was ^{22}Na (3 mCi, 1.5 mM) and as reference a low permeability compound (^{3}H-sucrose, 15 mCi) was used. The ratio of ^{22}Na to ^{3}H in the brain sample was divided by the corresponding ratio in the injected mixture (0.2 ml). This resulted in a ratio termed the relative extraction (E) of ^{22}Na compared with the ^{3}H reference compound. E cannot be converted to a quantitative measure of extraction but is appropriate to determine the effects of potential transport inhibitors[1].

Results

Changes in Serum Electrolytes and Osmolality Following AVP or ANF Treatment

The administration of the hypoosmolar fluid load caused small decrease in the serum sodium content and reduced the serum osmolality significantly, but did not alter the serum potassium content ("water intoxication"). The central administration of ANF or AVP alone did not change the ion content or the osmolality of the serum. Thus, changes in the brain water and electrolyte content could not be ascribed to a primary systemic effect of the hormones on the blood.

Changes in Brain Water and Ion Contents After AVP Treatment

Administration of a fluid load + 50 mU AVP or 0.5 µg DDAVP increased the brain water content significantly (1.3%), as did 50 mU AVP or 0.5 µg DDAVP alone ($p < 0.05$). There was no significant difference in water accumulation between animals treated with AVP or DDAVP + fluid load and those receiving AVP or DDAVP alone. The contents of sodium and potassium decreased after the administration of 50 mU AVP or 0.5 µg DDAVP by about 8% and 4% respectively. The water load slightly decreased the sodium content of the brain (by about 5%).

Changes in Brain Water and Ion Contents After ANF Treatment

Administration of hypoosmolar fluid brought about a water accumulation (by 1.3%) and slight decreases in sodium and potassium due to the water intoxication. Administration of 0.2 or 2.0 µg of ANF prevented this water uptake, and led to a statistically significant sodium loss from the brain (3.8%) ($p < 0.05$). Similar syn rANF administration to rats not treated with hypoosmolar fluid load caused no significant change in the water, potassium and sodium contents of the hemispheres.

^{22}Na Permeability Studies

The value of E was 1.45 in control animals. Pretreatment with 2 µg of syn rANF into the ventricle 10, 20, or 40 minutes before the carotid injection of the isotopes slightly reduced the E (1.30) but the decrease of E did not reach a significant level. If 2 µg ANF was administered into the carotid artery prior to injection of the isotopes, a significant increase in the relative extraction was observed ($E = 1.85$, $p < 0.05$).

Discussion

Central administration of AVP (50 mU) or DDAVP (0.5 µg), with or without an accompanying water load, brought about a 1–1.3% water accumulation. Changes in the ion contents were inconclusive as concerns the mechanisms of water accumulation. In a different oedema model, Reeder *et al.* likewise confirmed that pharmacological doses of central AVP facilitate the production of vasogenic oedema induced by cold injury[11]. The mechanism may be explained by the increased water permeability of the brain microvessels, as detected by Raichle and Grubb using the carotid bolus technique, external detection and positron-emitting tracers (^{15}O). Intraventricular administration of 0.2 or 2 µg syn rANF prevented the water accumulation elicited in rat brain by a systemic hypoosmolar fluid load, and led to a statistically significant sodium loss from the nervous tissue, while the potassium content remained unaltered. Similar syn rANF administration to rats not treated with the fluid load caused no significant change in the water, potassium and sodium contents of the hemispheres. In this experiment a primary systemic action of centrally released syn rANF with ensuing secondary changes in the brain ion and water homeostasis appears unlikely, as the serum osmolality and sodium and potassium concentration remained unaltered. Thus, a central influence of the intraventricularly administered hormone upon the water and ion balances of the nervous tissue can be hypothetized. The significant loss of sodium content may reflect the primary role of the volume regulation of the nervous tissue, *i.e.* the loss of extracellular "osmols" such as Na^+, in response to the hypoosmolar environment. These data lend further support to the concept that a central neuroendocrine system regulates brain ion and volume homeostasis. AVP antagonists and ANF analogues may have exciting clinical implications in the treatment of brain oedema.

References

1. Bradbury MWB (1979) The concept of a blood-brain barrier. J Wiley and Sons, Chicester New York Brisbane London
2. Brownfield MS, Kozlowski GP (1977) The hypothalamo-choroidal tract I. Immunohistochemical demonstration of neurophysin pathways to the telencephalic choroid plexus and cerebrospinal fluid. Cell Tissue Res 178: 111–127
3. Buijs RM, Swaab DF, Dogterom J, Van Leeuwen FW (1978) Intra- and extrahypothalamic vasopressin and oxytocin pathways in the rat. Cell Tissue Res 186: 423–433
4. Dóczi T, Szerdahelyi P, Gulya K, Kiss J (1982) Brain water accumulation after the central administration of vasopressin. Neurosurgery 11: 402–407
5. Dóczi T, László FA, Szerdahelyi P, Joó F (1984) The role of vasopressin in brain oedema formation: Further evidence obtained from the Brattleboro Diabetes Insipidus Rats with subarachnoid haemorrhage. Neurosurgery 14: 436–440
6. Dóczi T, Szerdahelyi P, Joó F, Bodosi M: Regulation of brain water content and electrolytes: The possible role of central atrial natriuretic hormone (ANF). Neurosurgery (in press)
7. Jójárt I, Joó F, Siklós L, László FA (1984) Immunoelectronhistochemical evidence for innervation of brain microvessels by vasopressin-immunoreactive neurons in the rat. Neurosci Let 51: 259–264
8. Palluk R, Gaida W, Hoefke W (1985) Minireview: Atrial natriuretic factor. Life Sci 36: 1415–1425
9. Raichle ME, Grubb RL (1978) Regulation of brain water permeability by centrally released vasopressin. Brain Res 143: 191–194
10. Raichle ME: Hypothesis: A central neuroendocrine system regulates brain ion homeostasis and volume. In: Martin JB, Reichlin S, Bick KL (eds) Neurosecretion and brain peptides. Raven Press, New York, pp 329–336
11. Reeder RF, Nattie EE, North WG (1986) Effect of vasopressin on cold-induced oedema in cats. J Neurosurg 64: 941–950
12. Rennels ML, Gregory TF, Blaumanis OR, Fujimoto K, Grady PA (1985) Evidence for a "paravascular" fluid circulation in the mammalian central nervous system, provided by a rapid distribution of tracer protein throughout the brain from subarachnoid space. Brain Res 326: 47–63
13. Rodriguez EM (1976) The cerebrospinal fluid as a pathway in neuroendocrine regulation. J Endrocrinol 71: 407–443
14. Rosenberg AG, Kyner WT, Fenstermacher JD, Patlak CS (1986) Effect of vasopressin on ependymal and capillary permeability to tritiated water in cat. Am J Physiol 251: F485–F489
15. Sorensen PS, Gjerris F, Hammer M (1984) Cerebrospinal fluid vasopressin and increased intracranial hypertension. Ann Neurol 15: 435–440
16. Sorensen PS, Vilhardt H, Gjerris F, Warberg J (1984) Impermeability of the blood-cerebrospinal fluid barriers to l-deamino-8-D-arginine-vasopressin (DDAVP) in patients with acquired communicating hydrocephalus. Europ Clin Invest 14: 435–438
17. Steardo L, Nathanson JA (1987) Brain barrier tissues: end organs for atriopeptins. Science 235: 470–473
18. Wang BC, Share L, Goetz KL (1985) Factors influencing the secretion of vasopressin into the cerebrospinal fluid. Fed Proc 44: 72–77
19. Wood JH (1982) Neuroendocrinology of cerebrospinal fluid: peptides, steroids, and other hormones. Neurosurgery 11: 293–305

Address for correspondence: Dr. T. Dóczi, Department of Neurosurgery, Central Research Laboratory, University Medical School, Szeged, Hungary.

Acta Neurochirurgica, Suppl. 43, 189–192 (1988)

An Increment in Unilateral Carotid Blood Flow Produces Cerebral Aneurysms in Rats*

J. M. Roda, F. Alvarez, A. L. Garcia-Villalon, M. R. Ruiz, M. Gutierrez, and **M. Garcia Blazquez**

Servicio de Neurocirugia and Laboratorio de Microcirugia (Hospital la Paz), and Departamento de Fisiologia, Facultad de Medicina, Universidad Autonoma de Madrid, Madrid, Spain

Summary

To study the effect of haemodynamic stress on cerebral aneurysm development, three groups of adult Wistar rats were used: 1) One group with ligation of the right common carotid artery; 2) One with an end-to-side common carotid artery anastomosis and 3) Another group that was sham-operated. Blood flow through the left common carotid artery was measured with an electromagnetic flow transducer (0.85 mm in diameter) in anesthetized animals. Before the experimental intervention, the mean carotid blood flow was 2.8 ± 0.14 ml/min. Following carotid ligation the blood flow in the patent carotid artery increased 25 ± 5% after 7 ± 0.3 days and 37 ± 18% after 149 ± 23 days; whereas following carotid anastomosis, carotid blood flow increased 102 ± 16% after 9.6 ± 2 days and 103 ± 35% after 169 ± 15 days. The sham operation did not significantly affect carotid blood flow. Control values for mean arterial pressure (102 ± 3 mmHg) and blood gases ($PaO_2 = 106 \pm 3.4$ mmHg, $PaCO_2 = 38 \pm 1.2$ mmHg and pH = 7.32 ± 0.02) were similar in the three groups of rats and did not change significantly over time after the experimental procedure. Necropsy was carried out 149 ± 15 days after the experimental intervention. No cerebral aneurysms were found in the rats with carotid ligation or which had been sham-operated, however, three of the eight rats ($p < 0.05$) with carotid anastomosis presented single aneurysms in the anterior cerebral-anterior communicating arterial complex. These results indicate that haemodynamic stress correlates with cerebral aneurysm development and suggest that aneurysms may, in fact, be induced by high, unilateral carotid blood flow which provokes cerebrovascular injury.

Keywords: Cerebral aneurysm; carotid blood flow; carotid anastomosis; carotid ligation.

Introduction

Among the several experimental models used to produce cerebral aneurysms[4–8, 11, 12, 17], haemodynamic stress has been proposed as a determinant factor in the development of vascular lesions observed[3, 15]. However, the haemodynamic changes induced by the experimental procedures have not been documented.

Clinical[9, 13] and experimental[7, 8] observations suggest that hydraulic imbalance in the circle of Willis is one of the major ethiological factors in the development of intracranial aneurysms. We have recently reported that end-to-side common carotid anastomosis is more effective in causing cerebral aneurysms in hypertensive rats than unilateral common carotid ligation[1]. These findings suggest that carotid anastomosis induces higher haemodynamic stress on the circle of Willis than does carotid ligation.

The present study compares the effect of end-to-side common carotid anastomosis with those of unilateral common carotid ligation on cerebral aneurysm development in normotensive rats. In addition, we have also measured the common carotid blood flow (with an electromagnetic flow probe) before and after the experimental interventions, in order to document the changes in carotid blood flow.

Material and Methods

Three groups of adult Wistar rats weighing 300–390 g were used: 1) Nine rats with ligation of the right common carotid artery; 2) Eight rats with an end-to-side (right to left) common carotid anastomosis and 3) Four rats which were sham operated. Anaesthesia was induced and maintained with ether inhalation. Systemic arterial pressure was monitored throughout the experiment in one of the femoral arteries and arterial blood samples were taken for PaO_2, $PaCO_2$ and pH determinations.

Blood flow through the left common carotid artery was measured with an electro-magnetic flow transducer (0.85 mm in diameter) placed around the vessel in the neck and connected to a Biotronex transducer. Blood flow determinations were made before, between one to fourteen days, and between 80 to 240 days after the experimental interventions (in the case of the group 2 measurements after

* Supported by the Grant from the Spanish Social Security Research Fund FISS 916/84.

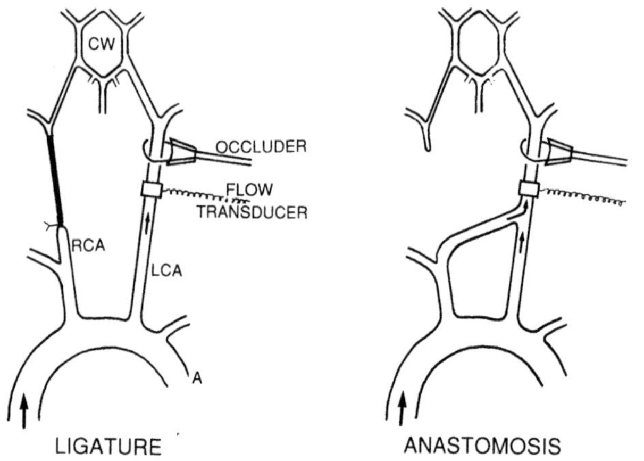

Fig. 1. Schematic representation of the Circle of Willis (*CW*) and carotid arteries of the rat illustrating the experimental preparations (*left side* = ligature of the right common carotid artery (*RCA*); *right side* = end-to-side carotid anastomosis) as well as placement of electromagnetic flow transducer and occluder on the left common carotid artery (*LCA*). Arrows, direction of blood flow. *A* aorta

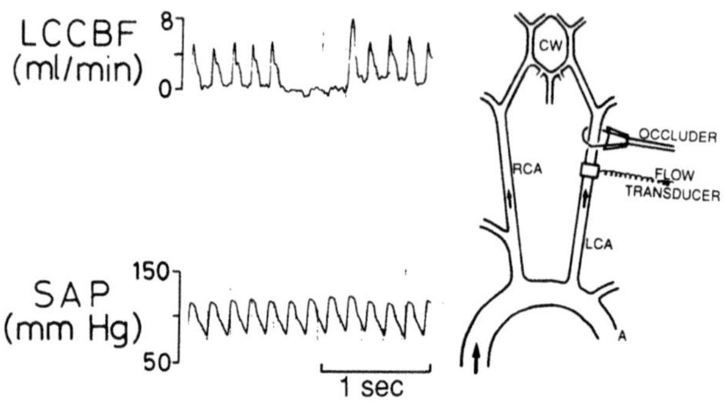

Fig. 2. On the left, representative tracings of pulsatile blood flow of the left common carotid artery (*LCCBF*) and systemic blood pressure (*SAP*) in an anesthetized rat under control conditions. Deflection in the blood flow recording corresponds to occlusion of the left common carotid artery

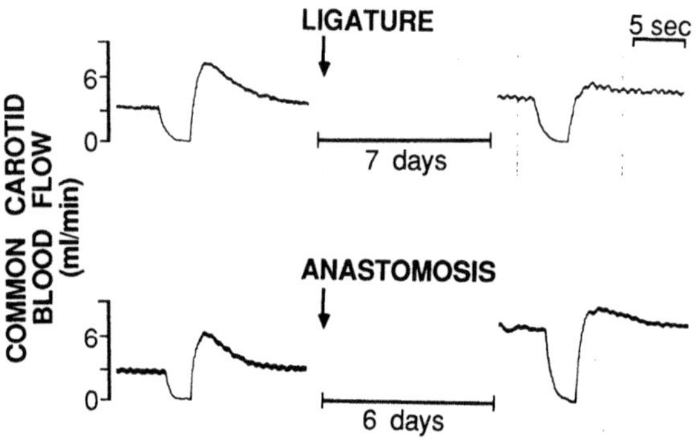

Fig. 3. Effects of contralateral carotid artery ligature or end-to-side carotid anastomosis on mean left common carotid blood flow

the surgical intervention were made distal to the anastomosis) (Fig. 1).

The animals were fed on standard laboratory diet "ad libitum". At the end of the experiments the rats were anesthetized and perfused with 50 ml of 1% heparinized normal saline serum through the abdominal aorta. Three ml of a formalin-containing solution of contrast medium (micropaque) were injected in the abdominal aorta and an angiography was obtained. The brains were removed and fixed with 10% formalin. Three weeks later the arteries at the base of the brain were dissected and examined with an operating microscope. The intracranial aneurysms observed were removed, embedded in paraffin and sectioned. The sections were stained with haematoxilin and eosin, P.A.S., van Gieson and Masson trichromic and examined with a light microscope.

Results

Under control conditions the mean carotid blood flow was 2.8 ± 0.14 ml/min (ranging between 2.7 and 4.2 ml/min) (Fig. 2). After carotid ligation the blood flow in the patent carotid artery increased 25 ± 5% after 7.1 ± 0.3 days and 37 ± 18% after 149 ± 23 days. After carotid anastomosis the carotid blood flow increased 102 ± 16% after 9.6 ± 2 days and 103 ± 35% after 169 ± 15 days (Fig. 3). The sham operation did not significantly affect carotid blood flow. Control values for mean arterial pressure (102 ± 3 mmHg) and blood gases ($pO_2 = 106 \pm 3.4$ mmHg, $pCO_2 = 38 \pm 1.2$ mmHg and pH = 7.32 ± 0.02) were similar in the three groups of animals and did not change significantly over time after the experimental procedure.

Post-mortem angiography evidenced occlusion of the right common carotid artery in the rats with carotid ligation and functioning anastomoses in the rats with carotid anastomoses. The arteries forming the circle of Willis were filled by the left common carotid artery and no aneurysm was clearly visualized. Using the operating microscope no cerebral aneurysms were found in the rats with carotid ligation or which had been sham-operated. However three of the eight rats ($p < 0.05$) with carotid anastomosis had single aneurysms in the anterior cerebral-anterior communicating arterial complex. Histologic examination of the aneurysms showed deficient arterial wall structure and the internal elastic lamina had completely disappeared. A proliferative phenomenon of the myo-intima consisting of vessel wall enlargements of varying intensity was observed. In one of the cases, isolated siderophages were found within the adventitia. Neither inflammatory infiltrate nor fibrinoid necrosis were visible (Fig. 4).

Discussion

Clinical[10, 15, 18] and experimental[8, 12, 16] observations suggest that haemodynamic stress is a main factor in

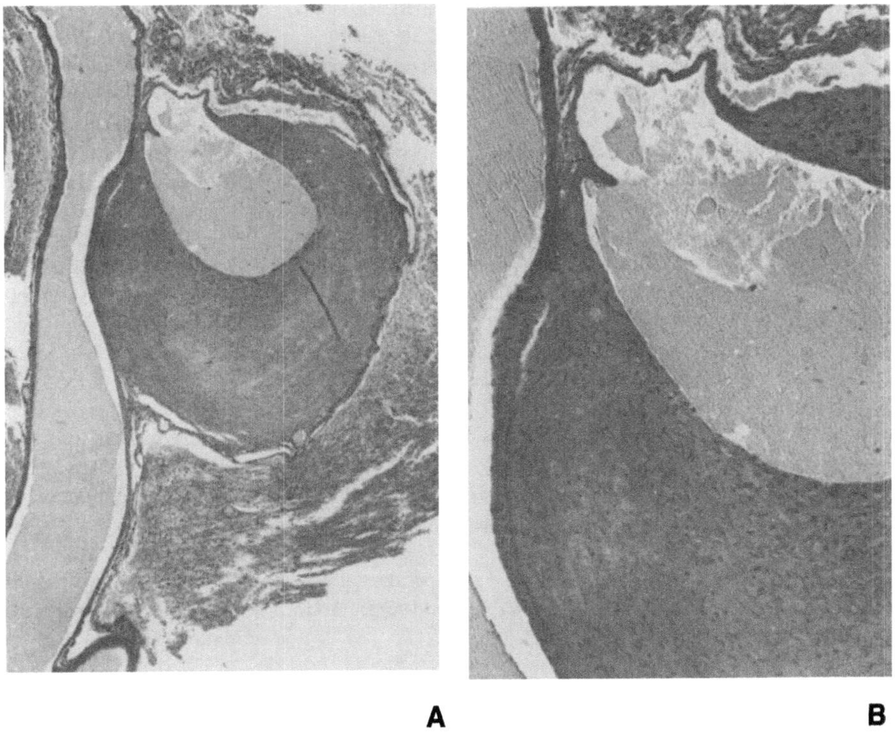

Fig. 4. A) Arterial aneurysm showing concentric cellular proliferation of the intimal. H & E 40 ×. B) Microscopic detail of intimal proliferation. Neither inflammatory infiltrate nor fibrinoid necrosis are seen. H & E 100 ×

cerebral aneurysm development. However, the haemodynamic changes caused by stress have not been documented. The effect of pulsatile flow and the haemodynamic forces of axial stream impingement on the apex of bifurcations are considered to have great influence in the development of intracranial aneurysms[2, 14]. The existence of media defects at the apex of the bifurcations, where the force of the central stream is higher, could explain the congenital origin of cerebral aneurysms[3]. Another hypothesis proposes that the production of aneurysms might be favoured by haemodynamic imbalance in the circle of Willis together with degenerative changes in the arterial wall at the site of bifurcations[16]. These forces would contribute to a focal degeneration of the internal elastic lamina. A third hypothesis attributes aneurysm production to a combination of developmental and degenerative changes[16].

In the present paper we have demonstrated that the increase in carotid blood flow in rats after end-to-side common carotid anastomosis is consistently higher than the increase found after contralateral common carotid ligation. These haemodynamic changes correlate well with the development of cerebral aneurysms in the anterior cerebral-anterior communicating arterial complex. This observation suggests that high haemodynamic stress in the cerebral vascular complex of the rats with carotid anastomosis causes cerebrovascular injury. The change in blood flow direction, the increase in haemodynamic stress and the effect of the pulsatile blood flow in the apex of the bifurcations in this anterior arterial complex would act as factors which induce aneurysms. Our results therefore suggest that persistent high, unilateral carotid blood flow induces cerebrovascular injury and so produces aneurysms in cerebral vessels. We propose that haemodynamic imbalance in the circle of Willis could be of relevance in the origin of cerebrovascular alterations. The number and size of aneurysms observed in the present study is lower than that reported in previous experimental models[1, 7, 12] but this is probably due to the absence of hypertension and/or the non-administration of lathyrogen in the present model, in contrast to other models.

References

1. Alvarez F, Roda JM (1986) Experimental model for induction of cerebral aneurysms in rats. J Neurosurg 65: 398–400
2. Ferguson GG (1972) Physical factors in the initiation, growth, and rupture of human intracranial saccular aneurysms. J Neurosurg 37: 666–677
3. Forbus WD (1930) On the origin of miliary aneurysms of the superficial cerebral arteries. Bull Johns Hopkins Hosp 47: 239–284

4. Hashimoto N, Handa H, Hazama F (1978) Experimentally induced cerebral aneurysms in rats. Surg Neurol 10: 3–8
5. Hashimoto N, Handa H, Hazama F (1979) Experimentally induced cerebral aneurysms in rats: Part II. Surg Neurol 11: 243–246
6. Hashimoto N, Handa H, Hazama F (1979) Experimentally induced cerebral aneurysms in rats: Part III. Pathology. Surg Neurol 11: 299–304
7. Hashimoto N, Handa H, Nagata I, Hazama F (1980) Experimentally induced cerebral aneurysms in rats. Part V. Relation of haemodynamic in the circle of Willis to formation of aneurysms. Surg Neurol 13: 41–45
8. Hassler O (1965) Experimental carotid ligation followed by aneurysmal formation and other morphological changes in the circle of Willis. J Neurosurg 20: 1–7
9. Jaffe ME, Mc Henry LC Jr (1968) Cerebral aneurysms following spontaneous carotid occlusion. Neurology 18: 1012–1014
10. Miyasaka K, Wolpert SM, Prager R (1982) The association of cerebral aneurysms, infundibula, and intracranial arterio-venous malformations. Stroke 13: 196–203
11. Nagata I, Handa H, Hashimoto N (1979) Experimentally induced cerebral aneurysms in rats: Part IV. Cerebral angiography. Surg Neurol 12: 419–424
12. Nagata I, Handa H, Hashimoto H, Hazama F (1980) Experimentally induced cerebral aneurysms in rats: Part VI. Hypertension. Surg Neurol 14: 477–479
13. Portnoy HD, Avellanosa A (1970) Carotid aneurysm and contralateral carotid stenosis with successful surgical treatment of both lesions. Case report. J Neurosurg 32: 476–482
14. Sekhar LN, Heros RC (1981) Origin, growth, and rupture of saccular aneurysms. A review. Neurosurg 2: 248–260
15. Stehbens WE (1963) Aneurysms and anatomical variation of cerebral arteries. Arch Pathol 75: 45–64
16. Stehbens WE (1983) Ethiology and pathogenesis of intracranial berry aneurysms. In: Fox JL (ed) Intracranial aneurysms, vol I. Springer, New York Berlin Heidelberg Tokyo, pp 358–395
17. Suzuki S, Robertson JT, White RP, Stadlan EM, Popoff N (1980) Experimental intracranial aneurysms in rats. A gross and microscopic study. J Neurosurg 52: 494–500
18. Wilson G, Riggs HE, Rupp C (1954) The pathologic anatomy of ruptured cerebral aneurysms. J Neurosurg 11: 128–134

Address for correspondence: J. M. Roda, M.D., Servicio de Neurocirugia, Hospital La Paz, Paseo de la Castellana 261, 28046 Madrid, Spain.

Acta Neurochirurgica, Suppl. 43, 193–197 (1988)

Occlusion of Experimental Artery Aneurysms by Intrasaccular Injection of Fibrin Sealant

J. R. Moringlane[1], **R. Grote**[1], **F.-J. Vonnahme**[2], **P. Mestres**[3], **G. Harbauer**[4], and **C. B. Ostertag**[1]

[1] Department of Stereotactic Neurosurgery, Neurosurgical Clinic, [2] Department of Pathology, [3] Department of Anatomy, [4] Department of Experimental Surgery, University of the Saarland, Homburg/Saar, Federal Republic of Germany

Summary

Aneurysms were produced by grafting a vein pouch onto the cervical carotid artery of rabbits following the removal of an elliptical piece of the arterial wall. The diameter of the opening was 3–4 mm and the maximum height 8 mm. Through direct puncture of the aneurysm a clot of the fibrin sealant Tissucol was injected into the aneurysm. The sequential morphological changes were studied by light microscopy. One case was examined after 3 weeks by scanning electron microscopy. Complete resorption of the fibrin sealant was observed. The aneurysm cavity was filled with a dense connective tissue covered by a layer of newly formed endothelial cells.

Keywords: Aneurysm occlusion; fibrin sealant; experimental aneurysms.

Introduction

In recent years, several authors have used isobutyl-cyanoacrylate (IBC) to occlude cerebrovascular malformations in laboratory animals and in human subjects[5, 10, 11, 16, 23]. Although some authors reported only mild reactions to the adhesive[4], IBC has been accused of provoking chronic inflammatory reactions and necrosis of the vessel wall[20, 22, 23]. Polymerization is accompanied by the development of heat. Occlusion of arteriovenous malformations with IBC was found to be incomplete[11]. Recently, attention has been drawn to a potential carcinogenic effect of IBC[15]. It thus seemed necessary to look for other embolizing agents, which have the advantage of instantaneous clotting – avoiding embolization of the parent artery – but do not provoke the adverse effects ascribed to IBC. The fibrin sealant Tissucol (Immuno GmbH/Heidelberg, FRG) is widely used as a surgical tissue adhesive, also for the haemostasis of bleeding from small vessels, and for vascular anastomosis[12].

The present study was undertaken to investigate the histological changes induced by a clot of the fibrin sealant following its injection into experimental aneurysms.

Materials and Methods

Thirty-eight aneurysms were produced in 34 rabbits weighing 2.5–3 kg. In 4 animals aneurysms were produced on both carotid arteries.

Animals were anaesthetized with 0.5 ml Rompun and 1–2 ml 1 : 1 mixture of NaCl/Ketavet, the latter being repeated if necessary. Animals were placed in a supine position with dorsal extension of the neck. The neck was shaved and, following surgical disinfection, a midline incision was made from the mandibular level to the sternum. Through a small stoma a cannula was inserted into the trachea. The largest branch of the external jugular vein was ligated and a segment of the vein with a ligature at one end to form a pouch was removed. Under the operating microscope the adventitia of the carotid artery at the site chosen for the aneurysm was removed. The artery was subsequently clamped. A lens-shaped biconvex piece of the arterial wall was removed. Through the opening the arterial lumen was washed with a heparinized saline solution. The vein pouch was grafted onto the defect of the arterial wall by single 10-0 sutures or a continuous suture on each side. After the clamps had been taken from the artery the blood stream could be observed through the wall of the aneurysm. This was documented by micro-doppler recording or by arteriography after puncture of the carotid artery proximal to the level of the aneurysm (Fig. 1). Aneurysms with an opening of about 3–4 mm in diameter and a maximum height of 8 mm ("sausage-shaped") with inclinations of approximately 45, 60 and 90 degrees to the axis of the blood stream were produced. An additional ligature shortened the pouch, producing spherical aneurysms. Since the object of the study was to investigate the sequential changes in an aneurysm occluded by the injection of a fibrin clot, the production of the aneurysm and the treatment were carried out in one stage.

With a sharp 23 G injection needle adapted to the injection set containing the components of the fibrin sealant Tissucol (aprotinin, $CaCl_2$, factor XIII, and thrombin at a concentration of 500 IE/ml), the aneurysms were punctured near the base. Using gentle and continuous pressure a clot, that was smooth enough to be advanced through the needle but compact enough not to spill into the carotid artery was injected toward the fundus. Occluded aneurysms ceased to pulsate, the wall became pale. The exclusion of the aneurysms

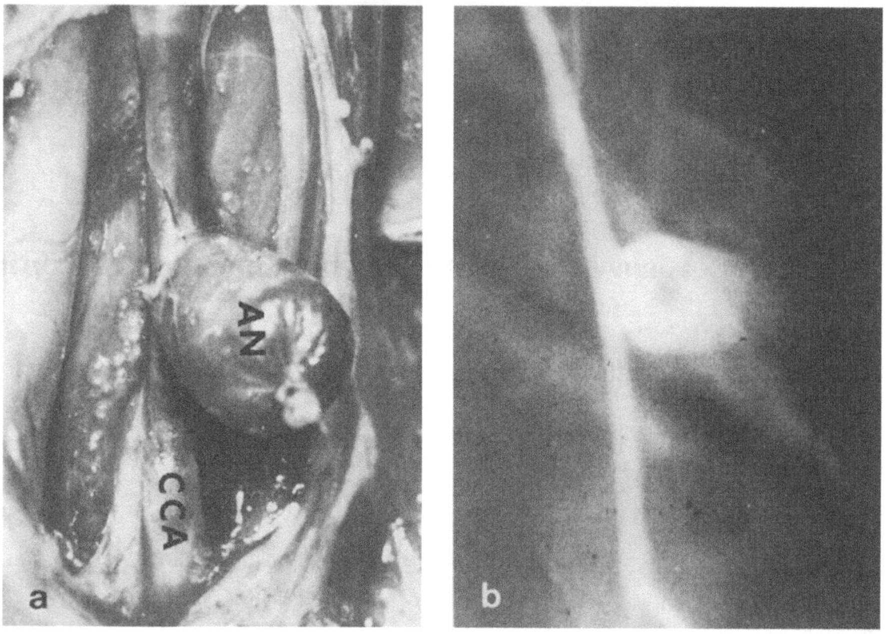

Fig. 1. a) Photograph of an aneurysm (*AN*) produced by grafting a vein pouch onto the common cervical carotid artery (*CCA*) of a rabbit. b) Angiographic study of the cervical carotid artery with the aneurysm

from the circulation was documented by microdoppler or by angiography.

The skin was closed with single sutures and the animals were returned to their cages. The animals were observed daily (eating habits, motor function, response to handling).

A total of 38 aneurysms were produced in 34 rabbits. Manipulation of the aneurysm after occlusion or faulty injection caused an embolization of the carotid artery in 4 cases. One animal developed a wound infection and was excluded from the study. Thirty-three aneurysms were available for examination. Animals were sacrificed at intervals varying from a few hours to 3, 7, 14, 21 and 28 days after treatment; in 1 case 5 weeks after thrombosis. Following repeat micro-doppler recording and/or angiography to verify the patency of the carotid artery, the aneurysms were removed with a distal and a proximal segment of the artery that was fixed in formaldehyde and embedded in paraffin. Sections were stained with haemtoxylin eosin, van Gieson and trichrome methylene blue stains.

One aneurysm removed on the 21st day after treatment was fixed in a phosphate buffered solution containing 1% formaldehyde and 3% glutaraldehyde with a resulting pH of 7.4[13], and in fresh fixative overnight at a temperature of 4 °C. The luminal surface of the carotid artery and the opening of the aneurysm were exposed, then the block of tissue was postfixed in ascending series of ethanol. The samples were dried in the critical point chamber and coated with gold with a sputtering apparatus (Balzers SCD 030). The specimens were examined under a CAmScan scanning electron microscope.

Results

On the third postoperative day fresh granulation tissue was present around the aneurysm and the carotid artery. On the 7th postoperative day the aneurysm could be easily separated from the surrounding tissue but from the 14th day on, the aneurysms became embedded in granulation tissue from which they had to be dissected by microsurgical preparation. To avoid injury to the wall of the aneurysm, the layer of scar tissue next to the aneurysm was not separated from it. Except for a few totally thrombosed aneurysms, angiography generally showed a partial filling of the basal portion of the aneurysms.

Light Microscopic Findings

The fibrin material was present in the aneurysm up to the 3rd postoperative day. Aneurysms examined on the 3rd day showed a leucocytic reaction extending from the wall of the aneurysm to the lateral portion of the fibrin clot.

At 7 days after treatment there was a small remnant of the fibrin clot. Granulation tissue was observed in the wall of the aneurysm and in the entire space previously occupied by the sealant. The residual cavity of the partially occluded aneurysms was not invaded by granulation tissue.

Two weeks after occlusion the aneurysms were surrounded and filled with granulation tissue in their outer zones, while the inner portions showed advanced tissue organization. A layer of new endothelial cells delineated the residual aneurysm cavity or bridged the opening of the completely obliterated aneurysms (Fig. 2).

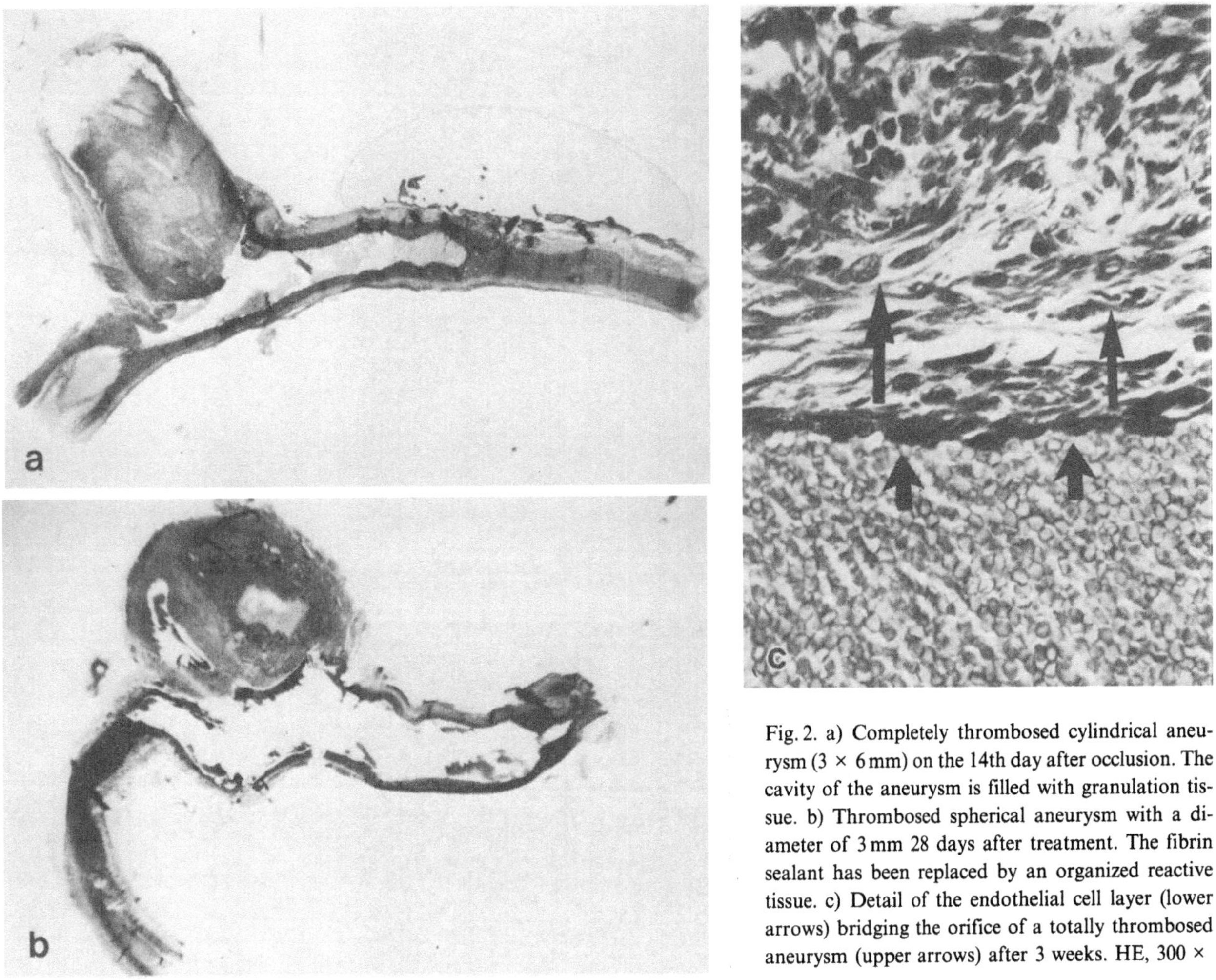

Fig. 2. a) Completely thrombosed cylindrical aneurysm (3 × 6 mm) on the 14th day after occlusion. The cavity of the aneurysm is filled with granulation tissue. b) Thrombosed spherical aneurysm with a diameter of 3 mm 28 days after treatment. The fibrin sealant has been replaced by an organized reactive tissue. c) Detail of the endothelial cell layer (lower arrows) bridging the orifice of a totally thrombosed aneurysm (upper arrows) after 3 weeks. HE, 300 ×

Three weeks after aneurysm occlusion the cavity was filled with completely organized reactive tissue including abundant capillary proliferation. An endothelial cell layer covered the orifice of the residual aneurysm cavity.

Four weeks after treatment there again was organized tissue filling the volume of the aneurysm, except for the residual cavity which contained fresh blood and was delineated by an endothelial cell layer (Fig. 2).

Ultrastructural Findings

A residual cavity was present in the case studied 3 weeks after occlusion (Fig. 3). No remnants of the fibrin clot were found. The degree of organization of the reactive tissue filling the aneurysmal pouch was indicated by the presence of numerous collagen fibres having different size, and by capillary vessels. Between the collagen fibrils were numerous bodies of connective tissue cells. The endothelial cell layer covering the residual cavity was composed of cells with a polygonal shaped apical surface, well defined cell borders and numerous microvilli. A single cilium was commonly found in most of the cells.

While the luminal surface of the carotid artery showed a characteristic longitudinal relief of many folds parallel to the blood stream, the relief patterns were different in the vicinity of and within the residual cavity of the aneurysm. In these regions the endothelial cells, which were rhomboidal in shape with a slight protrusion corresponding to the cell nucleus, had a cobblestone arrangement (Fig. 3). In contrast to this regular arrangement the endothelial cells in the transitional zone between the lumen of the artery and the residual aneurysm cavity seemed to be aligned in different directions and obviously not correlated to the axial plane of the arterial and of the aneurysmal lumen. There was no evidence of any smooth muscle cells in the reactive subendothelial tissue.

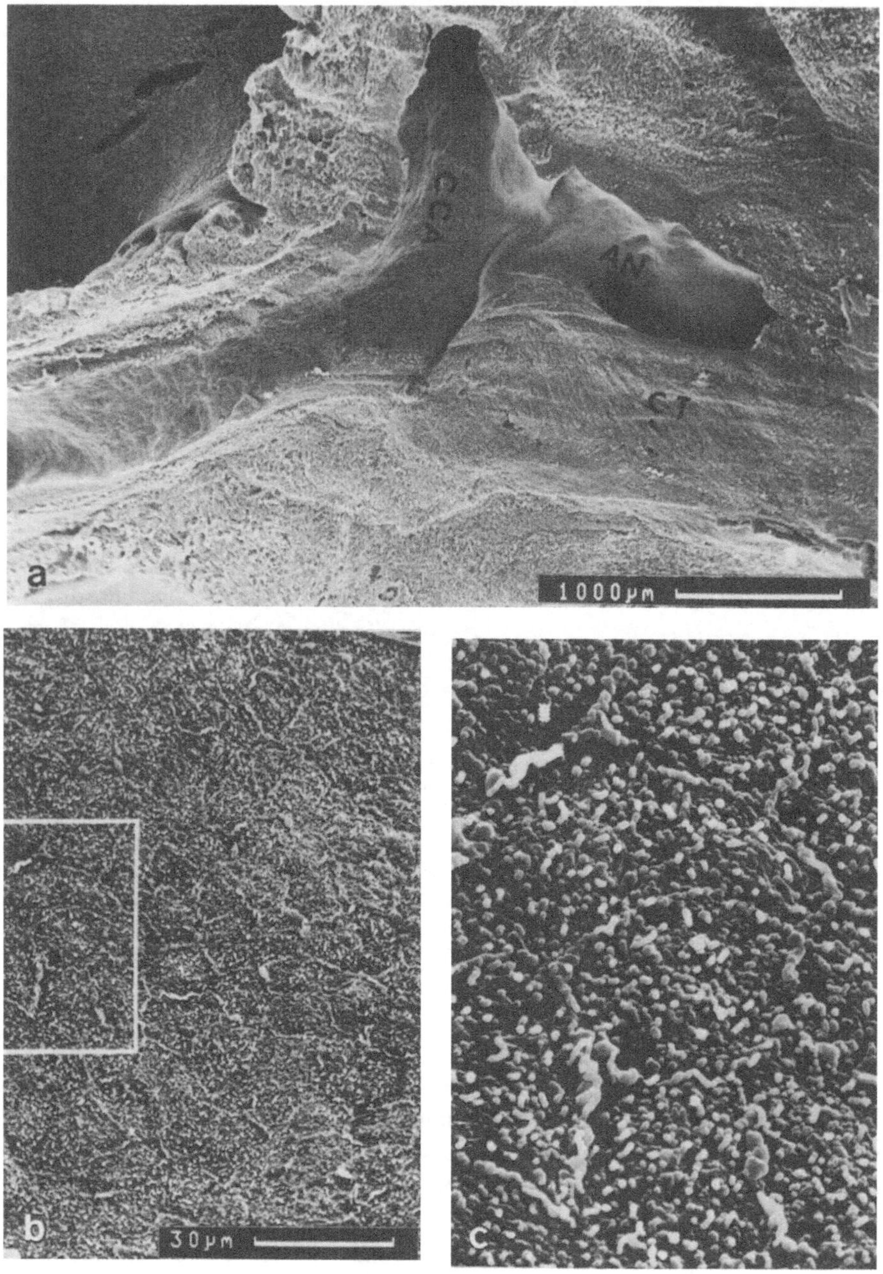

Fig. 3. SEM study of the endothelial cell layer covering the residual cavity of an aneurysm 21 days after treatment. a) Exposure of the lumen of the common carotid artery (*CCA*) and of the residual cavity of the aneurysm (*AN*) surrounded by connective tissue (*CT*). b) Relief of the endothelial cell layer: polygonal cells arranged in mosaic. c) 3 × magnification of 3 b) to show the cell surfaces of the new endothelial cells with microvilli and the cell borders

Discussion

Experimental aneurysms of intracranial arteries have been produced by non-surgical methods in rats, *e.g.* by beta-aminopropionitrile feeding combined with ligation of a common carotid artery and induced hypertension[7]. The ligation of carotid arteries in rabbits provokes the formation of intracranial artery aneurysms[8]. The aneurysms, however, are too small to allow the study of treatment modalities.

Experimental aneurysms are easily produced by microsurgical modifications of the method described by German and Black in 1954[6], which consists of grafting a vein pouch onto an incision of the arterial wall. In the present study the resection of an elliptical piece of the arterial wall was preferred to a simple linear incision. The morphological aspects of aneurysms created with this technique have been found to be similar to those of natural aneurysms[18]. Haemodynamic properties of the aneurysm model and its long lasting patency have been documented[2, 14, 17, 19].

In this study experimental aneurysms produced by the microsurgical technique at the cervical carotid artery of rabbits were occluded by direct puncture and the intrasaccular injection of the instantaneously polymerizing fibrin glue Tissucol in one stage. The fibrin sealant was completely resorbed. It was replaced by granulation tissue delineated by an endothelial cell layer that bridged the orifice of completely thrombosed aneurysms, or covered the border of those partially thrombosed.

The neoformative character of these endothelial cells is suggested by the high number of microvilli in comparison with the carotid artery. The newly formed endothelium showed no pathological alterations; this observation was supported by the lack of fibrin and adhesive platelets. The different alignments and arrangements of the endothelial cells in the aneurysm cavity and in the zone of transition to the arterial intima could be due to the influence of the perivascular reactive tissue in conjunction with haemodynamic factors in this area.

The injection of a thrombin solution resulting in the formation of a thrombosis has been reported as a method for the occlusion of otherwise non-treatable aneurysms of the aorta, of an accessory hepatic artery, of iliac, femoral and peroneal arteries[1, 3, 21]. Although our results are purely tentative, the physical properties of the instantaneously clotting fibrin sealant seem to be advantageous. The fibrin sealant has also been successfully used for the obliteration of high-flow carotid-cavernous fistulas[9].

References

1. Berguer R, Schneider J, Wilner HI (1978) Induced thrombosis of inoperable abdominal aortic aneurysm. Surgery 83: 425–429
2. Black SPJ, German WJ (1960) Observation on the relationship between the volume and the size of the orifice of experimental aneurysms. J Neurosurg 17: 984–990
3. Cope C, Zeit R (1986) Coagulation of aneurysms by direct percutaneous thrombin injection. AJR 147: 383–387
4. Cromwell LD, Freeny PC, Kerber CW, Kunz LL, Harris AB, Shaw CM (1986) Histologic analysis of tissue response to bucrylate-pantopaque mixture. AJR 147: 627–631
5. Debrun GM, Varsos V, Liszczak TM, Davis KR, Heros RS, Zervas NT (1984) Obliteration of experimental aneurysms in dogs with isobutyl-cyanoacrylate. J Neurosurg 61: 37–43
6. German WJ, Black SPW (1954) Experimental production of carotid aneurysms. N Engl J Med 250: 104–106
7. Hashimoto N, Handa H, Hazama F, Nagata I (1984) Animal model of cerebral aneurysms: pathology and pathogenesis of induced cerebral aneurysms in rats. Neurol Res 6: 33–40
8. Hassler O (1963) Experimental carotid ligation followed by aneurysmal formation and other morphological changes in the circle of Willis. J Neurosurg 20: 1–7
9. Isamat F, Ferrer E, Twose J (1986) Direct intracavernous obliteration of high-flow carotid-cavernous fistulas. J Neurosurg 65: 770–775
10. Kerber CW, Cromwell LD, Zanetti PH (1985) Experimental carotid aneurysms: Part 2 Endovascular treatment with cyanocrylate. Neurosurgery 16: 13–17
11. Klara PM, George ED, McDonnell DE, Pevsner PH (1985) Morphological studies of human arteriovenous malformations. Effects of isobutyl-2-cyanoacrylate embolization. J Neurosurg 63: 421–425
12. Kletter G, Matras H, Dinges HP (1978) Zur partiellen Klebung von Mikrogefäßanastomosen im intrakraniellen Bereich. Wiener Klin Wschr 90: 415–419
13. Mestres P, Rascher K, Garfia A (1980) Bedeutung und Perspektiven des REM in Neuroanatomie und experimenteller Neuropathologie. Beitr elektronenmikroskop Direktabb Oberfl 13: 265–276
14. O'Reilly GW, Utsunomiya R, Rumbaugh CL, Colucci VM (1981) Experimental arterial aneurysms: modification of the production technique. J Microsurg 2: 219–223
15. Samson D (1986) Isobutyl-2-cyanoacrylate. Neurosurgery 19: 885
16. Sheptak PE, Zanetti PH, Susen AF (1977) The treatment of intracranial aneurysms by injection with a tissue adhesive. Neurosurgery 1: 25–29
17. Stehbens WE (1973) Experimental production of aneurysms by microvascular surgery in rabbits. Vasc Surg 7: 165–173
18. Stehbens WE (1985) The ultrastructure of experimental aneurysms in rabbits. Pathology 17: 87–95
19. Varsos V, Heros RC, Debrun G, Zervas NT (1984) Construction of experimental "giant" aneurysms. Surg Neurol 22: 17–20
20. Vinters HV, Debrun K, Kaufmann JCE, Drake CG (1981) Pathology of arteriovenous malformations embolized with isobutyl-2-cyanoacrylate (bucrylate). J Neurosurg 55: 819–825
21. Wessler S, Reiner L, Freiman D, Reimer S, Lertzman M (1959) Serum-induced thrombosis. Studies of its induction and evolution under controlled conditions. Circulation 20: 864–874
22. White RI, Strandberg DVM, Gross GS, Barth KH (1977) Therapeutic embolization with long term occluding agents and their effects on embolized tissues. Radiology 125: 677–687
23. Zanetti PH, Sherman FE (1972) Experimental evaluation of a tissue adhesive as an agent for the treatment of aneurysms and arteriovenous anomalies. J Neurosurg 36: 72–79

Address for correspondence: Dr. J. R. Moringlane, Department of Stereotactic Neurosurgery, Neurosurgical Clinic, University of the Saarland, D-6650 Homburg/Saar, Federal Republic of Germany.

Acta Neurochirurgica, Suppl. 43, 198–204 (1988)

Local Pathological Findings in Experimental Dorsal Root Entry Zone Lesions Performed by Mechanical Section, Laser and Radiofrequency. A Comparative Study

J. Broseta[1], G. García-March[1], Mª. J. Sánchez-Ledesma[1], J. Gonçalves[1], J. Anaya[1], A. Torregrosa[2], J. Urbano[2], F. Collía[3], Mª. D. Ludeña[3], and M. Merchán[3]

[1] Departamento de Neurocirugía, Hospital Clínico Universitario, Salamanca, [2] Departamento de Neurocirugía, Hospital Gómez Ulla, Madrid, [3] Departamento de Ciencias Morfológicas, Facultad de Medicina, Salamanca, Spain

Summary

Dorsal root entry zone lesions are nowadays accepted as a first choice of treatment to alleviate certain types of chronic pain though it is still controversial to decide proper procedure for lesioning. Based on this last argument in 15 mongrel dogs different types of lesions using mechanical microsection, radiofrequency and laser were done. Under general anaesthesia a C-3 to C-7 laminectomy was performed in all animals and after open the dura mater the posterior nerve rootlets and posterolateral sulcus were identified using mangification. In 5 dogs, a longitudinal 1 mm deep incision from C-4 to C-6 spinal cord segments coinciding with the sulcus was carried out with a microknife; in other 5, a sequence of thermocoagulations produced by a radiofrequency current lower than 35 mA was done at same places and level with intervals of 2 mm; and in the rest of animals, an alike incision in depth, location and level was realised by carbon dioxide laser. Four days and three months after lesioning 2 and 3 animals of each group were sacrified and spinal cord specimens submitted for histological and ultrastructural studies. Low-power microscopic examination showed that all lesioning methods were able to produce a reasonable well delimited necrotic area involving the whole dorsal root entry zone structures, slightly more diffuse with the laser. Acute perilesional changes as well as chronic ones were nevertheless more evident in mechanical and radiofrequency lesions, over all regarding oedema, perivascular haemorrhage, intraluminal thrombosis and ischaemic alterations.

Keywords: Radiofrequency; carbon dioxide laser; dorsal root entry zone lesion; pain.

Introduction

In the dorsal root entry zone (DREZ) is located the leading segmentary control for modulation of the nociceptive afferent pathways. This includes the central portions of the dorsal rootlets, the Lissauer's tract and the first four laminae of the medullar dorsal horn. It thus seems rational to approach this area surgically for treating chronic pain. In fact DREZ lesions have been nowadays definetively incorporated in the surgical armamentarium as the first choice treatment to alleviate certain specific types of resistent pain such as those resulting from brachial or lumbosacral plexus avulsion[1, 8, 12], postherpetic neuralgia[3, 11], paraplegia[7, 12], multiple sclerosis[12] and cancer involving peripheral nerve structures[1, 12].

Although DREZ lesioning was initially performed using radiofrequency thermocoagulation[6], other procedures such as mechanical[13, 14] or laser[4, 9] sectioning of the area were further introduced, all them reporting similar fair results in their respective clinical series. Selection of one or other lesioning method is still controversial and therefore decided in accordance to the particular surgeon's usages, experience and disposal disregarding possible pathological side-effects of each method in the lack of individual or comparative experimental studies.

The goal of this study is to present the comparative histological and ultrastructural findings after mechanical, radiofrequency and carbon dioxide laser DREZ lesions produced in a series of 15 dogs.

Material and Methods

Experimental DREZ lesions were performed in 15 adult mongrel dogs of both sexes weighting 13 to 21 kg. They were anaesthetized with an intramuscular injection of 5 mg/kg of body weight of ketamine hydrochloride and then endotracheally intubated and artificially ventilated. Anaesthesia was maintained with 30 mg/kg body weight of pentobarbital and 0.05 mg/kg body weight of atropine through a saline intravenous infusion.

Following sterile techniques a complete laminectomy from C-3 to C-7 was carried out and the dura mater opened in all animals. With an operative microscope the right posterolateral sulcus and dorsal nerve rootlets were identified. Then, in 5 dogs of the series, DREZ lesions were made by mechanical section using a microknife. This tool was positioned in the sulcus at C-4 and introduced to a

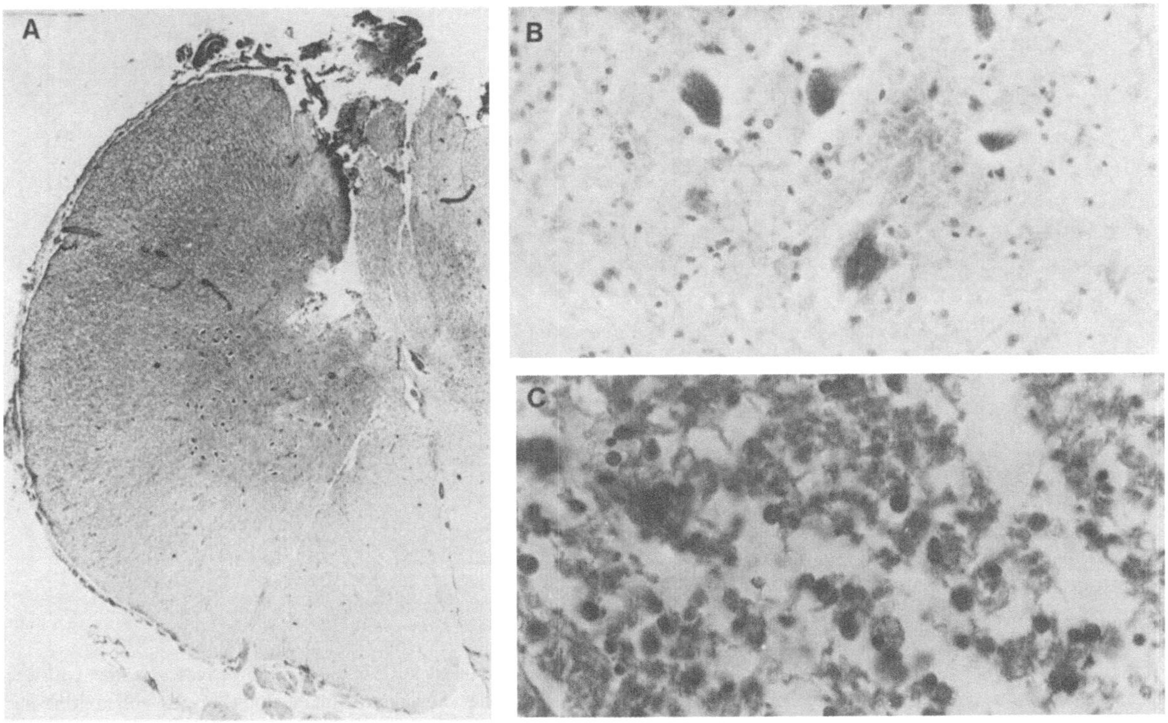

Fig. 1. Histological findings in mechanical DREZ section: A) low-power photomicrograph on acute DREZ lesion performed with the microknife showing necrotic area location. H & E × 1.6. B) perilesional dorsal horn neurons surrounded by acute secondary haemorrhage after lesion. Luxol-fast blue × 25. C) perilesional acute inflammatory infiltration with abundant macrophages. H & E × 40

depth of 1 mm taking as reference a trace previously carved on the blade. Under magnification the sulcus was caudally cut to C-6 in the same depth and also involving the local nerve rootlets.

In 5 other dogs, DREZ lesions were made by radiofrequency. A specially designed 0.5 mm epoxilite insulated stainless steel electrode with an active tip of 0.5 mm was employed for this purpose. Using the surgical microscope it was inserted 0.5 mm deep at the rostral level of the sulcus angled 30° in lateromedial direction and later the probe connected to a radiofrequency generator*. Thermocoagulations were made at 25 to 30 mA for 10 seconds with the electrode stabilized in this position. Through the microscope a small whitened coagulated area arround the tip could be observed. The procedure was repeated at intervals of 3 mm down along the longitudinal extension of the dorsolateral sulcus to C-6.

In another five dogs, the lesions were produced by a carbon dioxide laser** with a 3 W output power and a duration of 3 seconds at an optical spot size of 1 mm. The beam was angled 30° mediolaterally and focussed under magnification at C-4 level on the sulcus such was followed down to C-6. Following these parameters a longitudinal lesion on the DREZ of 1 mm wide and 0.5 mm deep was obtained.

In all cases the dura and surgical wound were closed after lesioning. Intravenous steroids were given following operation and for 4 days postoperatively.

Two animals of each group were sacrified by intravenous injection of potassium chloride 4 days after surgery and the rest 3 months later. The cervical spinal cord was removed, fixed in 10% formalin solution, embedded in paraffin and prepared for histological examination with haematoxylin-eosin, Luxol – fast blue and Van Gieson stain techniques. Small samples of dorsal root entry zone, dorsal columns and centromedullary area were immersed in 3% glutaraldehyde and phosphate buffer and later processed for ultrastructural studies.

Results

Despite the shortness of the experimental series some differences concerning morbidity and mortality were observed related with the DREZ lesioning system. On one hand, the postoperative clinical condition in animals with mechanical and laser DREZ lesions was significantly better than those with radiofrequency thermocoagulations, presenting less neurological deficits and handicaps. On the other hand, during chronic follow-up 2 out of the remaining dogs belonging the radiofrequency group died 10 and 24 days respectively after surgery; 1 dog with mechanical section died 7 weeks after lesioning and no postoperative mortality was observed in the laser group. Although these data suggest differences in the validity of the diverse lesion-

* Model RFG-55 Radionics manufactured by Radionics Inc., Massachusetts.

** Model 733 Sharplan CO_2 laser manufactured by Laser Industries Ltd., Tel Aviv.

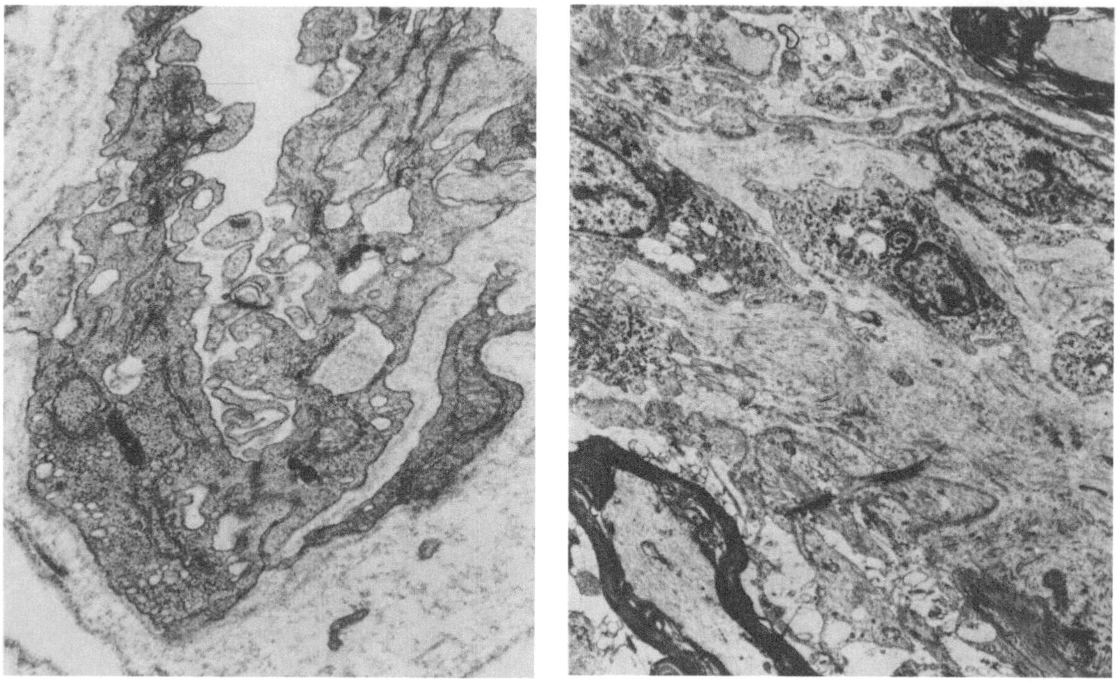

Fig. 2. Electron micrographs showing perilesional relevant findings in mechanical DREZ lesion: Left: perivascular area with acute oedema, macrophage infiltration and collagen fibers. × 2,000. Right: dorsal horn capillar showing irregular endothelial cells with multiple micropinocitic vesicles and middle electron density vacuolar material presenting on intact basal membrane and certain perivascular oedema. × 15,000

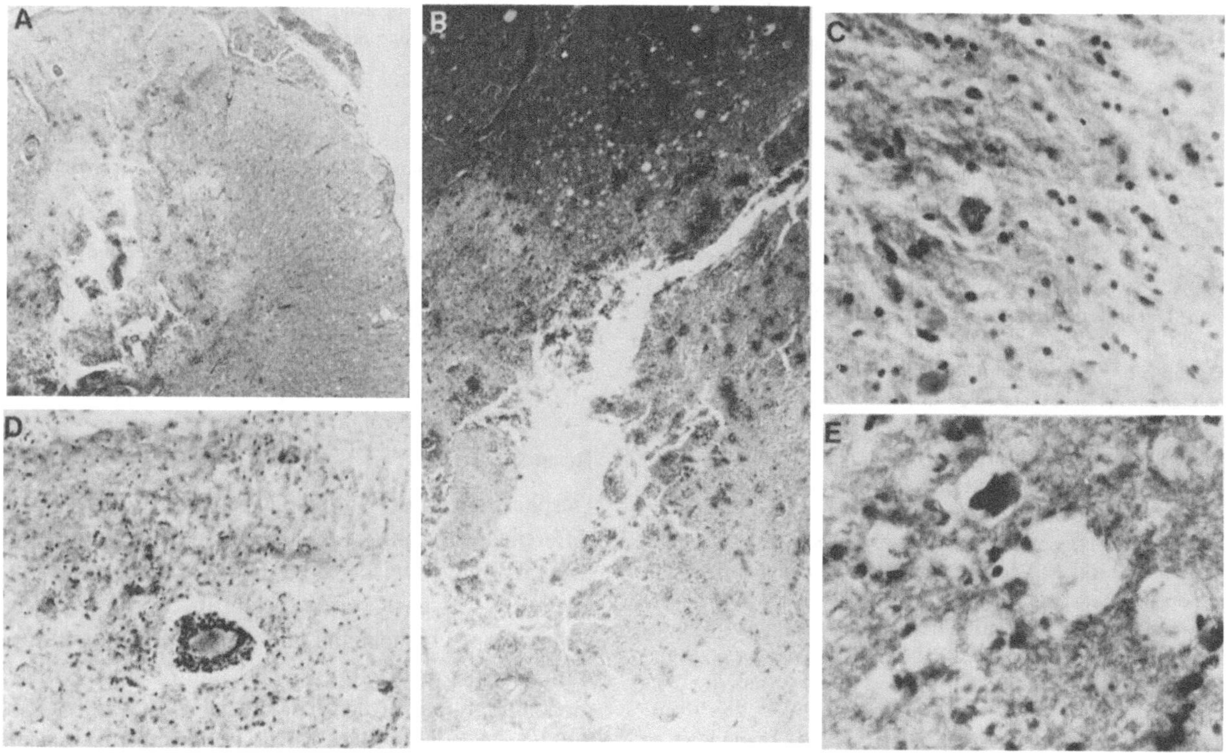

Fig. 3. Pathological observations in radiofrequency DREZ thermocoagulation: A) low-power photomicrograph illustrating acute necrotic lesion and its partial extension to dorsal columns. H & E × 1.6. B) acute DREZ thermolesion cavity and perilesional neighbouring necrotic detritus. H & E × 2.5. C) chronic degenerative neuronal bodies located in perilesional centromedullary area. H & E × 25. D) spino-cerebellar tract thrombosed capillar with perivascular haemorrhage. Luxol-fast blue × 10. E) perilesional dorsal column fibers degeneration with inflammatory infiltration. Luxol-fast blue × 25

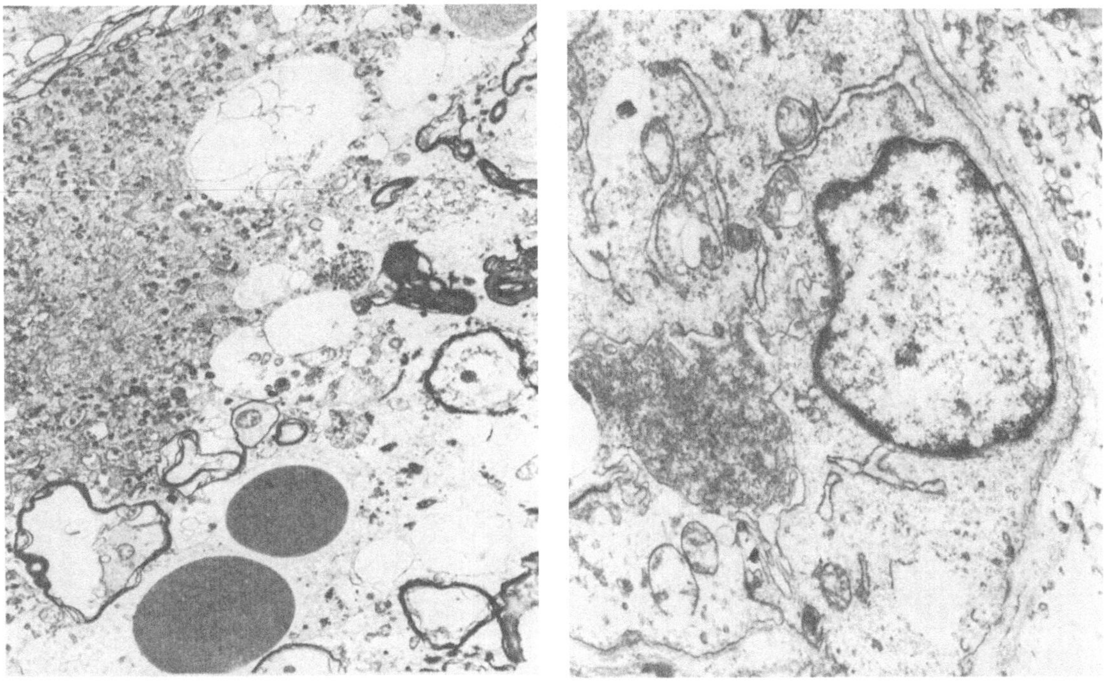

Fig. 4. Ultrastructural micrographs in perilesional areas after radiofrequency DREZ thermocoagulation: Left: perilesional cellular detritus showing degenerative fibers sprouts with macrophagic and red blood cells infiltration. × 3,000. Right: dorsal horn capillar with endothelial hyperplasia and intraluminal thrombosis. × 10,000

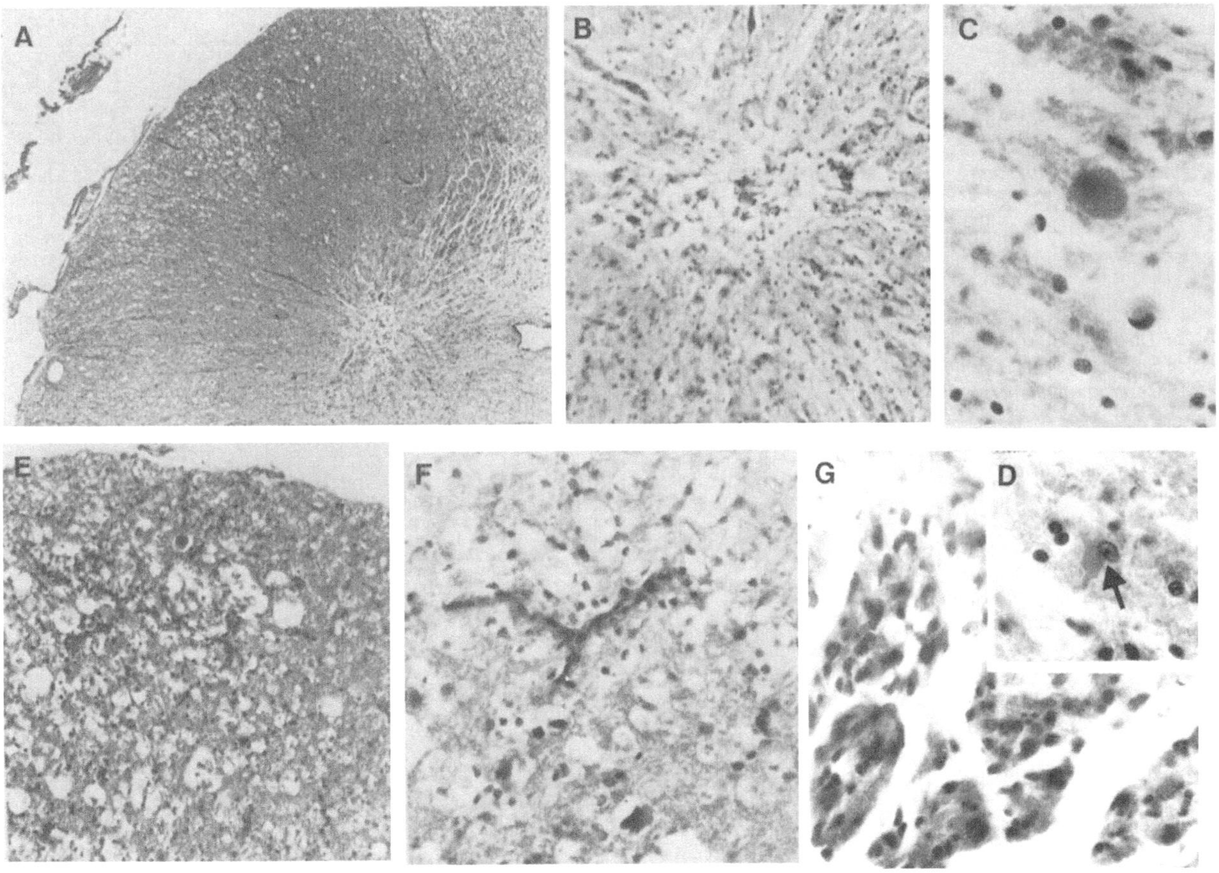

Fig. 5. Photomicrographs showing histological alterations after laser DREZ lesion: A) low-power acute DREZ lesion and extension. H & E × 25. B) acute necrotic field with inflammatory infiltration. H & E × 10. C) dorsal horn chronic degenerative neuronal bodies. H & E × 40. E) spinocerebellar fibrillary degeneration. Luxol-fast blue × 10. F) spinocerebellar fibrillary degeneration. Luxol-fast blue × 25. G) affected dorsal nerve root let showing and increasing of cell proliferation as regenerative indicator. H & E × 40

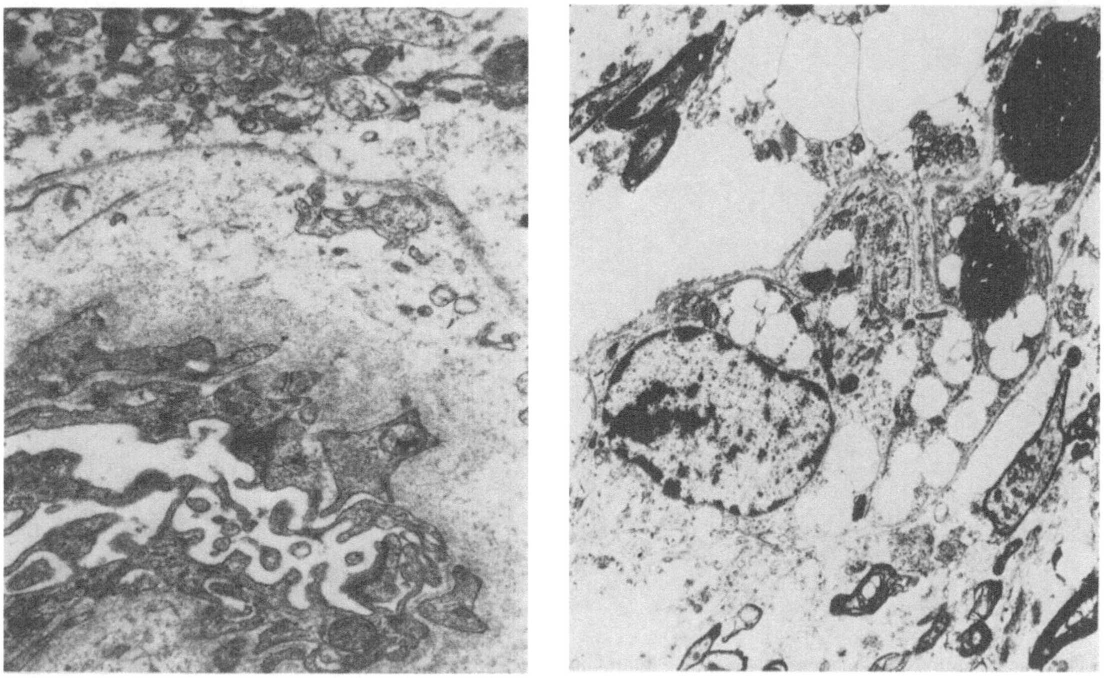

Fig. 6. Electron micrographs illustrating perilesional pathological details in chronic laser DREZ lesions: Left: dorsal horn capillar with scarce endothelial alterations and light oedema. × 15,000. Right: chronic perivascular intracellular vacuolization and extracellular small cavitation, intracellular calcium deposits and necrotic cellular detritus closed to the necrotic area. × 3,000

Table 1. *Comparative Pathological Findings in Experimental DREZ Lesions*

Pathological observations	*Mechanical section*	*Radiofrequency*	*CO_2 laser*
Necrotic lesion			
	Fair	Fair	Fair
DREZ location	+ + +	+ + +	+ + +
Dorsal column extension	+	+ +	+ +
Spinocerebellar extension	+ +	+ +	+ +
Centromedullar extension	+ +	+ + +	+
Acute perilesional changes			
Edema	+ + +	+ + +	+
Intracellular vacuolization	+ +	+ +	+ +
Tisular cavitation	-	-	-
Perivascular hemorrhage	+ +	+ + +	+
Intraluminal thrombosis	+ +	+ +	-
Inflammatory infiltration	+ + +	+ + +	+ +
Chronic perilesional changes			
Fibrillar degeneration	+	+ +	+ +
Neural degeneration bodies	+	+ +	+ + +
Reactive gliosis	-	-	+
Ischaemic alterations	+ +	+ + +	-
Connective reaction	-	-	-

-, absent; +, light; ++, moderate; +++, severe.

ing techniques here tested they are not at all conclusive when working in experimental conditions and with few animals.

Low-power microscopic examination of affected spinal cord sections showed that all methods reproduced a well defined damage area on the DREZ, more delimited when using mechanical and radiofrequency lesions and less precise with laser.

Histological features of the lesions were very alike in all groups disregarding the lesioning system and in the acute stage consisted in a necrotic area containing cellular detritus that usually involved the central portion of the dorsal rootlets, the Lissauer's tract and partially the dorsal horn, encircled by an irregular perilesional ring that often involved the neighbouring fibre tracts and represented a zone with demyelination, oedema, vascular alterations and neuronal dystrophies with inflammatory infiltration and abundant macrophages. In chronic stages this lesion appeared as a clean cavity surrounded by regenerative nervous tissue though showing multiple degenerative neuronal and fibrillary bodies. DREZ thermocoagulations seemed to extend to dorsal columns, spinocerebellar tract and centromedullary area more frequently and with a higher severity than mechanical or laser lesions.

While tissue cavitation distal to the lesion was not observed in any of the lesioning methods, intracellular vacuolization in the perilesional area was constant in all. Perivascular haemorrhage, intraluminal thrombosis and oedema shown in the acute stage as well as ischaemic alterations in chronic ones were less evident in laser sectioning than in lesions produced by the other techniques. Nevertheless fibrillary and neuronal degeneration close to the lesion were more intense in the former than in the latter. Neither reactive gliosis nor connective reaction were detected in the different lesions.

Figs. 1, 3 and 5 illustrate the main histological findings observed in the different experimental groups of the series and Figs. 2, 4 and 6 show the ultrastructural differential features.

Table 1 presents the comparative pathological observations in the distinct experimental DREZ lesions.

Discussion

Pathological findings on human DREZ lesions are seldom reported and always concern radiofrequency thermocoagulation. In one of these cases treated by our group[1], where a DREZ thermolesion was used to control malignant pain from a Pancoast's carcinoma involving the brachial plexus, pathological study showed proper location of the necrotic area in the DREZ with little extension to neighbouring structures and histological features conciding with other reports in literature[10].

However, histological observation on the effect of different lesioning systems on the DREZ in experimental conditions are more controversial. Levy *et al.*[5] compared experimental DREZ lesions in the cat performed by laser and radiofrequency thermocoagulation, concluding that the former creates a more precise and uniform necrotic area with less surrounding oedema and tissue disruption than the latter that affects more frequently the spinal tracts. In this paper, laser DREZ lesions are defended based on the short time required for lesioning, the light affection of surrounding regions and the very shallow penetration of the system. Laser aiming and focusing errors are the inconveniences adduced against this procedure. Watts[15] also defended the use of laser in this type of surgery since he observed more precise lesions and less pain recurrence. On the contrary, Young[16] opts for radiofrequency lesions even accepting that thermocoagulations can extend to other unwanted structures throughout laser will also do this. Nashold[8] also warns against laser application in DREZ surgery due to the possibility of producing centromedullary cavitation.

Our experimental results confirm that DREZ laser lesions have less extension and pathological side-effects in the perilesional area than mechanical section and much less than radiofrequency thermocoagulation, agreeing on this point with Levy *et al.*[5] and Power *et al.*[9]. However, our radiofrequency lesions were done without temperature monitoring and electrode stabilitation as Cosman *et al.*[2] recommended to achieve consistent results. Our study failed to find the centromedullary syrinxes reported by Nashold[8] when using laser on the DREZ.

Despite these conflicting experimental results, the three lesioning systems tested in this study present quite satisfactory long-lasting analgesia as shown by Sindou *et al.*[13, 14], and pathological differences are not so evident. Thus, selection of one or other lesioning system probably will be once more dependent on the surgeon's facilities and abilities more than the pathological and pathophysiological background.

References

1. Broseta J, García-March G, González-Darder J, Roldán P, Masbout G, Barberá J, Barcia-Salorio JL (1987) Lesión de la zona de entrada de las raices dorsales en el tratamiento del dolor crónico. Cir Esp 41: 431–439

2. Cosman ER, Nashold BS, Ovelman-Levitt J (1984) Theoretical aspects of radiofrequency lesions in the dorsal root entry zone. Neurosurgery 15: 945–950
3. Friedman AH, Nashold BS (1984) Dorsal root entry zone lesions for the treatment of postherpetic neuraliga. Neurosurgery 15: 969–970
4. Levy WJ, Nutkiewicz A, Ditmore QM, Watts C (1983) Laser-induced dorsal root entry zone lesions for pain control. J Neurosurg 59: 884–890
5. Levy WJ, Gallo C, Watts C (1985) Comparison of laser and radiofrequency dorsal root entry zone lesions in cats. Neurosurgery 16: 327–330
6. Nashold BS, Ostdhal RH (1979) Dorsal root entry zone lesions for pain relief. J Neurosurg 51: 59–69
7. Nashold BS, Bullitt E (1981) Dorsal root entry zone lesions to control central pain in paraplegics. J Neurosurg 55: 414–419
8. Nashold BS (1984) Current status of the DREZ operation: 1984. Neurosurgery 15: 942–944
9. Powers SK, Adams JE, Edwards MSB, Boggan JE, Hosobuchi Y (1984) Pain relief for dorsal root entry zone lesions made with argon and carbon dioxide microsurgical lasers. J Neurosurg 61: 841–847
10. Richter HP, Schachenmayr W (1984) Is the sustantia gelatinosa the target in dorsal root entry zone lesions? An autopsy report. Neurosurgery 15: 913–916
11. Samii M (1981) Thermocoagulation of the sustantia gelatinosa for pain relief (preliminary report). In: Siegfried J, Zimmermann M (eds) Phantom and stump pain. Springer, Berlin Heidelberg New York, pp 156–159
12. Samii M, Moringlane JR (1984) Thermocoagulation of the dorsal root entry zone for the treatment of intractable pain. Neurosurgery 15: 953–955
13. Sindou M, Mifsud JJ, Boisson D, Goutelle A (1986) Selective posterior rhizotomy in the dorsal root entry zone for treatment of hyperspasticity and pain in the hemiplegic upper limb. Neurosurgery 18: 587–595
14. Sindou M, Pelissou I, Goutelle A (1987) Anatomical bases of neurosurgical spinal cord procedures for treatment of chronic pain. Peripheral Nerve Repair and Regeneration (in press)
15. Watts C (1985) Letter to the editor: DREZ lesions using lasers. J Neurosurg 63: 150
16. Young RF (1986) Letter to the editor: Laser versus radiofrequency lesions of the DREZ. J Neurosurg 64: 341

Address for correspondence: J. Broseta, M.D., Departamento de Neurocirugía, Hospital Clínico Universitario, E-37007 Salamanca, Spain.

Acta Neurochirurgica, Suppl. 43, 205–209 (1988)

The Quality and Extension of Nerve Fibre Regeneration in the Centrocentral Anastomosis of the Peripheral Nerve

J. Barberá, J. Gonzalez, J. L. Gil, Mª A. Sanjuán, F. García, and **A. Lopez**

Department of Neurosurgery, Faculty of Medicine, University of Cadiz, Cadiz, Spain

Summary

Centro central anastomosis (CCA) is the connection created by means of a nerve autograft placed between the fascicles of the proximal stump of a sectioned nerve. We have used this technique in the laboratory and in the human to avoid the formation of a terminal neuroma after amputation of a limb. The aim of the present study is to know what is the quality and how long is the progression of the regenerating nerve fibers coming from a fascicle and growing inside the nerve graft and into the other central fascicle.

56 male, adult, Sprague-Dawley rats were operated on. The 4 rats of Group I were used for control: the right sciatic nerve was exposed and dissected in its four branches. One piece of 10 mm was taken of each of the sural and the peroneal branches for hystological and hystomorphometric studies. In the 16 rats of Group II, after the section of the sural and peroneal branches a terminoterminal suture between their proximal stumps was made. In the 16 rats of Group III a CCA with an interposed nerve graft of 7 mm was achieved. 5,5 and 6 rats of Groups II and III were sacrified at 10, 40 and 365 days after the operation and the entire graft was taken for histological analysis. In the 20 rats of Group IV, 60 days after CCA, the graft was sectioned at its center to produce wallerian degeneration of nerve fibers. 10 days after this second operation the animals were sacrified and the nerve specimens were taken for histological study.

The histological and morphometric examination showed that the terminoterminal suture between the central fascicles does not avoid the neuroma. The CCA avoids the neuroma and the interposed graft reaches an almost orthotypic miscroscopic and morphometric structure. In the Group IV we noted that the regenerative sprouting is arrested inside the graft and the wallerian degeneration does not extend beyond the second suture, even though it is permeable.

The axonal regenerative sprouting inside the graft produces an orthotypic reconstruction of the peripheral nerve without a significant neuroma but the nerve fibers do not progress beyond the suture between the graft and the other central nerve stump. The speculative causes of this arrested growth are discussed.

Keywords: Nerve fiber regeneration; nerve graft; neuroma; stump pain.

Introduction

The centrocentral anastomosis (CCA) is the connection created by a nerve autograft sutured between the fascicles of the proximal stump of a sectioned nerve. This technique has been used in the laboratory and with human patients achieving good results in the treatment of pain due to the amputation neuroma[2, 7, 8, 9, 13]. The aim of the present work was to know the histological characters of the nerve fiber regeneration inside the interposed nerve graft.

Material and Methods

We operated on 56 male, adult, Sprague-Dawley rats weighing 250–300 gr, separated into four groups. In all cases, under general anaesthesia, the right sciatic nerve was exposed through a posterior incision in the thigh and the nerve trunk was dissected in its three terminal branches: sural, peroneal and tibial. After the operation the surgical wound was closed in two planes.

In *Group I*, with 4 rats, a piece of nerve 10 mm long was taken from the sural and peroneal branches for histological and histomorphometric studies of the normal nerve of the rat. In the 16 animals of *Group II* the sural and peroneal branches were sectioned 4 mm below their origin in the sciatic nerve. After that a terminoterminal suture with 10-0 ethilon was made between the proximal stumps of the sectioned branches. In the 16 animals of *Group III* the sural branch was sectioned 4 mm below its origin, and the peroneal one was sectioned at two levels, 4 and 11 mm below its origin at the sciatic trunk. In this way we obtained a free piece of the peroneal nerve that was used as an interposed graft to connect the proximal stumps of the peroneal and sural branches, so creating a CCA between them. 5, 5, and 6 animals of Groups II and III were reoperated on 10, 40 and 365 days after the first operation. The sciatic nerve was exposed again and its sutured branches were taken off for histological, histomorphometric and ultrastructural studies. The 20 animals in *Group IV* were operated on in the same way as Group III with one major difference: in 10 rats the sural nerve was used as the graft donor, and in the other 10 animals, the peroneal nerve was the donor. All these rats were reoperated on 60 days after the first operation. The CCA was exposed and sectioned at the center of the graft. 10 days later a new operation was performed, taking off the sural and peroneal branches with their own part of the graft to study the wallerian degeneration due to the immediate previous section (Fig. 1).

In all the experiments only the sural and peroneal branches were

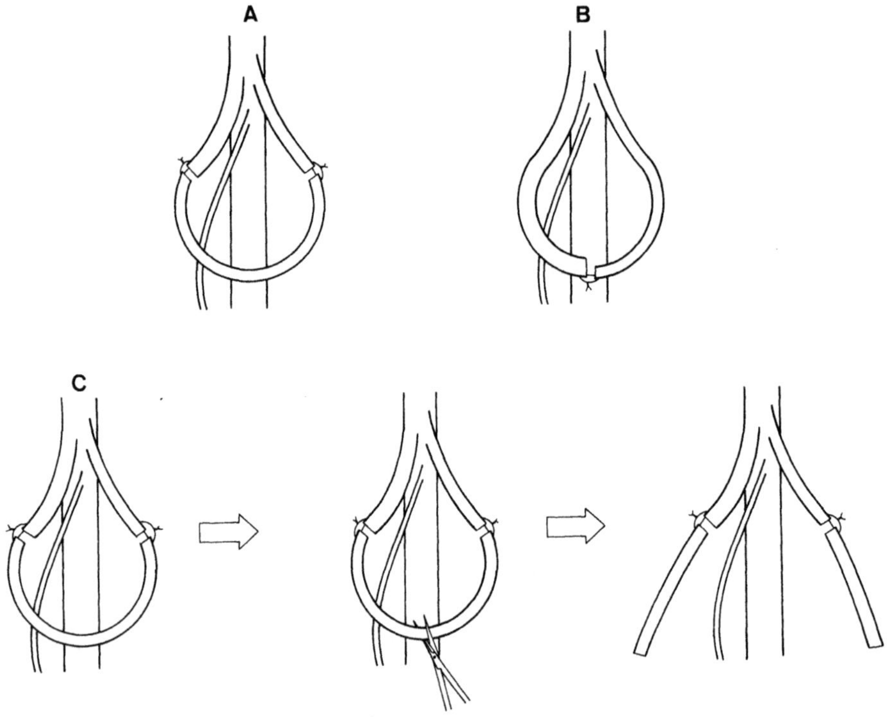

Fig. 1. Scheme of the experiment. A) Centrocentral anastomosis (Group III). B) Terminoterminal suture between both proximal ends (Group II). C) Sectioned nerve graft after 60 days of the CCA (Group IV)

sectioned in order to avoid the self-mutilation of the limb that follows the whole section of the sciatic nerve in rats, in accordance with the recommendations made by the International Association for Study of Pain[6].

Results

The *histological examination of the normal nerve* showed its typical characters. The morphometric study was carried out on a total of 1773 counted myelinated fibers with a ratio of 48 fibers per field. The myelinated fibers occupied an area of 57.36% as compared with the unmyelinated ones which occupied 1.93%. The average diameter of the myelinated fibers was 8.77 µm with a range between 1 and 20 µm and a maximum number of fibers between 6–7 and 9–10 µm, following a bimodal distribution. The average bulk of the myelin sheath was 1.22 µm, with a distribution range between 0 and 4 µm.

In Group II the histological study of the suture, 10 days after the operation, showed the axonal degeneration and rupture of the myelin sheath with phagocytic activity of the Schwann cells. Initial regenerative processes were also observed at the two proximal ends of the nerves. The space between these was occupied by a large number of Schwann cells and endoneural fibroblasts distributed in a disorganized way. After 40 days a large proliferation of endoneural connective tissue appeared, including nerve fibers in an irregular form and of different types, and distributed also in a random way. At the point of the suture the myelinated fibers occupied an area of 6.60% and the unmyelinated fibers the 2.10%. After 365 days the suture point was occupied by a large proliferation of connective tissue inside which small nerve fibers were found irregularly distributed and with abnormal morphology. The growth of the regenerated fibers was abnormal, following aberrant directions.

In Group III, 10 days after the operation the longitudinal and transversal sections of the graft offered the typical signs of wallerian degeneration, characterized by the disruption of the myelin sheath configurating irregular images of a variable calibre; the elipsoids, made up of axonal remains and myelin degeneration were observed. At the sutures a large number of connective cells existed as well as fibroblasts and Schwann cells. Images of retrograde degeneration appeared in short extension in both proximal nerve ends. After 40 days the nerve regeneration was obvious, both at the suture area and inside the graft. Large groups of axonal sprouts appeared, small, round in contour, with thin sheaths of myelin and dispossed in an homogeneous distribution. In the longitudinal sections the parallel order of the fibers is what stood out. The ultrastructural study showed axonal compartmentation. At the suture region the myelinated fibers occupied an area of 12.63% and the unmyelinated were in an area of 1.97%. The serial sections of the graft showed a density of 108 myelinated fibers per field.

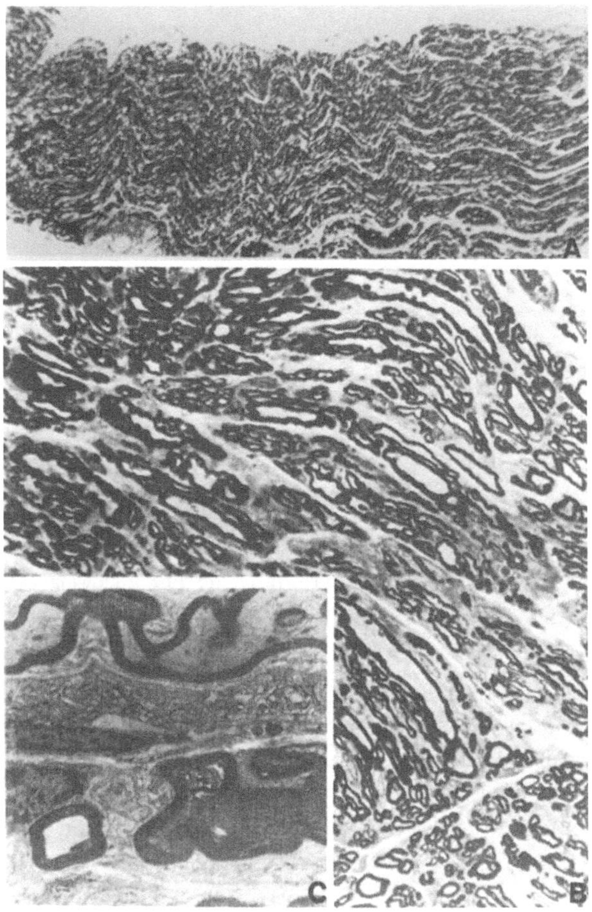

Fig. 2. Group III (CCA). 365 days of evolution. The pictures were taken from different grafts. A) Longitudinal section: the ordered disposition of the nerve fibers is seen. B) Transversal section: the fibers appear well structured, wrapped in their myelin sheath and surrounded by slight endoneurum. C) Ultrastructural aspect: the structure of the myelinated fibers have an almost orthotypic morphology

These fibers occupied an area of 21.06%; its average diametre was 3.99 µm with a distribution range between 1 and 13 µm. The average thickness of the myelin sheath was 0.37 µm with their values between 0 and 1.2 µm. The unmyelinated fibers occupied an area of 1.97%. One year after the operation the suture lines were difficult to identify histologically, the perineural sheath appeared repaired and there was a minimum proliferation of well organized endoneural tissue which included almost normal nerve fibers in the correct position. In the graft it was observed that the compartmentation had disappeared, allowing the free healthy nerve fibers to be distributed in an ordered way and included in a well organized connective tissue with parallely disposed collagen fibers (Fig. 2). The density was 71 myelinated fibers per field. The areas occupied by the myelinated and unmyelinated fibers were 40.58% and 1.77% respectively. The average diameter of the myelinated fibers was 6.92 µm with a distribution range between 2 and 20 µm. The average thickness of the myelin sheath was 0.83 µm with a distribution between 0 and 2.6 µm.

The whole analysis of the histological and histomorphometric studies allow us an overall vision of the evolution of the two groups. From the beginning, in group III there was a minimum fibrous reaction at the level of the suture which allowed the fibers to pass from the central nerve endings and be introduced into the graft. The disposition of the nerve fibers on crossing the suture area and inside the graft was fairly ordered, having great parallelism and being regular in form, remaining included in a conjunctive endoneural tissue, also well ordered. Their evolution followed an appropriate regenerative process which ended by offering a structure very similar to that of the animals in the control group. The morphometric studies provided facts that support the evolution of the regenerative process. The final number of fibers in the graft was high with a suitable proportion between the myelinated and unmyelinated fibers: After their growth the fibers followed a process of maturation demonstrated by an increase in diameter which reached, after a year, an average of 6.92 µm in the myelinated fibers as opposed to an average diameter of 8.77 µm to be found in a normal nerve. The thickness of the myelin increased at the same time as the axonal diameter, reaching by the end of a year an average of 0.83 µm as against the 1.2 µm which corresponds to the thickness of the myelin in a normal nerve. The distribution of these values was very wide as it is in the normal nerve. Summing up, the results obtained with the CCA are similar to whose obtained in a sectioned and sutured sciatic nerve. In Group II there was no suitable area for the growth of the regenerative axonal sprouts. The histological study showed the proliferation of endoneural material in which are included the regenerative tips growing in a disorganized and irregular way. At the level of the suture area the proportion between the myelinated and unmyelinated fibers was very inferior to normal.

In Group IV identical findings were obtained in the two models used (graft of the peroneal nerve or the sural nerve). In each case two areas were examined, the one corresponding to the graft and the other corresponding to the proximal stump in which this was implanted. The area corresponding to the proximal end of the nerve showed an orthotypic structure with all the characters observed in the control group; no in-

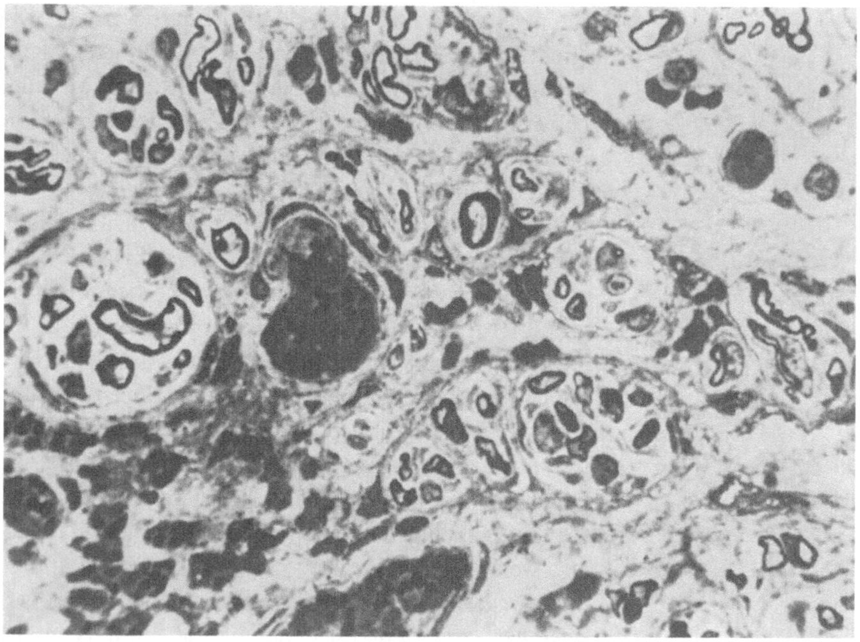

Fig. 3. Group IV: the different degree of maturation in the regenerative fibers is observed. Inside some regenerative fascicles there are fibers showing early phase of degeneration

dications of nerve fiber degeneration existed in any degree. However, the area that corresponded to the sectioned graft offered some interesting images. Along its whole length we could see a mixed structure of fibers in different stages of regeneration, some very advanced as corresponding to stages of maturation. Together with those were fibers in a recent degenerative phase. It was not strange to appreciate how, in the same regenerative fascicle appeared basal membranes which included, simultaneously, fibers in an advanced stage of regeneration and degenerative fibers (Fig. 3). So we could arrive at the conclusion that there was a double axonal direction via the graft, inside the same regenerative fascicles.

Discussion

The morphometric values obtained from the animals in the control group are similar to those of other authors[4, 10, 15, 16, 19]. According to our results the structure of the graft in the CCA group shows an identical evolution to that which follows the regeneration after the terminoterminal suture of a sectioned nerve[11, 15]. The regenerative fibers growing from the proximal stumps into the intercentrocentral graft are in an area with the best conditions for growth. Following this procedure the constant presence of the terminal neuroma which endangers the unrepaired nervous section is avoided and so the cause to which some stump pains are attributed[17] is also eliminated. Clinical experience[8, 13] and experimental models of chronic pain[2, 7] confirm these suppositions.

In contrast the terminoterminal suture of the two central ends provokes the close contact of the regenerative ends between them, producing a larger neuroma.

One apparently significant aspect which appears to be very interesting is the length and extension of the nerve regeneration proceeding from two different points, which occurs inside the nerve graft. The simultaneous presence, in the graft, of fibers in an advanced state of regeneration together with degenerative fibers lead us to believe that the first comes from the central end of the nerve homolateral to the fragment of graft analysed, while the second comes from the contralateral central stump. Our findings show also that the degeneration that follows the section of the intercentrocentral graft is unable to extend itself beyond the suture area. In none of the studied specimens were degenerative fibers observed in the central ends of the nerves. This is easy to explain because at these places there are no empty endoneural tubes to serve as a new support for the regenerative axons when these reach their second suture. However while studying the graft it has been observed that wrapped in a same basal membrane there exist fibers that proceed from different nerves. The obvious conclusion is that the regeneration after the anastomosis finishes in an area near to its second suture and that the fibers cannot penetrate in the central end that is different from the one they originally came from. This discovery seems to be a contradiction to what Carlstedt observed in 1983[5] when he sutured the dorsal root to the ventral one after their section, proving that the fibers of the ventral root were growing inside the dorsal root up to the vicinity of the central nervous system. The explanation depends upon whether the ventral fibers on his preparation found a clear way to go because the fibers of the dorsal root degenerated on being separated from their neuronal body in the ganglion. It is also interesting to observe that the impossibility of fibers continuing to grow in the other proximal end does not produce a reaction that distorts the morphology of the regenerative axons which occurs, under other circumstances, when this meets some obstacle which prevents it from progressing onwards[18]. It seems that growth in length spontaneously stops at this point even though the fibers have not reached their normal destination. In spite of that, the maturation of the regenerated fibers continues normally.

These findings can lead us to the following interpretation. The detention of the regenerative axonal growth, in length, at the level of the second suture, is owed to the absence of stimulative factors of regeneration to act on the axonal sprout[8]. In fact the presence in the other proximal stump of normal endoneural tubes occupied by normal axons created a physiological chemical atmosphere inside the nerve. Therefore no strange chemical agents appear that work on the isolated regenerative sprout and provoke the retrograde axonal flow responsible for the regenerative neuronal activity[3, 12]. The identity of the mechanisms of these factors are not known[14]. The maturation of the regenerative fibers is independent, whether they arrive or not at their distal effector. The axonal maturation would be due to the activity of the Schwann cells which surround the fiber[1].

The final conclusion we arrived at is that the CCA impides the presence of the terminal neuroma of the sectioned nerves. At the same time it allows an autolimited regeneration of mature fibers which protect them from the structural neuroma alterations that supply abnormal excitation which is the generator of pain.

References

1. Aguayo AJ, Bray GM (1984) Cell interactions studied in the peripheral nerves of experimental animals. In: Dyck PJ *et al* (eds) Peripheral neuropathy, 2nd ed, vol 1. WB Saunders, Philadelphia
2. Barberá J, Gonzalez J, Broseta J (1984) Evaluation of centrocentral anastomosis in autotomy following sciatic transection in rats. Pain [Suppl] 2: 444
3. Brimijoin S (1984) The role of axonal transport in nerve disease. In: Dyck PJ *et al* (eds) Peripheral neuropathy, 2nd ed, Vol 1. WB Saunders, Philadelphia
4. Brown MJ (1978) Special methods in the study of peripheral nerve. In: Asbury AK, Johnson PC (eds) Pathology of peripheral nerves. WB Saunders, Philadelphia
5. Carlstedt T (1983) Regrowth of anastomosed ventral root nerve fibers in the dorsal root of rats. Brain Res 272: 162–165
6. Committee for Research and Ethical Issues (1980). Ethical standards for investigation of experimental pain in animals. Pain 9: 141–143
7. Gonzalez J, Barberá J, Abellán MJ, Mora A (1985) Centrocentral anastomosis in the prevention and treatment of painful terminal neuroma. J Neurosurg 63: 754–758
8. Gorkisch K, Boese J, Vaubel E (1984) Treatment and prevention of amputation neuromas in hand surgery. Plast Reconstr Surg 73: 293–301
9. Lagarrigue J, Chavoin JP, Belahouart L, Scavazza R (1982) Traitement des nevromes douloreux par anastomose nerveuse en anse "piege a neurone". Neurochir 28: 91–92
10. Mira JC (1976 a) Etudes quantitatives sur la régénération des fibres nerveuses myélinisées. I. Variations du nombre et du calibre des fibres myélinisées dans les nerfs de rats normaux. Arch Anat Micr Morphol Exp 65: 209–229
11. Mira JC (1976 b) Etudes quantitatives sur la régénération des fibres nerveuses myélinisées. II. Variations du nombre et du calibre des fibres régénérées après un écrasement ou une section du nerf. Arch Anat Micr Morphol Exp 65: 255–284
12. Ochs S (1984) Basic properties of axoplasmic transport. In: Dyck PJ *et al* (eds) Peripheral neuropathy, 2nd ed, Vol 1. WB Saunders, Philadelphia
13. Samii M (1981) Centrocentral anastomosis of peripheral nerves: a neurosurgical treatment of amputation neuromas. In: Siegfried J, Zimmermann M (eds) Phantom and stump pain. Springer, Berlin Heidelberg New York
14. Selzer ME (1980) Regeneration of peripheral nerve. In: Sumner AJ (ed) The physiology of peripheral nerve disease. WB Saunders, Philadelphia
15. Sunderland S (1985) Nervios periféricos y sus lesiones. Salvat Editores, Barcelona
16. Thomas PK, Ochoa J (1984) Microscopic anatomy of peripheral nerve fibers. In: Dyck PJ *et al* (eds) Peripheral neuropathy, 2nd Ed, vol 1. WB Saunders Philadelphia
17. Wall PD, Scadding JW, Tomkiewicz MM (1979) The production and prevention of experimental anesthesia dolorosa. Pain 18: 267–278
18. Weiss P (1944) The technology of nerve regeneration: A review. Sutureless tubulation and related methods of nerve repair. J Neurosurg 1: 400–450
19. Weller RO, Cervós-Navarro J (1977) Pathology of peripheral nerves. Butterworths, London

Address for correspondence: J. Barberá, M.D., Department of Neurosurgery, Faculty of Medicine, University of Cadiz, Cadiz 11003, Spain.

Acta Neurochirurgica, Suppl. 43, 210–213 (1988)

Peripheral Nerve Autografts to the Rat Spinal Cord: a Study of the Origin of Regenerating Fibres Using Fluorescent Double Labelling

R. Pallini[1], **E. Fernandez**[1], **D. Minciacchi**[2], **E. Marchese**[1], and **A. Sbriccoli**[2]

[1] Institute of Neurosurgery, [2] Institute of Neurology, Catholic University, Rome, Italy

Summary

In adult Wistar rats a segment of the right sciatic nerve was grafted to the right dorsal funiculus of the spinal cord (SC). After survivals of 25–54 weeks a solution of the retrograde fluorescent cytoplasmatic tracer Fast Blue (FB) was injected into the SC and microgranules of the retrograde fluorescent nuclear tracer Diamidino Yellow (DY) were inserted into the graft. DY labelled neurones were found in the SC up to 30 mm both rostrally and caudally to the graft and were always single labelled. DY labelled cells were also found in the dorsal root ganglia (DRG) for 6–7 consecutive levels caudally to the graft. Ten to 30% of the DY labelled DRG cells were double labelled and also contained the FB tracer.

The present work using the multiple retrograde fluorescent tracing method demonstrates that peripheral nerve (PN) grafts implanted in the rat SC are reinnervated by axones arising from intrinsic SC neurones and DRG cells. The axones from intrinsic SC neurones seem to result only from the regrowth of damaged neurones which have lost their intraspinal projections. The regeneration from DRG cells seem to be sustained by both the regrowth of damaged axones and the preterminal and/or collateral sprouting.

Keywords: Spinal cord; regeneration; autograft; peripheral nerve; fast blue tracer; diamidino yellow tracer.

Introduction

Intraspinal grafts of peripheral nerve (PN) segments have attracted interest as a means for studying the influence of PN glia on mammalian spinal cord (SC) axonal regeneration. Recent studies have shown that the implanted peripheral glia has a high capacity to interact with the mammalian SC[5, 6, 13] and to modify its neuroglial environment providing more favourable conditions for the expression of the SC neuronal growth potentialities[1–3, 7]. Furthermore, intrinsic SC neurones and dorsal root ganglion (DRG) cells are capable of regenerating axones over long grafted PN segments[8, 9, 12, 15–19]. The intimate axonal mechanisms underlying these regenerative responses are less known. The present work directly addressed the question of whether the axones which regenerate in PN grafts implanted to the SC result from the regrowth of damaged axones which have lost their projections through the SC or from the preterminal and/or collateral sprouting of both damaged and undamaged axones which have retained their intraspinal projections. Therefore, a double retrograde fluorescent tracer strategy was employed, with one tracer placed in the SC and the other in the PN graft. According to our experimental model, single labelled neurones by the tracer injected in the graft should be an expression of the regrowth of damaged axones while double labelled neurones, should be an expression of preterminal and/or collateral sprouting.

Materials and Methods

Six female Wistar rats weighing 150–200 gr were anesthetized with Nembutal (4 mg/100 gr i.p.). In asepsis, using an operating microscope, a 20–25 mm long segment of the right sciatic nerve was re-

Table 1. *Experimental Data for Fluorescent Tracer Injections*

Case code	Survival after grafting (weeks)	Injection	
		Tracer	Dose
NCH 23	25	FB	2.0 µl
		DY	0.3 mgr
NCH 25	25	FB	1.6 µl
		DY	0.2 mgr
NCH 30	50	FB	2.0 µl
		DY	0.2 mgr
NCH 31	54	FB	1.5 µl
		DY	0.3 mgr
NCH 32	54	FB	2.0 µl
		DY	0.2 mgr
NCH 35	54	FB	1.5 µl
		DY	0.2 mgr

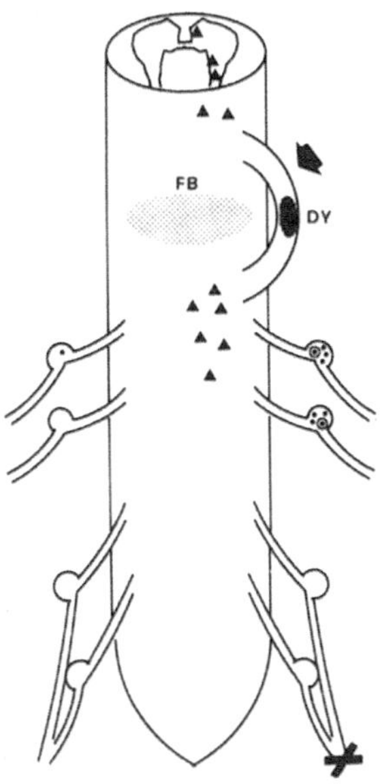

Fig. 1. Diagramatic drawing of the experimental procedure. A segment of the right sciatic nerve was implanted to the right SC dorsal funiculus. Twenty-five to 54 weeks after grafting the FB was injected into the SC and the DY was injected into the graft. In group B, before injecting the DY, the graft was sectioned near its rostral SC insertion (arrow). DY labelled neurons were found in the SC (triangles) and were always single labelled. DY labelled cells were also found in the DRG caudally to the graft (single circles), about 10 to 30% of these cells were also labelled by the FB tracer (double circles)

moved and implanted to the right SC dorsal funiculus. The SC was exposed through a T 8–T 9–T 10 laminectomy and two small myelotomies of the dorsal funiculus were performed at different rostrocaudal levels. The cut ends of the nerve segment were inserted into the myelotomies and retained by 10-0 perineurio-dural stitches. All the animals recovered from surgery without overt clinical deficits. After survivals of 25–54 weeks (see Table 1), the animals were reanesthetized and the SC and the graft were exposed. In 3 rats (group A) a 2% aqueous solution of the Fast Blue (FB) fluorescent cytoplasmatic tracer[4], was injected in the SC between the upper and the lower insertion of the graft using a Hamilton syringe; microgranules of the Diamidino Yellow (DY) fluorescent nuclear tracer[11] were inserted into the graft; the opening of the perineurial sheath of the graft was sutured to avoid diffusion of the tracer. In the remaining 3 rats (group B) the same injecting procedure was followed but, before inserting the DY the graft was sectioned near its rostral SC insertion and the free end of the injected stump was clipped. The injected quantities are listed in Table 1. Group B was designed to exclude double labelling which could result from hypothetic fibres that, after having crossed the injection area in the SC, penetrated the graft across its rostral SC insertion. Following the injecting procedure the animals were maintained from 6 to 8 days, then they were deeply anesthetized and transcardially perfused with saline followed by 10% buffered formalin (pH 7.2). The SC was soaked in phosphate buffered 30% sucrose (pH 7.2) for 12–24 hours. Serial 30 microns thick frozen sections were collected in distilled water. Alternate sections were counterstained with cresyl violet. The unstained sections were studied with a Leitz Ploempack fluorescence microscope utilizing Leitz filter-mirror system A and D which elicits the fluorescence of both the tracers.

Results

The results of the experiment are schematically summarised in Fig. 1.

In the SC, DY labelled neurones were found up to 30 mm from the insertions of the graft: in group A, both rostrally and caudally to the graft; in group B, only caudally to the graft. These intrinsic SC neurones were always single labelled and scattered in various laminae of the SC grey matter, but appeared more represented in laminae I to VI[14] (Fig. 2) both ispilaterally and contralaterally to the graft.

The results obtained in the DRG of both groups A

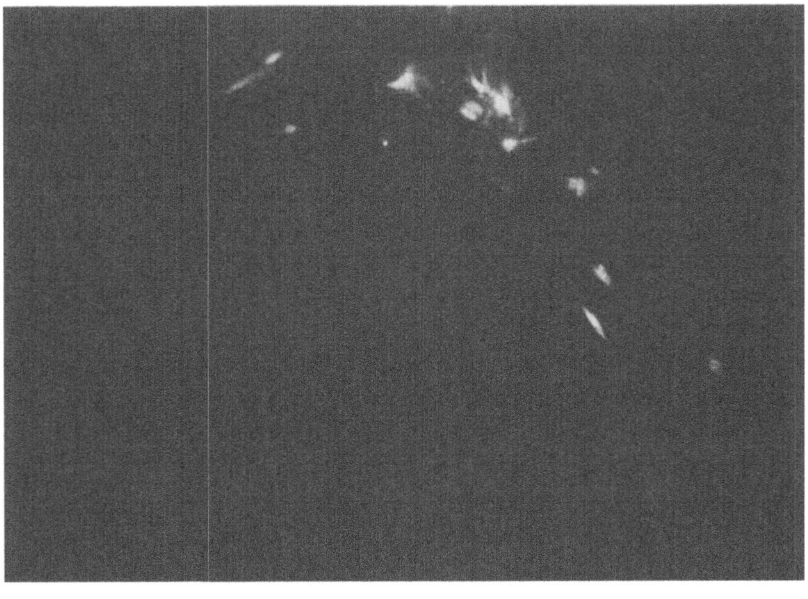

Fig. 2. Photomicrograph of DY labelled neurons in the dorsal horn of the SC. Magnification 100 ×

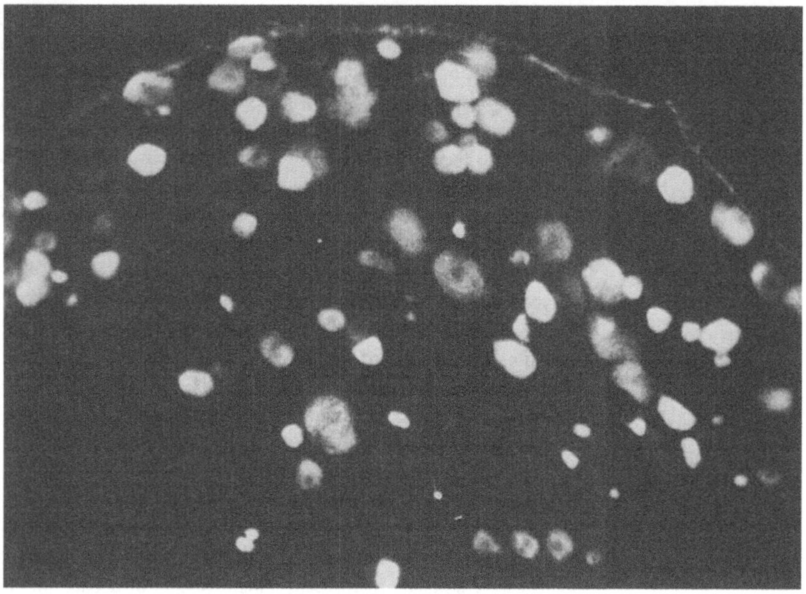

Fig. 3. Photomicrograph of a DRG obtained on the right side 2 levels caudally to the graft showing cells with cytoplasmatic (FB) fluorescence and cells with nuclear (DY) fluorescence. Magnification 100 ×

and B were similar. DY labelled cells were found in the right DRG at the level of the graft and in 6–7 consecutive levels caudal to it (Fig. 3) and in the left DRG at the level of the graft and in 1–2 levels caudal to it. The vast majority of these DY labelled cells were single labelled but 10 to 30% of this population was double DY-FB labelled.

Discussion

In the present study we have demonstrated with the multiple retrograde fluorescent tracing method that PN autografts to the rat SC are reinnervated by axones of intrinsic SC neurones and DRG cells.

The absence of double labelled neurones in the SC seem to indicate that the axones from intrinsic SC neurones, which regenerated into PN grafts, originate from the regrowth of axones damaged by the graft insertion; these axones have lost their projections through the SC.

The presence of both single DY labelled cells and double DY-FB labelled cells (10–30% of the entire DY labelled population) in the DRG caudal to the graft, suggest that the regeneration of axones from DRG cells into PN graft is chiefly due to the regrowth of damaged axones which have lost their projections in the SC and partly to the sprouting mechanism.

Other investigators have also suggested that sprouting from uninjured nerve cells is not an important source of axonal growth into PN grafts implanted to the central nervous system[10, 20]. Friedman and Aguayo[10] and So and Aguayo[20] have recently demonstrated by grafting PN segment into the olfactory bulb and the retina of rats that the regenerated axones arise as a regrowth of the severed fibres. The present study thus partly confirms these previous observations but also indicates that a possible source of axones within intraspinal PN grafts is represented by the sprouting of DRG cells. The sprouting process is probably represented by both the extension of preterminal collaterals from ascending primary sensory axones damaged at the grafting site and the collateral sprouting of uninjured primary sensory axones.

References

1. Aguayo AJ, Bray GM, Perkins CS, Duncan ID (1979) Axon-Schwann cell interactions in peripheral and central nervous tissue transplants. Soc Neurosci Symp 4: 361–383
2. Aguayo AJ, David S, Garth MB (1981) Influences of the glial environment on the elongation of axons after injury: transplantation studies in adult rodents. J Exp Biol 95: 231–240
3. Aguayo AJ, David S, Richardson PM, Bray GM (1982) Axonal elongation studied in peripheral and central nervous system transplants. In: Fedoroff S *et al* (eds) Advances in cellular neurobiology, vol 3. Academic Press, New York, pp 215–234
4. Bentivoglio M, Kuypers HGJM, Catsman-Berrevoets CE, Loewe H, Dann O (1980) Two new fluorescent retrograde neuronal tracers which are transported over long distances. Neurosci Lett 18: 25–30
5. Blakemore WF (1979) Remyelination in the CNS. In: Behan PO *et al* (eds) Progress in neurological research. Pitman, Bath, pp 12–25
6. Blakemore WF (1983) Remyelination of demyelinated spinal cord axons by Schwann cells. In: Kao CC *et al* (eds) Spinal cord reconstruction. Raven Press, New York, pp 281–291
7. Chi NH, Bignami A, Bich NT, Dahl D (1980) Autologous sciatic

nerve grafts to the rat spinal cord. Immunofluorescence studies with neurofilament and gliofilament (GFA) antisera. Exp Neurol 68: 568–580

8. David S, Aguayo AJ (1981) Axonal elongation into PNS "bridges" after CNS injury in adult rats. Science 214: 931–933
9. Fernandez E, Pallini R, Minciacchi D, Sbriccoli A (1986) Peripheral nerve autografts to the rat spinal cord: study on the origin and course of regenerating fibers. Acta Neurochir (Wien) 82: 57–63
10. Friedman B, Aguayo AJ (1985) Axonal regeneration from injured axons in the adult rat olfactory bulb. J Neurosci 5: 764–773
11. Keizer K, Kuypers HGJM, Huisman AM, Dann O (1983) Diamidino yellow dehydrochloride (DY · 2HCl); a new fluorescent retrograde neuronal tracer which migrates only very slowly out of the cell. Exp Brain Res 51: 179–191
12. Pallini R, Fernandez E, Minciacchi D, Sbriccoli A (1985) Peripheral nerve autografts to the rat spinal cord: HRP studies on the origin and the course of regenerated fibers. In: Hudson AR *et al* (eds) Proceedings 8th International Congress of Neurological Surgery, Toronto, p 248
13. Reier PJ, Stensaas LJ, Guth L (1983) The astrocytic scar as an impediment to regeneration in the central nervous system. In: Kao CC *et al* (eds) Spinal cord reconstruction. Raven Press, New York, pp 163–195
14. Rexed B (1964) Some aspects of cytoarchitectonics and synaptology of the spinal cord. Prog Brain Res 11: 58–92
15. Richardson PM, McGuinness UM, Aguayo AJ (1980) Axons from CNS neurons regenerate into PNS grafts. Nature 284: 264–265
16. Richardson PM, McGuinness UM, Aguayo AJ (1982) Peripheral nerve autografts to the rat spinal cord: studies with axonal tracing methods. Brain Res 237: 147–162
17. Richardson PM, Issa VMK (1984) Peripheral injury enhances central regeneration of the primary sensory neurons. Nature 309: 791–792
18. Richardson PM, Issa VMK, Aguayo AJ (1984) Regeneration of long spinal axons in the rat. J Neurocytol 13: 165–182
19. Sceats DJ, Friedman WA, Sypert GW, Ballinger WE (1986) Regeneration in peripheral nerve grafts to the cat spinal cord. Brain Res 362: 149–156
20. So KF, Aguayo AJ (1985) Lengthy regrowth of cut axons from ganglion cells after peripheral nerve transplantation into the retina of adult rats. Brain Res 328: 349–354

Address for correspondence: E. Fernandez, M.D., Istituto di Neurochirurgia, Università Cattolica, Largo A. Gemelli 8, I-00168 Rome, Italy.

CHICAGO

Park District Employee Retirement Guide 2025